LAKE BAIKAL

MONOGRAPHIAE BIOLOGICAE

EDITORES

W. W. WEISBACH
The Hague

P. van OYE
Ghent

VOL. XI

SPRINGER–SCIENCE+BUSINESS MEDIA, B.V.

LAKE BAIKAL AND ITS LIFE

BY

MIKHAIL KOZHOV
Professor of Zoology
State University of Irkutsk USSR

SPRINGER–SCIENCE+BUSINESS MEDIA, B.V.

ISBN 978-94-015-7390-0   ISBN 978-94-015-7388-7 (eBook)
DOI 10.1007/978-94-015-7388-7

CONTENTS

FOREWORD

Baikal is the deepest lake on earth and one of the most ancient. The pronounced endemism and specific wealth of its fauna and flora has attracted the keen interest of biologists and biogeographers all over the world.

A start on the Baikal studies was made more than 200 years ago, but they have been carried on with the greatest intensity in the last 30 to 40 years, and more than 1,000 scientific works devoted to it have appeared in this period. Hence there is an urgent need for a summary of the main results of more than 200 years' study of one of the most remarkable lakes of our planet, and this the author has endeavoured to provide. A zoologist and hydrobiologist himself, he has concentrated on the living world of the lake.

The author has for many years worked at Baikal as head of the Biologo-Geographical Institute and the Baikal Biological Station of Irkutsk University. In preparing this book for the press the author has received invaluable assistance from cartographer N. V. TYUMEN-TSEV, algologists N. L. ANTIPOVA and O. M. KOZHOVA, hydrobiologists G. L. VASILYEVA, G. J. SHNYAGINA, L. J. PROTASOVA and R. A. GOLYSHKINA, painters B. I. LEBEDINSKY, N. N. KONDAKOV and from V. B. KOCHETKOV, who has been of great help in preparing the English text. Very valuable advice and recommendations have been given to the author by L. A. ZENKEVICH (Moscow), B. G. IOGANZEN (Tomsk), Y. I. LUKIN (Kharkov), M. Y. BEKMAN (Institute of Limnology), G. G. MARTINSON (Institute of Limnology), N. I. LIVA-NOV (Kazan), V. V. IZOSIMOV (Kazan), A. L. LINEVICH (Irkutsk), G. G. ABRIKOSOV (Moscow), G. I. GALAZY (Institute of Limnology), and K. K. VOTINTSEV (Institute of Limnology), to all of whom he expresses profound gratitude.

The author is grateful to the late Prof. Dr. W. W. WEISBACH, Den Haag, and to Dr. W. Junk, Publishers, Den Haag, for having given him an opportunity of publishing this work in the Monogra-phiae Biologicae series, to Mr. ALLAN BRINDLE of the University of Manchester, who read the whole manuscript before it was sent to the printers and to Mr. K. J. PLASTERK of Dr. W. JUNK, Publishers for the proof-reading and his work in the preparation of the book and the indices.

Irkutsk, USSR.   MIKHAIL KOZHOV
March 1962.

INTRODUCTION

The unusual size and depth of Baikal and the originality and wealth of its fauna and flora have long attracted the attention of scientists the world over.

The first data on Baikal were published back in the 18th century by I. E. GMELIN 1751-52, P. S. PALLAS 1776, 1811, I. G. GEORGI 1775, and other explorers of Asia. But the foundations of the Baikal studies were laid by the remarkable investigations of B. DYBOWSKY, who worked here in 1868, 1870 and 1876, and also by the expeditions of Prof. A. A. KOROTNEV of Kiev University, who studied Baikal in 1900–1902. The same years witnessed important research in the morphometry, cartography and hydrology of the lake. Scientists in many countries took an active part in the studying of material collected by these expeditions and individual researchers.

After a certain interval the studies of Baikal were restarted in the period of the October Socialist Revolution and have continued ever since. An important part in these studies has been taken by the U.S.S.R. Academy of Sciences and Irkutsk State University. In 1918–20 the Academy of Sciences set up a permanent base for the study of the lake, situated 23 km to the north of the out-flow of the River Angara, in the area of the mouth of the River Bolshiye Koty. Soon afterwards this base was reorganised into the Biological Station and since 1921 it has been at the disposal of Irkutsk State University. Prof. V. N. YASNITSKY was its first director; since 1931 it has been headed by the author. In 1925, near the outflow of the Angara, the Academy established the per-manent Baikal Limnological Station, which till 1942 worked under the guidance of the distinguished limnologist G. Y. VERESHCHAGIN, who died in that year. Recently it has been reorganised into the Institute of Limnology of the Siberian Branch of the U.S.S.R. Academy of Sciences. In the Soviet period, studies of Baikal have also been conducted by many other scientific establishments of the U.S.S.R. and workers of the fishing industry.

As a result of these studies, Baikal has emerged before us, not only as a museum of living antiquities, which it was considered to be before, but also as a vast centre of autochthonous specialisation continuing at present as well. This process is the most remarkable feature of the living world of the great lake. Today, science has fathomed deeper than ever before the distant past of Baikal, and we can better visualise the history of its remarkable fauna. We know

now that over millions of years Baikal and the country around it have lived through many changes. They have seen the rise and change of mountain structures; the subsidence of extensive intermountain tectonic depressions connected with Baikal and their filling with deep waters; the transition from the warm, almost subtropical climate of the middle of the Tertiary period to the rigorous Quaternary climate and, finally, the moderately cold climate of the modern period, and the reshaping of the hydrographic network linking Baikal with the neighbouring biogeographical regions from which it could receive new immigrants. The outlook of the lake itself has also been changing gradually. A system of comparatively small lakes, embryos of Baikal, had developed into a single vast and colossally deep body of water whose trough is the deepest continental depression in the crust of the Earth.

All these phenomena were bound to have a decisive influence on the formation of a distinct world of organisms which gradually colonised the lake from the shores to the greatest depths. Life conditions in Baikal proved so favourable for some of the immigrants that they have thrived on an unprecedented scale there, forming a multitude of new species and genera. Today Baikal is already known to have more than 1,200 species of animals and up to 500 species of plants, two-thirds of which live only in its open waters. Among the 842 animal species inhabiting the open waters of the lake, 82% are not found anywhere else. There are 87 endemic genera and 11 endemic families and subfamilies in Baikal. The number of animal species in the lake is evidently not less than in the Black Sea and much greater than in the Caspian Sea. Elucidation of the time and ways of penetration of the ancient fauna into the Baikal basin and the laws governing its further evolution, which has brought about such striking results, is a primary task which will help in understanding the problems of world biogeography and the theory of evolution.

Elaboration of this problem has been greatly facilitated by recent palaeontological studies in the areas of old continental depressions in Central Asia which were once filled with extensive lakes. Some of them have been found to contain remnants of gastropods related to the modern Baicaliidae. These finds make it possible to judge with greater certainty of the ancient roots of the more enigmatic elements of the Baikalian fauna.

The problem of the history of Baikal has in itself provided an impetus to a profound study of the geological history and biogeography of the regions neighbouring on Baikal and also of the ancient hydrographical communications which existed in Central and North Asia. Among the results of recent geomorphological and biogeographical studies of special importance for the Baikal problem is the discovery of extensive and deep tectonic depressions of the Baikal system situated in the vicinity of the lake. The large residual lakes still covering the bottom of these depressions have been found to contain living remnants of the Baikalian fauna (KOZHOV, 1942, 1949).

Considerable additions to the knowledge of the systematic composition of the Baikalian fauna and its congeneric ties with analogous faunae of other countries have been made in the last 20 to 30 years. They include the discovery of such interesting groups as Bathynellidae, Hydracarina, benthonic Cyclopoida, etc. Large-scale studies have been carried out on the horizontal and vertical distribution of fauna and flora on the bottom and in the mass of water of the lake, as well as of such important factors of the water environment as temperature, chemistry, light intensity and types of soil peculiar to the separate zones and biotopes of Baikal.

Of great interest are the results of the study of seasonal phenomena in the life of Baikal's pelagic zone, vertical and horizontal movements of pelagic organisms, food relationships between widespread species, diurnal vertical migrations and seasonal and annual fluctuations in the crop of phyto- and zooplankton. These results may prove a major contribution to the limnology and ecology of hydrobiota.

In connection with the problem of Baikal, a study has been made of the hydrofauna of the drainage of the Angara and the Yenisei, and the lakes and rivers of the Baikal and trans-Baikal areas. It has been found that only a few Baikalian species live in some of the lakes connected with Baikal. But none of them ever penetrates very far up the affluents of Baikal. It is only down the Angara and the Yenisei that dozens of littoral species of the Baikalian fauna spread for thousands of kilometres, right up to the Arctic coast inclusive. En route they settle in big running-water lakes, including relict lakes (Taimyr, for instance). These facts help towards a deeper understanding of the part played by ecological conditions in the spread of fauna and of the impact of isolation on the specialisation process. They also indicate that unusual ecological conditions are responsible for the so-called immiscibility of the Baikalian fauna with the ordinary widespread Euro-Siberian fauna inhabiting the lakes and rivers of Siberia.

We have mentioned here only the main trends of the Baikal studies. The results of these studies have been printed in hundreds of works scattered in numerous publications, and the need has long since been felt for a work summing up the results of these studies. This book is designed to fill, as far as possible, this need and provide a concise review of the results of the many years of Baikal studies and also to outline their further tasks.

We hope that our work will further help to enhance the interest in Baikal, which presents, as it were, a gigantic natural laboratory

4

making possible a fruitful study of the evolution of aquatic organisms, the formation of peculiar endemic complexes, and other important biological and biogeographical problems.

I

GEOGRAPHY AND HYDROLOGY OF BAIKAL

1. Geographical Position (fig. 1)

Baikal, the world's deepest lake and one of the largest, stretches like a giant crescent 636 km long and up to 80 km wide amidst mountains on the north-eastern borders of Central Asia, at an altitude of 455.6 m above the Pacific level. Covering an area of 31,500 km², Baikal ranks seventh in the world after the Caspian and Aral seas, the North-American lakes Huron, Michigan and Superior and Lake Victoria in Africa.

In its depth (1,741 m), Baikal has no equals in the world. With its bottom lying at 1,285 m below ocean level, it is the deepest continental depression on our planet.*

The lake is situated in the centre of a vast mountain region bearing its name. In the south it is bounded by the Eastern Sayan mountain massif and its spur, the Khamar-Daban Range (fig. 2, 11). The highest point of the Sayan, the Munku-Sardyk (Eternally Snowcapped) complex of mountains, has an absolute altitude of 3,491 m and that of the Khamar-Daban, 2,400 m. The Munku-Sardyk carries small glaciers. At the foot of its southern slope extending into the confines of Mongolia lies the large and deep Lake Khubsugul (Koso Gol), to the south of which stretch the boundless Mongolian steppes. The linear distance between Baikal and Khubsugul is 200 km. Khubsugul belongs to the Baikal basin and communicates with it through the system of tributaries of the River Selenga. The Eastern Sayan Range gives rise to the Irkut, Kitoi, Belaya, Oka and other important tributaries of the Angara; many turbulent mountain rivers and streams flow into Baikal from the Khamar-Daban Range.

From the central massif of the Eastern Sayan, branch off the picturesque Tunka and Kitoi Alps (absolute altitude: 3,000 to 3,200 m). The Tunka Alps comprise the northern flange of the tectonic Tunka depression, which once contained a deep and large lake (fig. 110). Today it carries the meandering Irkut, which flows into the Angara near the city of Irkutsk, and numerous shallow lakes are scattered in it. Traces have also been preserved of the old bed of the Irkut, which in ancient times emptied directly into Baikal. The Tertiary lacustrine deposits in the Tunka depression are about 2,500 m thick.

---

* According to the latest acoustic soundings conducted by the Baikal Limnological Station of the U.S.S.R. Academy of Sciences, the maximum depth of Baikal is 1,620 m.

6

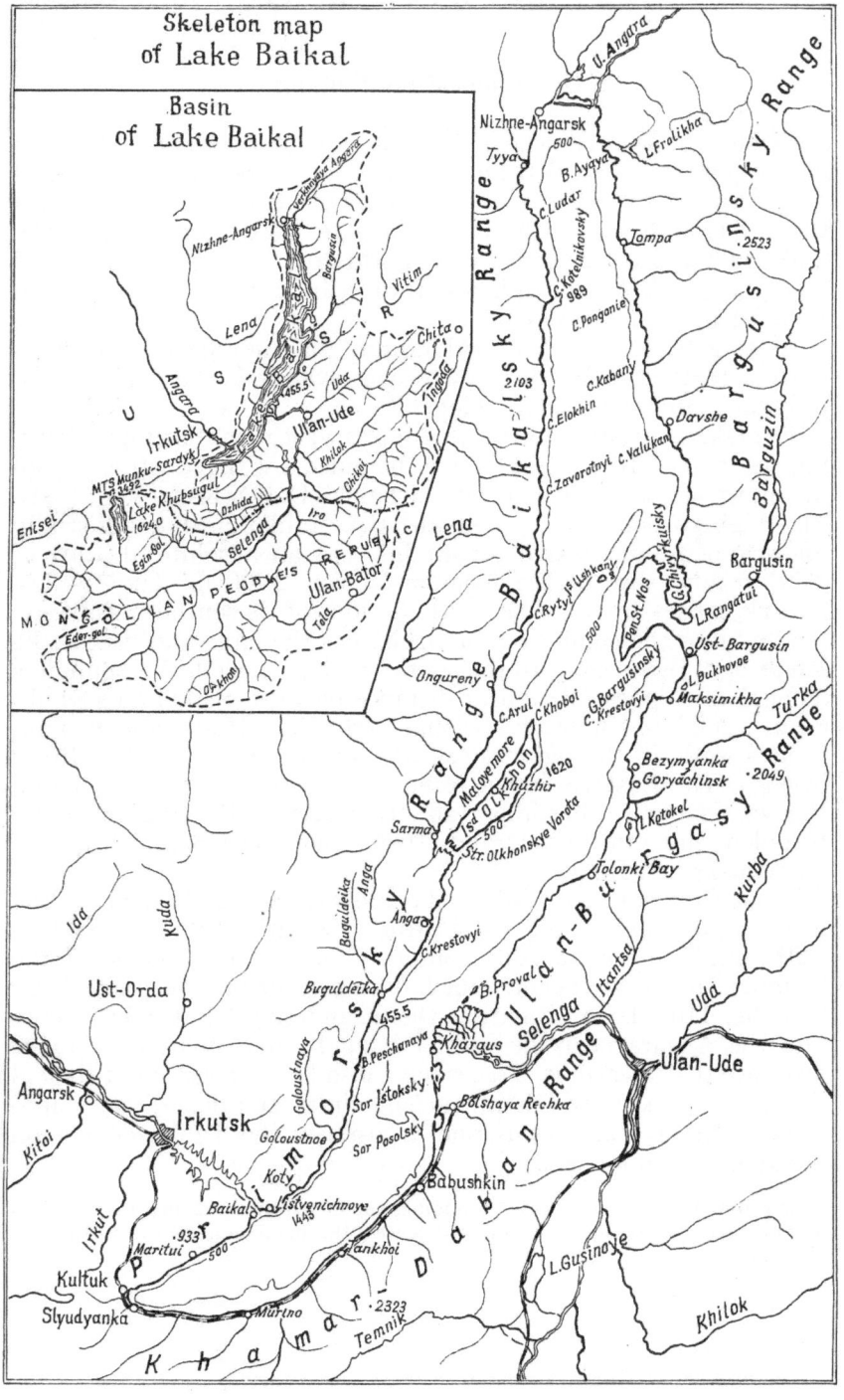

Fig. 1. Outline map of Lake Baikal and its basin.

The Khamar-Daban Range, which fringes Baikal in the south, consists of several mountain chains and spurs constituting the divide between the tributaries of the lake and the River Selenga. It slopes steeply towards Baikal, its snow-capped jagged bald peaks seeming to hang over the lake (fig. 2, 11). The Khamar-Daban is crossed by the Selenga valley and runs farther on along the north-eastern coast of Baikal as the Ulan-Burgasy Range which is lower than the Khamar-Daban and is dissected by the valleys of the tributaries of Baikal and the Selenga.

Fig. 2. Lake Baikal. Photo by N. TYUMENTSEV.

The Eastern Sayan, the Khamar-Daban, the Ulan-Burgasy and other ranges fringing South Baikal in the south and west are composed of Archean crystalline rocks: gneisses, mica and hornblende slates, amphibolites, marbles and very widely distributed Protero-zoic metamorphic schists. In many places they are broken by gra-nites and other plutonic pre-Cambrian, Paleozoic and younger rocks. Tertiary basalts lie on the tops and crests of the mountains; Quarternary basalts occur in the river valleys. Plutonic rocks rank first among others in the area occupied by them. Mesozoic continen-tal deposits, such as carboniferous sandstones, conglomerates and schistose clays, are found on the bottom of ancient tectonic depres-sions. The better known Tertiary lacustrine deposits are those of the terraces of the south-eastern coast of Baikal resting against the Khamar-Daban and lying at its very foot. A colossal mass of Ter-tiary and Quarternary sands and clays, 2,000 to 3,000 and probably more metres thick, has been found in the area of the Selenga delta.

The more gentle slopes of ranges and valley bottoms are in some

8

Fig. 3. The Bargusin Range in winter. Photo by O. GUSEV.

places covered by glacial deposits in the form of strongly scoured moraines.

The whole of this part of the South Baikal area is characterised by great vulcanicity, especially in the vicinity of the Sayan mountains, the Tunka depression and the Selenga delta. The Eastern Sayan Range still contains remnants of dead volcanoes with well defined craters.

To the east of the Ulan-Burgasy along the coast of Baikal stretch

several ranges of which the Ikat is the highest. The northern slopes
of this range comprise the left (southern) flange of the vast Barguzin
tectonic depression. In the Tertiary period this depression, too,
contained an extensive lake. Today its bottom, which is inclined
towards Baikal, is cut by the River Barguzin flowing into the lake in
its middle part. The right (northern) flange of the Barguzin depression is formed by the Barguzin Range (fig. 3). Its main ridge, reaching 2,724 m above ocean level, stretches at a distance of 10 to
40 km from the north-eastern coast of Baikal formed by the foothills
and scarps of this range. The Barguzin tectonic depression is up to
200 km long and 20 to 30 km wide. In its lower part it develops
directly into the depression of the Barguzin Gulf of Baikal (fig. 16).
The bottom of the Barguzin depression is filled with a mass of
Quaternary and, underneath, evidently Tertiary lacustrine deposits
analogous to those of the Selenga delta area. The total thickness of
these deposits reaches 1,700 m (LOGACHOV, 1958).

In the Quaternary period the elevated sections of the Barguzin
Range were covered by glaciers, which crept down the valleys towards Baikal, piling up heaps of debris before them. The north-
eastern coasts of Baikal adjoining the slopes of the Barguzin Range
are in many places composed of moraines forming broad terraces
and in some places descending under the level of the waters of the
lake.

To the north-east of the Barguzin Range lie the South-Muya and
North-Muya ranges separated by the extensive tectonic Muya-
Chara depression (fig. 111) which is analogous to the Barguzin
depression and which in ancient times also formed the bottom of
a large lake or a system of such lakes. This depression, like the ranges
flanking it, crosses the valley of the Vitim, the biggest right tributary of the Lena, and extends far to the north-east into the drainage
of the Chara, a right tributary of the Olekma in the Lena basin.
The north-western slopes of the South Muya Range serve as the left
flange of the large tectonic Tsipa depression (fig. 111), which is up to
200 km long and 25–30 km wide. Numerous lakes lie on the bottom
of this depression. The biggest of them are still inhabited by some
faunal species identical with or very close to indigenous Baikalian
species, such as the polychaete *Manayunkia baicalensis* NUSB., one
species of the Baikalian fish genus *Asprocottus* and species of the
genus *Paracottus*. The polychaete *Manayunkia* has also been found
in big running-water lakes of the Muya-Chara depression. This will
be dealt with in greater detail in Chapter VI of this book.

The coasts of the northern extremity of Baikal are formed by the
foothills of the Upper Angara Range (absolute altitude: up to 2,000 m),
which constitutes the divide between the Upper Angara, a big
affluent of the lake, and the Vitim, a tributary of the Lena. The
Upper Angara Range bounds on the right flange of the extensive

tectonic Upper Angara depression (fig. 111), along the bottom of which the Upper Angara flows into Baikal. The left flange of this depression is formed by the slopes of the Barguzin and North Muya

Fig. 4. Cliffs along the western shore. Drawing by B. LEBEDINSKY.

ranges. The Upper Angara depression is a natural northern continuation of the northern trough of Baikal.

The north-western coast of Baikal is composed of the steep precipitous slopes of the Baikal Range rising to 2,673 m above ocean level (fig. 8a) and the south-western coast, up to the outflow of the Angara, by the foothills of the comparatively low Primorsky Range. The Baikal Range comes very close to the waters of Baikal, forming very picturesque rocky cliffs (fig. 4–8). Its western slopes descend towards the Lena tableland, the drainage of which is directed already towards the Lena. The Lena takes its source on the northern slopes of the Baikal Range, in its central section 7 km from the lake.

The slopes of the Baikal Range still bear traces of the work of glaciers, such as hanging valleys and cirques with snow at the bottom. In summer the vivid white patches of snow stand out boldly against the background of grey rocks and debris. The slopes are cut by deep gorges with noisy rivers rushing down them (fig. 5). These rivers meander in the sea of stones, washing out their own debris and old glacial moraines. Often they disappear under debris and

Fig. 5. A creek on the Baikal Range slope. Photo by O. GUSEV.

reach Baikal as subterranean streams or drop in waterfalls from sheer cliffs of considerable height.

The geological composition of the mountain structures fringing the northern half of Baikal, as throughout the Baikal area, is characterised by the predomination of Archean and partly Proterozoic crystalline rocks. An outstanding part is played everywhere by plutonic rocks.

Fig. 6. The Khobot Cape on the western shore of Baikal. Drawing by B. Lebedinsky.

Fig. 7. A cedar on a rock on the western shore of Baikal. Drawing by B. LEBEDINSKY.

The mountain frame of Baikal is broken only in a few places by broad through valleys of affluents of the lake, such as those of the Selenga, Barguzin, Upper Angara, Turka, and others. Low-lying deltas and sand and rocky beaches are formed in the mouths of these rivers.

Fig. 8. The Storozhevoi Cape, with the Kolokolnya (Belltower) Cape in the background. Drawing by B. LEBEDINSKY.

The geological history of the Baikal mountain region will be discussed in Chapter VI devoted to the history of the lake.

Fig. 8a. The slopes of the Baikal mountain-range. Photo by M. Kozhov.

2. Climate of the Baikal Area

The climate of the country around Baikal is distinctly continental. The winter is long, cold and dry, the summer short and relatively hot, with considerable precipitation. January is the coldest month of the year and the mean temperature then in various parts of the coast of Baikal ranges from –17 to –25° C, the absolute minimum being –37 to –40°. In regions lying far from Baikal the winter is still colder. For instance, in the city of Ulan-Ude, situated 75 km to the south of the lake, the mean temperature in January is –25 to –26°, in some days dropping to –50°. In the Eastern Sayan Range the winter temperature often falls to 55 to 60° below zero. Table I shows the mean monthly temperatures of the air at Baikal and in Ulan-Ude.

July is the warmest month in the Baikal area but August is the warmest on the actual coasts of the lake. The mean summer temperature of July in the trans-Baikal area (Ulan-Ude) is 19 to 20° above zero, the absolute maximum reaching 38° (PREOBRAZHENSKY et al., 1959; TKACHUK et al., 1957). In the coast of Baikal the temperature of the warmest month, August, varies between 13° and 14°; the maximum is 20° and on specially warm and calm days, 25 to 30°.

As appears from these data, the waters of Baikal noticeably influence the climate of its shores: January and December are about 10° warmer and June-July 7° colder than in Ulan-Ude and in the city of Irkutsk lying 60 km to the north of Baikal, where the temperature of the air in June reaches 25 to 30°, while on the coast of Baikal at the same time it does not exceed 15 to 18°.

Table I.

Mean Monthly Temperatures of the Air (in °C) on the Coast of Baikal (the Southern and Central Parts, Listvyanka, Goryachinsk) and in Ulan-Ude (the Trans-Baikal Area).

Locality	January	February	March	April	May	June	July	August	September	October	November	December	Average annual	Amplitude
Coast of the southern part of Baikal, Listvyanka	—15.4	—15.7	—10.1	—1.0	5.0	10.1	12.7	13.9	8.7	1.5	—5.3	—11.9	—0.6	29.6
Goryachinsk (coast of Central Baikal)	—18.7	—18.2	—10.9	—2.2	3.7	9.1	12.9	13.5	8.3	1.5	—6.7	—12.6	—1.7	32.2
Ulan-Ude (the trans-Baikal area)	—25.7	—21.2	—11.8	0.5	8.5	17.0	19.7	16.9	8.8	—0.6	—12.5	—22.3	—1.9	45.4

* The data on Listvyanka have been borrowed from the works of L. S. BERG (1955) and those on Goryachinsk and Ulan-Ude, from the works of V. G. TKACHUK, N. V. YASNITSKAYA & G. L. ANKUDINOVA (1957).

In the first half of December the frosts in Irkutsk reach –20 to –25°, while on the coasts of Baikal the temperature does not fall more than 12 to 15° below zero and above the water, not more than 7 to 8° below zero. But this influence of Baikal on the climate of the coast does not spread farther than the dividing ridges fringing it.

The country around Baikal is noted for the clearness of the sky and transparency of the air, especially in January-February, and also in spring and early summer. In the number of sunny days it is not inferior to the southern coast of the Crimea (BUYANTUYEV, 1955). Winter and spring in the Baikal area and on the coasts of the lake are characterised by long periods of calm, with a deep-blue colour of the sky, bright sun and pure and transparent air. On sunny days the coastal ranges with their queerly jagged crests, grim abrupt promontories and foothills covered by the dark-green taiga blend with the water mirror of the lake into a beautiful picture enveloped by a transparent violet haze at dawn and sunset.

Annual precipitation in the Baikal area averages 300–400 mm, with only 30–35 mm falling in winter, but it reaches 800–1,200 mm on the Khamar-Daban slopes and 960 mm in the Baikal Range (PREOBRAZHENSKY et al., 1959). In some sections of the coast, however, annual precipitation does not exceed 160–170 mm. These include, for instance, the Olkhon Island and the adjacent regions of the western coast. Precipitation reaches its maximum in July and August.

The thickness of the snow cover in various parts of the Baikal area ranges from 15–20 cm to 130–140 cm. Snow is particularly abundant on the mountain slopes facing Baikal on the eastern coast (fig. 3). In regions with a thin snow blanket eternal congelation of the ground is sustained. But permafrost is not spread throughout the Baikal area; it occurs only in separate patches (BUYANTUYEV, 1955).

A greater part of the Baikal area presents a "sea" of the Siberian taiga covering mountain slopes, dividing boggy hilly plateaus, valleys of rivers and streams, and even gorges. Only the crests of high ridges show jagged bare rocks and there are steep slopes of talus devoid of vegetation. Below the bare mountain zone appear plants represented by *Betula Middendorfii* TRAUT & MEY, *Pinus pumila* PALL. (RDL) and other shrubs, the berry-bearing *Vaccinium vitis idaea* L., *V. uliginosus* L., *V. myrtillus* L. and *Ledum palustre* L., mosses, lichens, etc. *Pinus pumila* is a fairly typical plant of the mountains of South Siberia. It fringes the upper limit of the woods, but hardly penetrable thickets of this plant can also often be seen on low sandy shores of Baikal, right to the very edge of the water, especially in its northern part.

The tree-line on the mountain slopes extends to various altitudes depending on local topography, exposure and humidity. In the Upper Angara and Barguzin ranges the woods reach 1,300–1,400 m above ocean level. The larches *Larix dahurica* (Turcz.) Traut and *L. sibirica* Ldb. occur at an altitude of 1,200–1,500 m, but on the slopes facing Baikal they do not rise above 1,000 m. In the Eastern Sayan the woods reach an absolute altitude of 1,500–2,200 m.

The mountain taiga of the Baikal area consists predominantly of *L. dahurica, L. sibirica, Pinus silvestris* L., *P. sibirica* (Rupr.) Mayr, *Abies sibirica* Ldb., *Picea obovata* Ldb., with an admixture of the birch *(Betula verrucosa* Ehrb. and *B. pubescens* Ehrb.), aspen *(Populus tremula* L.) and alder *(Alnaster glutipes* Jorn.). *Rhododendron dahuricum* L., honeysuckle *(Lonicera)*, rowan *(Pyrus)* and the above-mentioned *Ledum palustre*, berry-bearing plants, club-mosses and many other plants are usually very common in the undergrowth.

A plant very widely distributed in the Baikal mountains is *Bergenia crassifolia* L. (Fritsch), which is found from the coasts of Baikal to bare mountain tops but prefers well moistened northern slopes and deep creek valleys (N. A. Yepova, 1955). Large tracts of pine forests *(Pinus silvestris)* stretch on sandy soils and in inter-mountain areas. Meadows lie along the banks of big rivers, and in some regions large territories are taken up by steppes.

In the recent period in the undergrowth on the slopes of the ranges bounding the lake in the east several dozen species of mosses, lichens and ferns have been found presenting relics of the undergrowth of the heat-loving platyphyllous forests which were widespread in the Baikal area in the Tertiary period.

The autumn withering of foliate plants in the Baikal mountains is evident in September. At the end of September the birch and aspen growths covering sites of old forest fires on mountain slopes appear as bright-yellow patches against the dark-green background of the taiga; the thickets of honeysuckle and *Bergenia crassifolia* L. (Fritsch) show red in some places, and tracts of larch forests turn pale-green. In summer the terraces of the coasts of Baikal and gentle slopes and scarps are covered with a motley carpet of meadow and forest grasses with vivid patches of *Lilium tenuifolium* Fisch. with bright-red flowers, lilies with large red and yellow flowers *(Hemerocallis, Pseudolirion)*, wild Siberian hop *(Atragena sibirica* L.) and many other Siberian plants.

In spring, in May-June, bright-green foliage appears on the trees and shrubs and light-green needles on the larches. In this period, on thickly wooded slopes, one can see bright-orange thickets of flowering *Rhododendron dahuricum* which add a bright touch to the austere beauty of the coasts of the great lake.

The products of the taiga are not unimportant to the waters of Baikal, for they enrich them with organic substances. On the bottom of the lake we often come across accumulations of decaying leaves and needles and pollen of coniferous trees swept there by storms, and whole trees brought by freshets can be found on the bottom near river mouths.

3. Morphology of Baikal

Printed below are figures on the morphology of Baikal in its modern boundaries (after VERESHCHAGIN, 1949).

Altitude above the Pacific level	455.6	m
Length	636	km
Maximum breadth	79.4	km
Minimum breadth	25	km
Mean breadth	47.81	km
Area	31,500	km²
Length of coastline (without islands)	2,000	km
Length of coastline of islands	139.2	km
Maximum depth	1,741	m*
Mean depth	740	m
Volume of water mass	23,000	km³

The basin of the lake consists of three troughs separated by submerged elevations.

The southern trough, situated to the south of the Selenga delta, includes the entire southern part of the lake. It is cordoned off from the central trough by a submerged elevation with a complicated relief, lying opposite to the Selenga delta. The maximum depth of the southern trough, after VERESHCHAGIN, is 1,473 m**, whereas the depths above the elevation do not exceed 400 m. It covers an area of 6,890 km².

The extensive region formed by this underwater elevation and the banks opposite to the Selenga delta and on both sides of it is called the Selenga shallow in the literature.

In the south the central trough of Baikal is bounded by the Selenga shallow and in the north by a massive submerged elevation (the Akademichesky Range of VERESHCHAGIN), stretching from the Olkhon Island to the south-east obliquely across Baikal in the direction of the Ushkany Islands. With the help of rope fathoming near the Olkhon Island, several kilometres from the Ukhan Promontory, G. Y. VERESHCHAGIN found a depth of 1,741 m (according to acoustic soundings in 1961, the maximum depth there equals 1,620 m), whereas the depth over the Akademichesky Range is not more than 300–400 m. The area of the central trough is 11,295 km².

The northern trough embraces the whole of the northern part

---

* During acoustic soundings in 1961 the maximum depth of Baikal was found to be 1,620 m.
** 1,419 m, according to acoustic soundings in 1961.

of Baikal. Its maximum depth is 890 m and the area, **13,315 km²**.

In recent years the morphology of Baikal has been studied by geophysical methods (V. V. FEDYNSKY, 1951). These studies have given reason to suppose that the southern and central troughs of Baikal present a single morphological whole and that the elevation separating them (the Selenga shallow) consists of thick sedimentary Tertiary and Quaternary deposits. V. V. FEDYNSKY writes in this connection:

"The single deep depression of the southern trough of Baikal with the maximum depth near the mouth of the Selenga is in its deepest part filled, as it were, by a colossal underwater sandy-argillaceous levee. This levee, formed as a result of a recent tectonic subsidence accompanied by many centuries of the working of the Selenga, has for a long time been regarded as an intermediate intermountain dike dividing South Baikal into two parts".

A characteristic feature of the relief of all three troughs is their asymmetrical structure, the underwater gradient along the eastern coast being, as a rule, much gentler than along the western coast. This asymmetry is also peculiar to the underwater elevations, the western slopes of which are gentler than those on the opposite side.

The littoral along the western coast of all three troughs (with depths less than 20 m) is very narrow; it usually ranges from 20 to 200 m, the gradient sharply increasing after 2–3 m or slightly greater depths, often to an angle of 60–80°. In such places by rocky shores

Table II.

Area of Depth Zones of Baikal (after KOZHOV & TYUMENTSEV, 1960).

Depth in m	Area in 1,000 ha	In % to total bottom area
0— 5	94	3.00
5— 10	56	1.77
10— 20	78	2.47
0— 20	238	7.24
20— 70	192	6.08
70—250	180	5.71
0—250	600 } 3150	19.03 } 100
From 250 to the bottom	2,550	80.97

the water directly washes the bases of sheer coastal cliffs. Along the eastern coast and especially opposite to the mouths of big rivers the gradient is less steep, but even at these places at a depth of 10–15 m the bottom also steeply plunges down. Table II gives the area of the bottom at different depths.

Thus, the zone of depths to the 250 m isobath occupies about

600,000 hectares, i.e., only about a fifth, and the zone of depths of 0–20 m not more than one-fourteenth, of the total area of the lake.

The underwater slopes of the Baikal depression retain distinct traces of the ancient relief of the coastline. Up to depths of 500–600 m one often detects submerged mouths of rivers and creeks, ancient valleys and coastal promontories deeply extending into the lake. But the shores also bear vivid traces of a higher level of water, and in some places terraces have been well preserved 2–3, 6, 15–20, 40–50, 80–100 metres and higher above the level of the lake. The Ushkany Islands in the central part of the lake have up to 11 such terraces and step-like rocky scarps with pebbles at the foot. The highest of them lies at more than 200 metres above the modern level of the lake. All this points to a complicated history of the Baikal basin, which will be discussed in greater detail in the chapter on the history of the lake.

The coast of Baikal is comparatively weakly indented. The biggest gulfs, Barguzin and Chivyrkui, are situated on the eastern coast in the central part of Baikal. They are separated from each other by the Svyatoi Nos Peninsula presenting a system of two ridges rising to 1,315 m above the level of the lake. The peninsula is 50 km long and 20–22 km wide, and a broad and low ridge connects it with the mainland.

Close to the western coast of the central part of Baikal lies its biggest island, Olkhon (fig. 9, 91, 92), the length of which is 71.7 km, the average breadth 10.5 km and the maximum breadth close to 14 km. The island has an area of about 730 km². It consists of a mountain range the highest point of which is more than 800 m above the level of the lake. The island is practically devoid of vegetation, only its northern part being covered by a coniferous forest. Its eastern coast is very steep and precipitous. The western coast is not so steep, but it also has many high abrupt scarps.

The extensive stretch of water lying between the Olkhon Island and the western continental coast is called the Maloye More (the Minor Sea). It is 70 km long and 18 km across in its broadest northern part. In the south it is connected with the lake by the Olkhonskiye Vorota Strait about 2 km broad, 7 km long and up to 30–40 m deep. The total area of the Maloye More is about 800 km².

Among other islands, mention should be made of an archipelago of four islands called Ushkany (fig. 82). They lie in the central part of Baikal 7 km from the Svyatoi Nos Peninsula. The Greater Ushkany Island has an area of 9.4 km² and rised to 211 m above the level of Baikal. In many places the picturesque abrupt shores of the islands are formed by marble rocks. The rows of large marble blocks rising from the water along the coasts are favoured by the Baikalian seal. The summits and slopes of the islands are covered by forests consisting chiefly of larches.

Fig. 9. The Deva Cape on the northern extremity of Olkhon Island. Photo by M. Kozhov.

Fig. 10. The Baikal Biological Station of the Irkutsk University in the area of the Bolshiye Koty. Photo by N. Tyumentsev.

Besides big gulfs, there are several dozen bays. Some of them are very deep, gently sloping and poorly protected against the prevailing winds (fig. 10); others deeply indent the coast and are shallow and very picturesque. Such are, for instance, the Anga (fig. 88) and Peschanaya bays on the western coast of Baikal. In the area of the Peschanaya Bay (fig. 86) the thickly wooded slopes of the Primorsky Range descend to the water in steep scarps. The products of the weathering of rocky scarps and bare ridges accumulate along the shores, forming broad sandy beaches. In the south and north the bay is bounded by rocky promontories rising majestically over the water and called "Kolokolni", or Bell Towers (fig. 8, 86).

The regions of the mouths of big affluents contain extensive shallow lagoons popularly called "sors". The water level in the sors is regulated by that in the lake. They are connected with Baikal by gullets of varying width. The sors are formed as a result of the inter-action of the surf of the lake and river currents filling the mouths with silt, sand and other material. The biggest sor, called Proval, has an area of about 18,500 ha and a maximum depth of 5–6 m. It lies to the north of the Selenga delta (fig. 15). Before January 1861 a bogged lowland with small lakes lay in its place. During a strong earthquake at the beginning of January 1861 all this lowland subsided and was inundated with the waters of Baikal (FITINGOFF, 1865). What has remained above the water is only a long and narrow strip of land which separates the Proval from Baikal today. This sandy spit is broken in many places and in years with a high water level it is fully covered with water.

The bottom soils of Baikal vary widely. Their properties depend on the distance from the shore and river mouths, the composition of the shore and the relief of the bottom, the direction of prevailing currents, and so on.

According to L. M. KNYAZEVA (1954), the following main types of soil predominate in South Baikal: stones, shingle, gravel, sands and silts (coarse- and fine-aleuritic, slightly diatomaceous and dia-tomaceous). Far from the shores, beyond 400–600 m depths, the predominant part is played, as a rule, by silty soils enriched with valves of diatom algae, particularly *Melosira* and *Cyclotella* (PATRI-KEYEVA, 1959; O. KOZHOVA, 1959 b). Along the shores tongues of silty soil extend into submerged creek valleys. In many places the steep slopes of the basin of the lake are formed by rock but slightly covered with a thin layer of silt and sand. Opposite to the mouths of the Selenga, Upper Angara, Kichera and Turka the bottom is sandy-silty; near the shores it is sandy. Pebbles, boulders and rough-ly rounded stones and fragments of rocks predominate at small depths along rocky shores. Opposite to low sandy shores the gentle slope of the bottom to considerable depths is covered with more or less silted sands.

Fig. 11. The River Snezhnaya. The Khamar-Daban Range is in the background.
Photo by V. LAMAKIN.

In many places, especially in the Maloye More, the central part of
Baikal and in the regions of underwater elevations sections of the
bottom strewn with coarse detritus (gravel, pebbles and even large
blocks) occur at considerable depths, often at 100–200 m and deeper.

The thickness of the silty deposits in the deep sections of the
Baikal depression is not known. Probably it is very great. In the
Selenga delta region, as has been noted above, the deposits of sand
and silt washed out by the river are up to 3,000 m thick. This delta
far protrudes into Baikal and is close to 40 km broad.

Some details of the distribution of soils in various sections of the
lake will be given in the review of the distribution of benthos.

4. Affluents and Drainage

More than 300 rivers and rivulets flow into Baikal from a vast
catchment area of 540,000 km². The rivers Selenga, Upper Angara,
Kichera, Barguzin, Turka and Snezhnaya (fig. 11) are especially
noteworthy. All big affluents rise in the eastern and northern parts
of the catchment area. The rivers and streams flowing from the
mountains of the western coast are very short; before emptying into
Baikal many of them disappear in screes and in winter are frozen to

Fig. 12. A hot spring in the Baikal taiga in winter. Photo by O. GUSEV.

the bottom, forming extensive ice bodies. Only after heavy and prolonged rainfalls in summer do they turn into tempestuous mountain streams carrying all kinds of material from the destroyed wooded banks into Baikal: stones, pebbles, clay, whole trees and their fragments and sometimes carcasses of dead animals. In such periods the waters of Baikal over large distances from the shores turn turbid with vast amounts of material washed out by the rivers.

The Selenga, Baikal's biggest affluent, rises in the spurs of the Khingan Range in Mongolia. Its total length is 1,480 km (1,591 km according to other data). It drains North Mongolia and the eastern part of the trans-Baikal area. The Egin-Gol, one of its tributaries, flows out of the big Mongolian Lake Khubsugul (Kosogol), which has already been mentioned above.

The Upper Angara takes its source in the North-Baikal highland. It is 640 km long and flows first in a narrow mountain valley and then along the broad Upper Angara tectonic depression, emptying into the northern extremity of Baikal. The wide pre-delta section of this depression directly adjoins the trough of North Baikal.

The Barguzin, 370 km long, flows into the Barguzin Gulf of Baikal. Its valley is structurally analogous to the valley of the Upper Angara. After emerging from the mountains the river flows towards Baikal along the broad Barguzin tectonic depression.

The Turka, 171 km long, empties into Baikal in the central part of the eastern coast.

All affluents deliver into Baikal up to 60 km^3 of water annually, with the Selenga accounting for 50%, the Upper Angara 15% and the Barguzin 7%. Besides, Baikal receives more than 9 km^3 of water owing to atmospheric precipitation on its surface (table III).

The coasts of Baikal and mountain slopes abound in hot springs which do not freeze even in the severest frosts. One of them is shown in fig. 12. In such springs or in their immediate vicinity one can sometimes see wintering ducks; and tertiary relicts of plants and some invertebrates whose life is connected with water can also be observed most often there.

All water brought into Baikal throughout the year, with the exception of 9 km^3 expended on evaporation, is carried out by the Angara (fig. 13, 13a). Cutting through the granite massif of the Primorsky Range, the Angara flows in a powerful river more than 1 km broad out of the south-eastern part of Baikal and rushes its cold pure waters to the north. After covering 1,853 km it merges with the Yenisei, which at the point of confluence carries half the amount of water carried by the Angara. The point of confluence lies almost 360 m lower than the level of Baikal. With the erection of big hydropower stations on the Angara, Baikal has begun to play the part of a giant reservoir feeding the entire system of the Angara hydropower chain.

The level of Baikal varies in the course of the year. The peak level is attained in September-October, sometimes in August; the lowest mark is reached in April-May. In years with scant precipitation the difference between the lowest spring and highest autumn levels constitutes 60–80 cm, and in years with maximum precipitation, 120–140 cm (VERESHCHAGIN, 1949; LOPATIN, 1954).

The mean perennial level of Baikal, as already noted, stands at 455.6 m, but there are years with exceptionally high or low levels. The highest level in the last century, observed on October 2, 1869, was 457.14 m; the lowest, 454.92 m, was registered in April 1904.

Fig. 13. The outflow of the Angara in summer. Photo by M. Kozhov.

Fig. 13a. The outflow of the Angara in winter. Photo by M. Kozhov.

According to research done by G. I. GALAZY (1956) by means of a botanical method, in 1785 the level of Baikal was 30 cm and in the period between 1395–1405, 50–60 cm higher than in 1869, judging by beach ramparts and the trees that have survived since then. G. I. GALAZY points out that in many parts these ramparts are overgrown with arboreal vegetation, with many trees being 450 to 550 years old. It can be supposed that the amplitude of secular fluctuations in the level of Baikal in the last 500–600 years has reached 3 metres.

Table III.

Water supply and loss in Lake Baikal.
Mean Perennial Estimates for the years 1901—1955.
(after AFANASYEV, 1960).

Sources of Supply	Layer in mm	Volume in km³	%	Losses	Layer in mm	Volume in km³	%
Precipitation	294	9.26	13.2	Outflow	1,933	60.89	86.8
Inflow-superficial	1,834	57.77	82.4	Evaporation	294	9.26	13.2
Inflow-underground	99	3.12	4.4				
Total Supply	2,227	70.15	100.0	Total losses	2,227	70.15	100.0

Under the impact of winds and changes in atmospheric pressure the level of the waters of the lake in various sections changes periodically. During strong winds the level falls near the leeward shores and rises at the opposite shores. In South Baikal the maximum wind-induced fall of water level at the western coast reaches 17 cm and the maximum rise 14 cm. A maximum wind tide of 17 cm has been observed for North Baikal and 20 cm for the southern extremity of the lake. Oscillations of the level of water during strong winds and a sharp change in atmospheric pressure occur also during the sub-ice period (POMYTKIN, 1960).

The wind-induced oscillations of the lake level and the changes in pressure give rise to seiches. According to observations by SOLOVYOV (1925), the seiches vary in periodicity and amplitude. In the northern part of the lake the amplitude does not exceed 10–12 cm and in the southern part reaches 14–15 cm. The periods of oscillations of seiches differ widely. G. Y. VERESHCHAGIN writes that a seiche with a period of oscillations of 4 hours 51 minutes is common in South Baikal.

5. Water Circulation and Temperature

Baikal is a very turbulent lake, especially in autumn, when the number of stormy days exceeds that of calm days. The winds induce powerful horizontal currents and a vertical circulation embracing the entire vast mass of water, down to the bottom is established. North-easterlies and south-westerlies are the prevailing winds in Baikal. The north-easterly begins in the morning and calms down at night, but in late summer and in autumn it is usually sustained for several days on end. During the north-eastern wind, called Barguzin, Baikal is especially beautiful. This wind blows, as a rule, on bright sunny days, making the foamy crests of high waves gleam and the sprays and foam opalesce in the sun. The rumble of the surf is heard far on the slopes of the mountains and in deep gorges. The south-western winds, also blowing along Baikal but in the opposite direction, are locally called kultuk. They, too, can be very strong and prolonged, especially in inclement weather in autumn.

The differences between the temperature of the air and the atmospheric pressure over Baikal and in the coastal regions give rise to constant winds alternately blowing from the shores or from the lake. Of particular constancy are cold breezes blowing in summer and winter in the evening and sometimes also at night from deep creek valleys and gorges. In summer they drive the heated superficial waters from nearshore shallows to open regions and cool them.

The greatest power is packed by the hurricane winds blowing across Baikal from the mountains of the north-western coast at a velocity of up to 30–40 m/sec. They have an especially strong influence on rise and fall in the water level and vertical movements of masses of water.

A recent research (SOKOLNIKOV, 1960) has revealed a rather complicated system of currents and vertical circulation both in the period free from ice and in the sub-ice period. There are cyclonic currents with a horizontal and vertical axis, and currents directed offshore in the superficial layers and shoreward in the lower layers, with their emergence to the surface. Often currents change direction in different periods of the day. Several independent closed systems of horizontal counter-clockwise circulations have been established. These horizontal circulations are overlapped by vertical currents of different speed and direction.

The process of the warming up of the waters is also greatly influenced by convection inversion of layers of different density, which is particularly intense in autumn before the establishment of homothermy and in spring during the spring warming. The influence of convection inversion evidently extends to 200–300 m depths and to the bottom in shallows.

All these phenomena have a strong bearing on the horizontal and vertical distribution of pelagic organisms.

Baikal is not thermally uniform (fig. 14). Depending on the relief of the bottom, the distance from the mouths of the big rivers, the indentedness of the coastline and the size of the shallows, the following regions can be distinguished in it:

30

1. coastal shallows cordoned off from open waters (sors, internal sections of bays and gulfs);

2. extensive open shallows lying opposite to the mouths of big affluents;

3. big gulfs (not the depths) and the Maloye More;

4. open waters far from river mouths and extensive shallows, including not only the deep-water part of the lake, but also the open coastal zone with a poorly developed (narrow) littoral not more than several tens or hundreds of metres wide.

The waters of these regions interact in various seasons of the

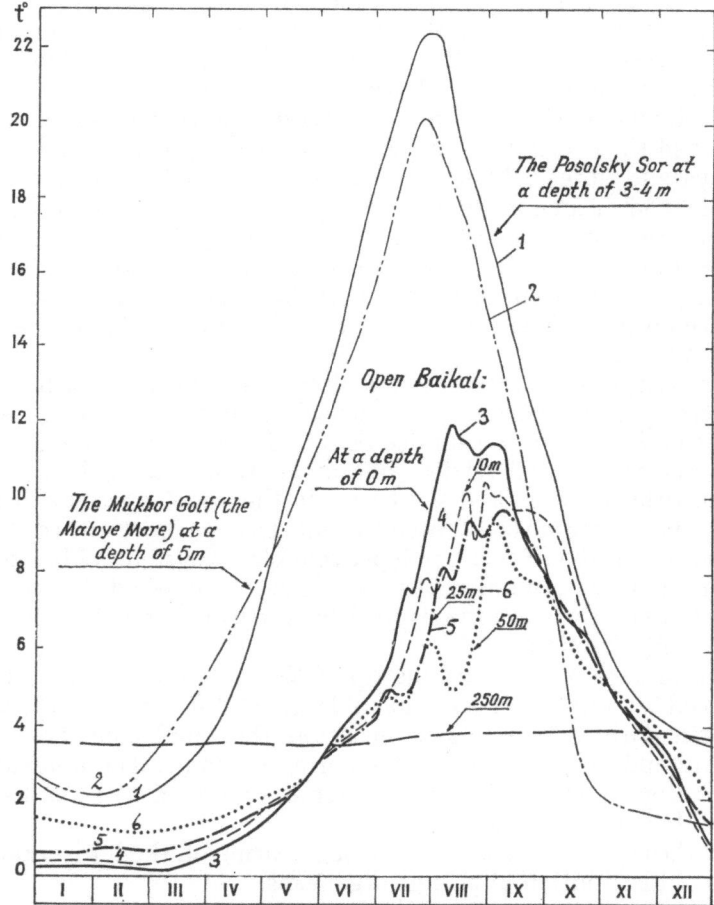

Fig. 14. Seasonal changes in the temperature of water in the sors, gulfs and open waters of Baikal. Original. 1-the Posolsky Sor; 2-the Mukhor Gulf (the Maloye More); 3—7-open Baikal: 3-at a depth of 0m, 4-10 m, 5-25 m, 6-50 m, 7-250 m.

year, thus making the establishment of any strict boundaries between them impossible.

We shall begin the review of the thermal regime of Baikal with sors and shallow depths of gulfs, which in the water regime differ but little from the ordinary eutrophic lakes of South Siberia. As an example, let us take the Posolsky Sor and the Proval Gulf (fig. 14, 15).

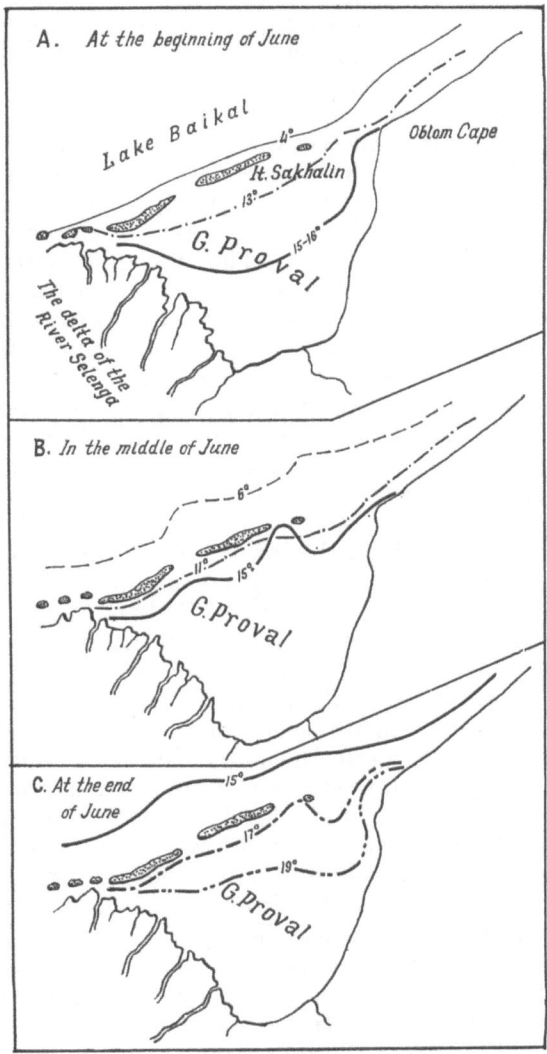

Fig. 15. Water temperature in the area of the Proval Gulf in June 1955. A-at the beginning of June; B-in the middle of June; C-at the end of June. After POPOWSKAJA, in litt.

The Posolsky Sor lies to the south of the Selenga delta. Its area is 3,500 hectares and its maximum depth 3.3 m, and several rivers flow into it. The water in it is renewed two or three times a year. It is connected with Baikal by a short shallow gullet about 200 m broad. The ice cover forms at the end of October and lasts till the beginning of May, breaking up at the mouths of the affluents in April.

After the ice break-up the water in the sor is rapidly warmed through to the bottom. Already towards the end of May the temperature of the superficial layer reaches 14–18.5° and that of the bottom layer, 13–15°C. In open Baikal the superficial temperature in the same period does not exceed 0.5–1.0°. In July and August the temperature of the entire mass of water in the sor attains the annual maximum of 19–23°. In the middle and later half of August a gradual cooling sets in. Early in September the temperature remains at a level of 14.5–16° and then drops sharply. At the end of October the sor becomes covered with ice. The thickness of the ice cover reaches 1.2–1.4 m, and therefore a considerable part of the shallow coastal belt freezes to the bottom. At the outset of the sub-ice period the bottom temperature remains comparatively high. In 1938–43, for instance, it was 4.5–4.8° and stayed at a 3.5–4° level even in the first ten days of April. During the sub-ice period the bottom layer retains a higher temperature thanks to the heat accumulated during the summer by the viscous soil rich in organic substances, the temperature of which does not drop below 4–5° throughout the winter (KOZHOV, 1947).

The outflow of water from the sor to Baikal through the gullet is maintained the year round. In spring and summer this water slightly warms up the nearshore superficies of Baikal.

The extensive shallow Proval Gulf, which has an area of 18,500 ha and is 4.5 to 5 m deep, is situated to the north of the Selenga delta and receives several arms of that river carrying vast amounts of suspended material which is responsible for the exceptional turbidity of the water in the gulf. In summer its transparency does not exceed 10–20 cm, becoming still worse after strong winds. During winds causing a rise in the water level, the waters of Baikal easily penetrate into the gulf through broad and very short channels. In quiet weather one can observe a slow permanent current directed from the pre-estuarine areas of the Selenga arms through the channels to Baikal. As in the Posolsky Sor, the waters of the Proval Gulf are rapidly warmed through after the break-up of the ice cover in the middle or later half of May (table IV). In the second half of July the temperature rises to 18–20° on the surface and 16–18° at the bottom.

In July the temperature remains at a level of 19–26° throughout the mass of water. In summer the water temperature in the vicinity

Table IV.

Water Temperature in the Proval Gulf in 1943 (after Kozhov, 1947)

Region	Total depth in m	March 29 Surface	March 29 Bottom	April 29 Surface	April 29 Bottom	June 16—24 Surface	June 16—24 Bottom	July Surface	July Bottom	Sept. 3—10 Surface	Sept. 3—10 Bottom	Sept. 15—16 Surface	Sept. 15—16 Bottom
South-western part of the gulf, 1—2 km from the shore	1—2	—	—	0.2	—	19.6	18.2	22—26	—	14.5	—	—	—
Middle of the gulf	4—4.5	—	—	0.2	2.5	19.2	16.7	22—26	—	14.7—15.0	14.0	—	—
Opposite to the Oblomovskaya Bayou	3—4.5	—	—	—	—	15.0	15.0	—	—	15.0	—	—	—
Off the Sakhalin Island	2—3	0.3	3.9	—	—	—	—	—	—	14.4	—	12.6	—
Lake Baikal near Cape Oblom	10—15	—	—	—	—	—	—	19.0	—	15.6	—	—	—

of the channels is usually 4° to 5° lower than in the middle of the gulf. In mid-August there appear signs of autumnal cooling. In the first half of September the temperature falls to 14–15°.

At the end of October and beginning of November the Proval Gulf becomes covered with ice. The water temperature in the sub-ice period is shown in table V.

Table V.

Temperature in the Central Part of the Proval Gulf in the Winter of 1925—26 (after ROSSOLIMO, 1957).

November 27		December 30		February 8		March 3		April 13	
Depth in m	t° C	Depth in m	t° C	Depth in m	t° C	Depth in m	t° C	Depth in m	t° C
0	0	0	0.40	0	0.60	0	0.40	0	0.40
3 (bottom)	1.0	1.3	2.00	1.5	1.80	—	—	—	—
		2.5 (bottom)	3.20	2.9 (bottom)	3.80	3 (bottom)	4.50	2.5	2.90

The regions opposite to the mouths of big rivers are characterised by the prevalence of shallows strongly influenced by fluvial waters. This influence is particularly pronounced opposite to the delta of the Selenga. In winter the cold fluvial waters penetrating into Baikal cool its open waters, but in April-May they are heated better than the waters of Baikal and warm them. Between June 10 and 20, when a homothermy of 4.3–3° is established in the deep-water area of the lake, the superficial temperature opposite to the Selenga delta reaches 8–11° (fig. 15). As the distance from the shore grows the temperature usually falls sharply to 4–5°. Later on, towards the end of June and in July, the temperature of nearshore waters opposite to the delta reaches 15–16° on the surface, whereas 2–3 km from the shore over considerable depths it does not exceed, as a rule, 11–12°. In August the surface water temperature opposite to the delta over a depth of about 4 m rises to 18–19°, and 3 km from the shore over a depth of more than 50 m, 14–15°.

In summer the warm waters of the Selenga shallow are carried by currents to the neighbouring deep-water regions and considerably raise the temperature of their water. A permanent current is known to be directed from the shallow to the southwest towards the outflow of the Angara. Spreading over the surface, the waters of the Selenga shallow gradually mix with the water of Baikal. The thickness of the layer of mixed waters in which an admixture of the Selenga waters can still be detected by chemical analysis does not exceed 5–10 m, but in some instances patches of these waters are observed to extend to a depth of 25–50 m (VOTINTSEV, 1960). When the southwesterlies prevail the Selenga waters are driven comparatively far away to the north of the delta, but in calm weather and during winds from the north-east the southward current restarts. The changes in the direction of the Selenga current often result in a rather complicated picture of the horizontal distribution of temperatures in the central part of Baikal. Some peculiarities of this distribution will be described in the chapter devoted to life in the mass of water.

The influence of fluvial waters on the thermal regime of the adjacent regions of the lake is vividly seen also in the Barguzin Gulf. A diagram of the warming through of the waters of the gulf in 1932 is given in fig. 16 (KOZHOV, 1934b). The pre-estuarine area of the River Barguzin is shallow, the 20 m isobath passing there at a distance of 4–6 km from the shore. The Barguzin empties into the gulf in the central part of its eastern shore. The fluvial waters incite a current in the gulf directed from the mouth of the river into the open lake along the Svyatoi Nos Peninsula. During winds from the south in summer this current shifts towards the coast of the Svyatoi Nos and flows along it in a band of 2 to 3 km wide and 2 to 4 m deep. Accordingly with the flow of fluvial waters, in June the temperature of the superficial layers of water in the gulf along the Svyatoi Nos coast, where the depth already reaches 300–500 m and more, is 9° to 10° higher than in open Baikal. For instance, on June 20, 1932, the temperature of the surface layer there reached 11–16°, whereas a 3.4–4° homothermy prevails at this time in the range of the gulf and in open Baikal.

Toward the beginning of July the water temperature throughout the gulf's shallow becomes more or less evened up, reaching 15–20° on the surface. In the outer part of the gulf it does not exceed 11–12° and in the depth of the gulf, 6–8°. In early August the temperature of the superficial layers climbs to 17–22° in shallows, 17–18° at a depth of 5 m and 12–14° at 10 m. In the open outer part the superficial layers are also warmed up to 15–16° and the bottom layers at 25 m to 5–6°. At the end of August and beginning of September the water temperature from the surface to a depth of 10 m is to a considerable degree evened up throughout the gulf at a level of 12–15°.

A permanent current of fluvial waters has also been established in the northern part of Baikal. It is directed from the mouths of the Upper Angara and the Kichera southward along the north-western coast. In summer it has a marked warming-up effect on the region lying along the north-western coast of the lake.

A brief geographical survey of the Maloye More has already been given above. In its open parts the ice cover is formed between December 20 and 31 and in the shallow bays and gulfs, much

36

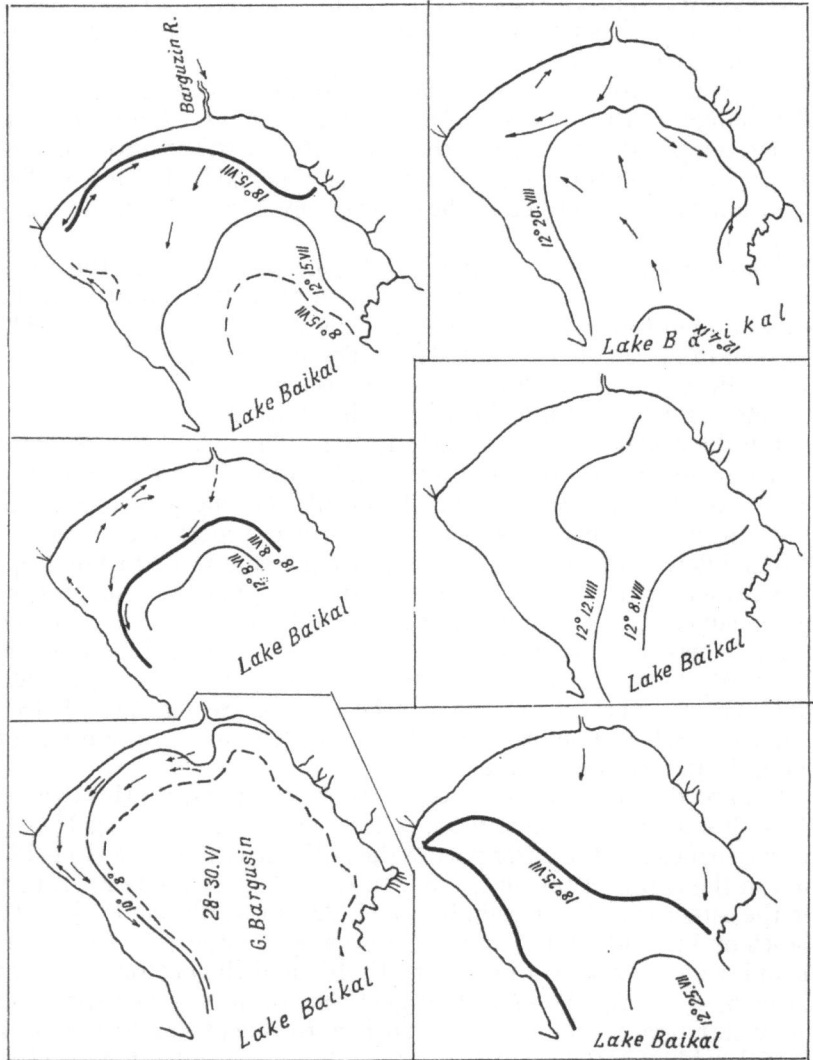

Fig. 16. Water temperature in the Barguzin Gulf in the summer of 1932. The arrows show the direction of the current. After Kozнov, 1934.

earlier. The Mukhor Gulf, for instance, freezes in late October. If the autumn is warm, the formation of the ice cover is delayed till the end of December, and in the northern part at times even into January. The ice cover in the southern and central parts of the Maloye More breaks up in the middle or between the 20th and 31st of May and in the northern part at the end of May. The ice break-up occurs still later if the spring is cold. For instance, at the end of May in 1951 the central and northern parts of the Maloye More were still under the ice cover, and in 1953 its southern part was freed from

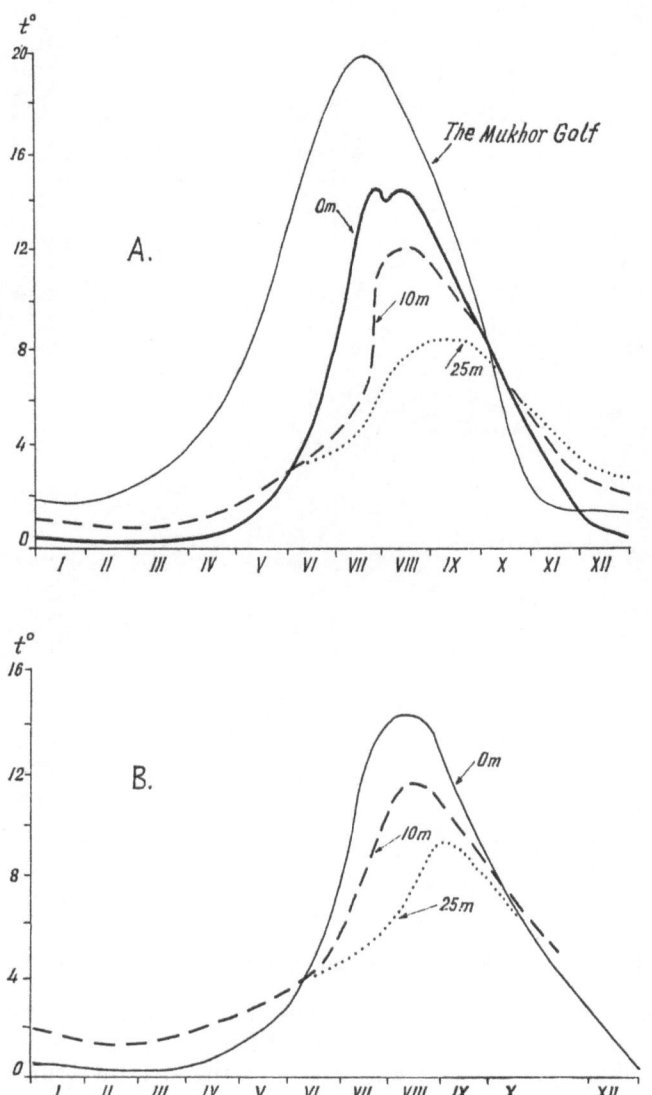

Fig. 17. Seasonal changes in water temperature in the Maloye More, after data by Kozhov. A-the southern part; B-the central part. After Kozhov, 1958.

Fig. 18. Hummocking along the shore during ice formation. Photo by M. Kozhov.

ice only at the beginning of June. The thermal regime of the Maloye More in its northern part does not differ essentially from that of open Baikal, but the thermal regime of the shallow southern part has its own peculiarities (fig. 17). In the Mukhor Bulf (the southern extremity of the Maloye More) the water temperature reaches 18–20° at the end of June, whereas in the same period in open Baikal the superficial layers are at a temperature of 4–5°. Towards mid-August the temperature throughout the Maloye More is more or less evened up and differs but little from that in the open lake. By the middle of September the superficial temperature falls to 10–12° everywhere. The cooling of the waters of the southern and central parts of the Maloye More proceeds faster than in open Baikal.

As has been noted, in open Baikal the inshore shallows 3 to 10 m deep are very narrow, with the exception of shallow regions opposite to river mouths. Due to this the open coastal waters are under a constant influence of the neighbouring deep-water areas while exerting but a faint influence on the thermal regime of the latter.

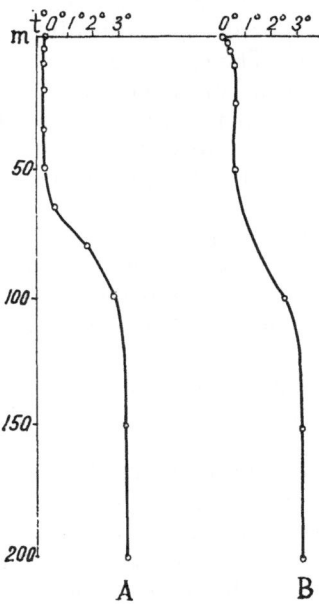

Fig. 19. Vertical distribution of temperature in Baikal's deep-water regions at the outset of the sub-ice period. A - in the area of the Ushkany Islands on January 9, 1915 (average temperature in the 0—200 m layer: 1.82° C); B - the Bolshiye Koty on January 14, 1932 (average temperature in the 0—200 m layer: 1.97° C). After Rossolimo, 1957.

Open Baikal freezes in the first half of January in its southern part and somewhat earlier in the northern part. The duration of ice-cover formation varies markedly with years depending on meteorological conditions. Often the just-formed young ice is broken by strong winds, then restored, to be destroyed again. Not infrequently this process lasts for many weeks. The ice cover is formed for good at an average temperature of 2–2.9° in the 0–250 m layer, with fluctuations from 1.60° to 2.93° (ROSSOLIMO, 1957) (fig. 18).

In February-March the thickness of the ice reaches 80–120 cm. The ice break-up in South Baikal occurs at the beginning of May and in the northern part at the end of May and in some years in June. Thus, Baikal carries the ice cover for 4 to 5 months.

The fluctuations in the temperature of the ice caused by ice-air heat exchange and heating by direct solar radiation are chiefly responsible for the formation of numerous cracks and breaks in the ice cover. Cracks evidently develop also due to strong winds causing sub-ice turbulence. Some cracks stretch for many kilometres and are up to a metre and more broad. During strong winds ice movements and formation of thrusts and hummock-building are observed along the cracks or, conversely, the cracks widen and form open-water spaces, which happens especially often in spring before the break-up of the ice cover. The cracking of ice and the appearance of hummocks and thrusts is accompanied by a gamut of weird sounds striking everyone who happens to be on the lake for the first time in winter. Now one hears a threatening rumble of distant ice ruptures, now thunderous strokes are heard quite near by and a wide crack suddenly opens before one's eyes; then there comes a long-drawn groan, produced by small local ruptures, or the noise and crackle of hummocking. This "symphony" attains particular force in the morning and evening at the beginning of winter.

Thermally, the mass of water of open Baikal divides into two vertical zones: the upper, or alternating (after VERESHCHAGIN, 1927a, 1933b), and the lower, or perennial. The boundary between them lies approximately at a depth of 250–300 m, but it is not permanent (fig. 19, 20, 21).

In November a reverse thermal stratification is established in the upper layer and is sustained there throughout the sub-ice period, lasting till the end of June or beginning of July. In the sub-ice period the temperature of the surface is close to zero and gradually grows to 3–3.5° with increase in depth. At a depth of about 200–350 m the temperature stops rising and its gradual fall begins, which continues down to the very bottom. Thus, in this period the maximum of temperature at 200–350 m depth stands at 3.5–3.6°. G. Y. VERESHCHAGIN called this maximum mesothermic. It can be observed, of course, only in the autumn-winter period and partially in spring before the start of the spring warm-up. The position of the mesothermic maximum is usually accepted as the boundary between the upper and lower zones of the mass of Baikal water, but it often lies lower than 300 m and is sometimes found at a depth less than 200 m.

While the reverse thermal stratification in the upper zone in the cold period is explained by the autumn-winter cooling of the upper layer, the slow fall in temperature below the mesothermic maximum

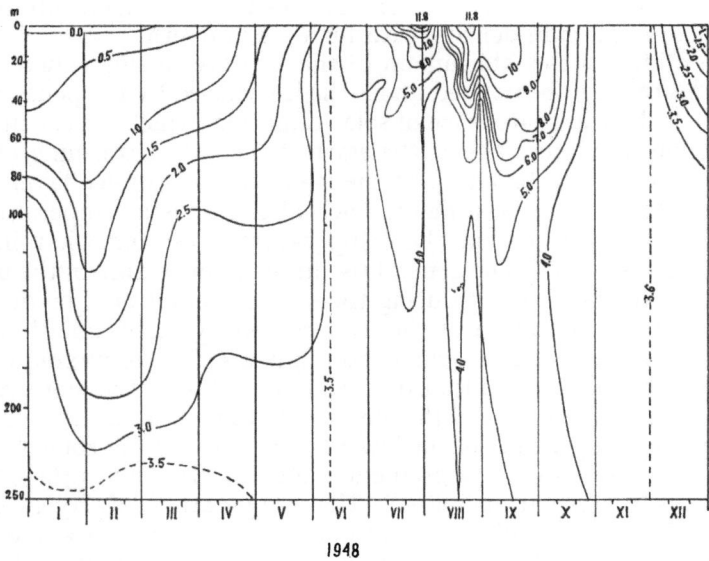

Fig. 20. Thermoisopleths in ° C in the southern part of Baikal (the Bolshiye Koty) in 1948 (a cold year). Original.

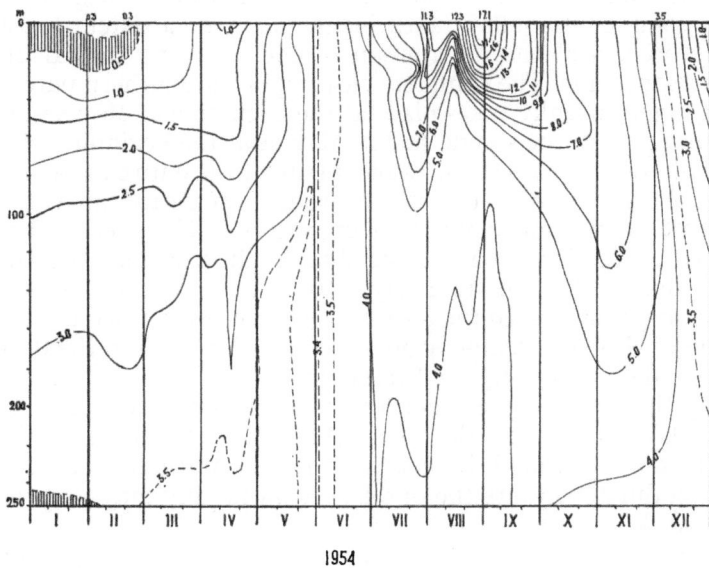

Fig. 21. Thermoisopleths in °C in the southern part of Baikal (the Bolshiye Koty) in 1954 (a warm year). Original.

towards the bottom depends on changes in the temperature of the water of maximum density with increase in depth.

Soon after the establishment of the ice cover a slight increase in temperature can be noticed under the ice, caused by the penetration through the transparent ice of solar radiation, which is very intense in Baikal in winter. This is the start of the spring warming of the water. Before the break-up of the ice cover the sub-ice layer from several centimetres to a metre thick already has a temperature of 0.4–1.7°. In this period the temperature rises somewhat in the lower layers as well. The general rise in the temperature of the upper layer up to 200 m thick during the sub-ice period may be due not only to the penetration of the heat caused by solar radiation into the depths by means of convection, but also to the mixing of the water of this layer with that of warmer deep layers the temperature of which is close to that of the mesothermic maximum. In the sub-ice period the temperature of the lower layers of the upper zone changes with a very small vertical gradient, while the gradient in the middle layers is somewhat increased. The temperature of layers lying below the mesothermic maximum does not remain constant, either. It changes vertically first within a range of several tenths of a degree, and at a depth of more than 400 m the thermal gradient per every 100 metres does not exceed, on the average, 0.05°.

The changes in temperatures in the mass of water in the sub-ice period show that the movement and mixing of waters continues even under the ice. For this reason the classical state of "winter stagnation," common in shallow lakes, is practically nonexistent in Baikal.

The process of the destruction of the ice cover in Baikal also takes a long time. The complete clearing of the southern part of the lake from ice usually takes place at the beginning of May and in the northern part in the second half of May or at the beginning of June. It is followed by a very slow warming up of the entire upper zone of the lake. At the outset of the warming up period an especially intensive convection circulation of waters is observed and wind-induced circulation also increases. As a result, a partial homothermy sets in first in the uppermost layer 10–20 m thick and then spreads to the deeper layers. In the 0–50 m layer it is established around the middle of May at a temperature close to 1–1.5° and in the 0–100 m layer at the end of May or beginning of June, at a 2.2–2.5° level. Finally the temperature is evened up throughout the mass of water of the lake. In South Baikal a relatively complete homothermy usually sets in around June 20 and in the northern part, in mid-July. In different years the onset of homothermy may vary within 2 to 3 weeks.

The period of summer warming up begins with the appearance of the first signs of direct thermal stratification. Already the beginning of summer warming up witnesses the establishment of epilimnion with increased temperatures and a small gradient of their fall; metalimnion, i.e. the layer of a sharp increase in gradient (thermocline); and hypolimnion, the layer of a gradual drop of temperatures with a very small gradient. In sunny calm weather in the period of summer warming up the metalimnion can be clearly located. We have observed on many occasions, at the end of July and beginning of August, a sharp temperature drop at a depth of only 4–5 m. In such periods the upper layer of 0–4 m may have a temperature of up to 12–15°, at 3–4 m the temperature will be up to 10–11°, while at 5–6 m only 4–5°. In August the metalimnion descends to 10–20 m and deeper. But the temperature layering is very often broken after strong winds causing rise and fall in the water level. In summer, after a certain period of calm days, the temperature of the upper layer often rises to 14–16°, but in the first hours after a strong north-western wind it drops to 4° along the leeward coasts, which is indicative of the emergence of deep cold waters to the surface.

At the end of the summer warming-up period the temperature layering of the upper zone becomes more stable than at the outset of this period, which is probably explained by the subsidence of the metalimnion below the zone of intense mixing.

Owing to the disturbances of layering the warming up of the upper zone proceeds very unevenly, by leaps and bounds. These disturbances are especially manifest near the shores.

As has been noted above, the surface temperature of open deep-water regions is influenced by currents from river mouths and extensive shallows. The zone of influence of the waters of the Selenga shallow takes up a large area spreading in both directions from the delta of the river. They warm up the superficial waters of the neighbouring deep-water regions in summer and cool them in autumn.

In the summer warming-up period the superficial layers of open Baikal reach a maximum of 14–16° in warm years and not more than 12–13° in colder years. In South Baikal this maximum falls on the end of August or beginning of September, after which the autumn cooling sets in. Stratification can still be observed in the upper zone at the outset of the cooling, but it is often broken. The depth where the metalimnion begins increases in this period to 30–50 m, its lower limit lying at about 100 m. The autumn homothermy grows gradually. Around the middle of October the temperature of the upper layer of 0–50 m in South Baikal evens up at 9–10°. Later on the layer of homothermy spreads deeper and deeper, with a simultaneous decrease in temperature. In South Baikal full autumn homothermy usually sets in in the course of the last ten days of November or at the beginning of December, at 3.2–3.7°, and in North Baikal a little earlier.

After the onset of autumn homothermy, the temperature of the upper layer goes on decreasing for four to six weeks, until the formation of the ice cover, and the reverse stratification characteristic of the entire late-autumn and sub-ice periods is established.

The amplitude of annual fluctuations in the temperature of the surface waters in South Baikal reaches 12–16° (from zero to 12–16°) with the maximum at the end of August and beginning of September and the minimum in January. In the 10 m layer the lowest temperature (0.2–0.5°) is reached in January and the highest (12–13°) in August and the first half of September. At 20–25 m the periods of the thermal minimum and maximum are about the same, while the amplitude decreases to 9–11°, in some years reaching 12–13°. At a depth of 50 m the maximum of temperature is usually attained later, in autumn, and the amplitude dwindles to 6–8° in warm years and 4.5–5° in cold years. At 100 m the lowest temperature (2.2–2.8°) is observed in April and the highest (6.0–6.5°) in autumn, with the amplitude not exceeding 4–4.3°. Finally, at a depth of 200–250 m, as has already been noted, the temperature does not go beyond the 3.5–4.5° limits the year round. Seasonal changes are sometimes observed deeper, to 300 m and more, but their amplitude is extremely small.

Table VI shows mean monthly temperatures of the superficial layers in South Baikal. We can see from it that January and February are the coldest months and August is the warmest month there.

G. Y. VERESHCHAGIN (1949) held, on the basis of isolated measurements, that at the very bottom of the abyssal region of Baikal the water temperature rises again within a few hundredths of a degree,

Table VI.

Mean Monthly Temperatures in °C of the superficial layers of water in the Maritui Area (after VERESHCHAGIN, 1949).

Months	I	II	III	IV	V	VI	VII	VIII	IX	X	XI	XII	Year
Temperature	0.17	0.01	0.11	0.41	1.95	4.29	9.32	12.75	10.18	7.98	4.64	2.41	4

which supposedly testified to geothermic warming of the bottom layers owing to the deep position of the bottom of Baikal in the lithosphere. But this supposition has to be confirmed by measurements with the use of highly sensitive instruments.

The amount of warmth in various sections and at various depths of Baikal is essential to its inhabitants. It is a powerful factor bearing upon their distribution. We have tried to count the average annual amount of warmth in day degrees by using the material on South Baikal available to us. The results of these computations are shown in fig. 22. What first strikes the eye is the important fact that the nearshore waters of the open lake (0–20 m depths) differ very little from the deeper zones in annual accumulated temperature. This means that the waters of open Baikal present, thermally, a single whole from the shores down to the greatest depths. At the same time they differ sharply from the sors and shallow sheltered

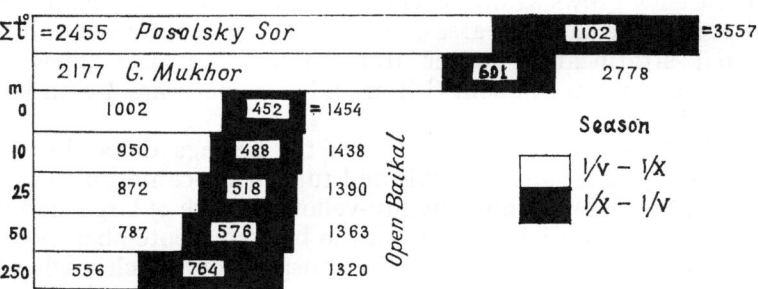

Fig. 22. Accumulated temperature in Baikal's sors, gulfs and open waters (in day degrees). Original.

gulfs, in which annual accumulated temperature is 2 to 3 times higher than in the open littoral of the lake. There are also sharp differences between shallow gulfs and open waters as regards the seasonal distribution of heat. This is certainly to be taken into account in discussing the problem of the evolution of the Baikalian fauna and its immiscibility with the fauna of sors and adjacent lakes.

The transparency of the open waters of Baikal and its seasonal changes are shown in fig. 23. In open deep-water regions the optimal transparency is observed from the end of November (the period of autumn homothermy) till January-mid-February. At that time the white disk can be seen up to a depth of 25–30 m and in regions re-

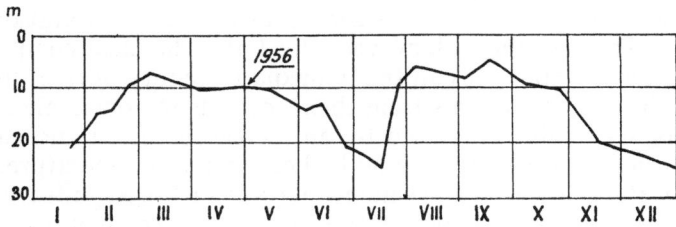

Fig. 23. Seasonal changes in Secchi disc transparency in the open waters of South Baikal in the area of the Bolshiye Koty, 1.5 km offshore. Original.

moved from the shores and river mouths to 40–41 m. In shallows and opposite to river mouths the water transparency is much lower than in deep-water regions. During the vernal development of algae the transparency gradually diminishes, and in March-May the white disc cannot be seen at 8–10 m even in deep-water regions. In years with particularly high crops of algae water transparency can fall to 6–8 m in April-May, i.e., still under the ice cover, and to 2–3 m in shallows. As the vernal algae die out and homothermy sets in, transparency grows again, reaching towards the end of June 20–25 m and even more in some areas. In summer, with the onset of direct thermal stratification, water transparency decreases anew, the summer minimum of about 6–8 m and in some years 4–5 m being reached in August.

In shallows, especially opposite to the Selenga delta, the water transparency after strong winds and turbulence can drop to 1–2 m and less, the water turning muddy-yellow because of large amounts of suspended material raised from the bottom. September sees the beginning of the autumn increase in transparency, which reaches the maximum towards December, i.e., the period of autumn homothermy.

6. Chemistry

The Baikal water is very poor in mineral constituents. The dry residue after evaporation does not exceed 95 mg/l in the superficial layers and 100 mg/l in deep waters. In their composition the waters of Baikal belong to the group of alkaline waters of low mineral content.

Table VII refers to the average chemical composition of the open waters of Baikal and, for comparison's sake, the analyses of the water of its affluents.

The water of the affluents somewhat differs chemically from the average for the lake in having a higher Ca, Mg, Fe, SO_4, Cl and HCO_3 content and being more easily oxidized. But these differences are insignificant. It should be noted that the chemical composition of affluent waters is highly variable owing to precipitation and freshets. These changes, however, affect the chemical composition of the Baikal waters only in the immediate vicinity of river mouths and in the belt of currents from them towards the open lake. These currents are easily recognised by an increased silicon and calcium content, lower transparency and higher summer temperature.

The water of the sors and shallow parts of sheltered gulfs receiving rivers has an increased content of Si, Fe, Ca and some other mineral components, especially during the sub-ice period. For example, the water of the Proval Gulf contains up to 22—27 mg/l of CaO, 7.1 mg/l of MgO and 1.3 mg/l of Fe. The water of the Mukhor Gulf in the Maloye More also has a higher mineral content than in the open lake (2.4—8.4 mg/l of SiO_2, 0.1—0.68 mg/l of Fe). The mineral

Table VII.

Average Chemical Composition of Water in Open Baikal and Its Affluents in mg/l.

	Open waters		Affluents	
	After VERE-SHCHAGIN (1949) for South Baikal*	After VOTIN-TSEV (1961)	After VOTIN-TSEV (1961)	After BOCHKA-RYOV (1959)
HCO_3'	63.5	63.5	79.3	72.8
SO_4''	4.8	5.2	6.7	7.0
Cl'	0.7	0.6—1.4	1.8	1.3
$Ca^{..}$	15.2	15.2	20.0	18
$Mg^{..}$	4.1	3.1	4.3	3.6
$Na^{.}$	3.9	3.8 }	5.1 }	4.6 }
$K^{.}$	2.3	2.0		
Si	SiO_2 1.5—5.5	Si 1.070	4.4	—
Al	traces	traces	—	—
Mn	—	0.0015	—	—
Fe, total	—	0.028	0.28	—
N	NO_3-0.19—0.62	N-0.045	—	—
P	PO_4-0.01—0.06	P-0.024	—	—
Oxidizability mg O_2 per litre	—	1.62	4.3	—
CO_2 free	0.44—5.28	1.49	—	—
O_2	14.4—9.6	11.64	—	—
N_2	22.4—16.8	—	—	—
Sum of ions	—	93.4	117.2	

* First figure stands for the superficial layers, second figure for deep waters.

composition of the water of big gulfs and the Maloye More is basically identical with that of the open waters of Baikal (BOCHKARYOV, 1935).

Indicated in table VIII are changes in the chemical composition of the open waters of Baikal with increase in depth (after VERE-SHCHAGIN, 1949). We see that chlorides, sulphates, bicarbonates, potassium and gaseous nitrogen are distributed evenly throughout the mass of water. Only the bottom layer about 1 m in depth has a markedly increased mineral content owing to the influence of the soil. The vertical distribution of these components is not subjected to seasonal changes, either. Stratification and sharp seasonal changes are manifest only in the content of such components vital for organisms as compounds of P, N, Si, Fe as well as O_2, CO_2 and some others. Let us note peculiarities in the vertical distribution and seasonal changes of these important chemical components of the

48

open waters of the lake, the dynamics of which have been studied in recent years by K. K. Votintsev (fig. 24).

P in phosphates. The maximum P content of elemental phosphorus of 30—40 mg/m³ in the 0—50 m layer is reached at the end of December and beginning of January. As a result of consumption by algae, whose vernal maximum of development is attained in

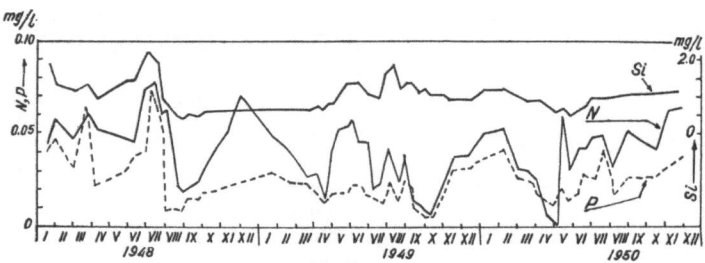

Fig. 24. Seasonal changes in the content of biogenous elements in the open waters of South Baikal in the area of the Bolshiye Koty in 1948—1950. After Votintsev, 1952.

April, the P content decreases to 13—18 mg/m³. The maximum of 20—30 mg/m³ is reached in June, during homothermy, and the summer minimum of 9—20 mg/m³ towards August. In autumn the P content increases again.

N in nitrates behaves likewise. In winter the nitrates content in the 0—50 m layer rises to 45—50 mg/m³ of elemental nitrogen, and in some years to 70—80 mg/m³. In spring it diminishes, down to analytical zero in years with a particularly high crop of diatoms. In June the nitrates content rises again, reaching 7—20 mg/m³ of nitrogen. Down to a depth of 500 m nitrite and ammoniacal nitrogen is not found in analytically determinable amounts.

The Si content undergoes especially sharp fluctuations in diatom-rich years. For instance, in 1950, after the winter maximum of 1.26 mg/l, it dropped to 0.34 mg/l. Following this the silicon content gradually grows, again reaching 1.21 mg/l towards August. In diatom-poor years the seasonal changes in the Si content are less vivid. Beyond the trophogenous layer the Si content increases gradually to 2.5 mg/l at 1,200—1,600 m.

The above-indicated seasonal changes in the content of biogenic elements are detected down to 100 m.

The Fe content in the surface layers also varies broadly in the course of the year from 0.010 to 0.020 mg/l in summer and spring and to 0.060 mg/l in autumn. With increase in depth it also grows. Votintsev (1961) considers that Fe compounds in the waters of Baikal are chiefly contained in colloids and coarse suspensions.

Considerable seasonal fluctuations are observed in the manganese

Table VIII.

Changes in the Chemical Composition of the Water of South Baikal with Increase in Depth in mg/l (after VERESHCHAGIN, 1949, midsummer)

Depth in m	O_2	Free CO_2	Oxidizability in O_2	Silicates SiO_3''	Nitrates NO_3'	Phosphates PO_4'''	Free N_2	HCO_3'	Sulphates SO_4''	Chlorides Cl'
0	12.44	2.6	1.11	1.60	0.30	0.031	17.27	63.9	5.18	0.78
100	11.35	2.9	1.03	2.42	0.31	0.025	20.05	63.9	5.27	0.78
200	11.32	3.0	0.87	2.73	0.43	0.034	20.08	63.6	5.46	0.71
300	11.24	3.0	0.89	3.64	0.44	0.037	—	64.1	5.37	0.71
400	11.01	3.3	0.91	3.11	0.40	0.039	20.27	63.7	5.27	0.71
500	10.71	3.9	0.96	3.64	0.38	0.042	20.30	63.3	5.66	0.75
600	10.55	3.7	0.77	4.10	0.46	0.047	—	63.7	5.46	0.68
700	10.44	3.8	0.79	4.95	0.46	0.049	20.30	63.5	5.18	0.64
800	10.31	3.9	0.78	5.09	0.46	0.049	20.30	63.7	4.90	0.78
900	10.27	4.2	0.79	5.00	0.47	0.049	20.30	63.5	5.09	0.82
1,000	10.20	4.1	0.90	5.09	0.47	0.049	—	63.5	5.27	0.75
1,100	10.18	4.2	0.87	5.61	0.49	0.062	—	63.1	5.46	0.71
1,200	10.16	4.2	1.00	5.40	0.52	0.061	20.30	63.8	5.56	0.75
1,300	10.12	4.2	0.68	4.78	0.53	0.062	20.30	63.7	5.56	0.75
1,400	10.10	4.3	0.75	5.40	0.54	0.063	20.24	63.4	5.27	0.71

content. VOTINTSEV (1961) has established that in autumn and winter the manganese content increases, reaching the annual maximum in April. In 1948 and 1949 this maximum stood at 0.0023 mg/l and kept around this value throughout the summer. VOTINTSEV believes that the seasonal changes in the manganese content are also to a certain degree connected with biological processes.

In sors and shallow parts of gulfs the content of biogenous compounds drops sharply during the vegetation of planktonic and bottom plants. For example, in June—July nitrogen compounds in the southern part of the Maloye More disappear almost completely and the silicon content falls sharply (down to 0.17 mg/l of SiO_2).

The total content of organic matter in the 0—250 m layer of the open waters determined by means of oxidizability ranges in different years from 4.0 to 5.5 mg/l of O_2 (bichromatic oxidizability). This corresponds to 3—4 mg/l of dry ash-free organic matter. The highest organic matter content is found in the upper trophogenous layer. Its minimum content in the trophogenous layer is observed in the sub-ice period and its maximum content in April and October, i.e., following the spring and summer maxima in the development of phytoplankton. The predominant part is played in Baikal by oxidation-resisting organic substances (humic complex), averaging 3.82 mg of O_2/l in the superficial layers and 3.27 mg of O_2/l in deep waters (1,200 m).

The waters of Baikal are rich in oxygen throughout their tremendous mass. In the upper layers saturation nears 100%, and during the maximum in the development of phytoplankton oversaturation to 120% and more is observed. Even in the bottom layers of the deep region of the lake saturation in O_2 does not fall below 75—80%.

Table IX.

Vertical Distribution of Oxygen in the Area of the Bolshiye Koty in 1952 (in mg/l).

Depth	March 19	April 2	ɔ August 29	September 12
0	12.76	12.90	10.83	11.14
25	12.94	12.98	10.97	11.30
50	11.55	11.70	10.67	10.86
100	10.78	10.83	10.78	10.96
250	10.59	10.51	10.43	10.42
500	10.48	10.46	10.30	10.30
600	10.04	10.10	—	—
750	—	—	10.20	10.17
800	9.92	9.87	—	—
1.000	9.74	9.77	10.14	10.08
1.200	9.62	9.71	10.03	10.00
1.350	9.56	9.63	—	—

Table IX shows the vertical distribution of oxygen in the southern part of Baikal in the area of the River Bolshiye Koty at different times of the year, after K. K. VOTINTSEV & A. V. SAMARINA (1957).

In the deep zone the O_2 content does not change noticeably in the course of the year, the rate of decrease being very small, constituting only 0.02—0.07 mg/l per 100 m and rarely rising to 0.1—0.4 mg/l. Still, insignificant changes can be detected both in the course of one year and in different years, which evidently depends on the character of water circulation.

Seasonal changes in the oxygen content are most clearly expressed in the photosynthesis zone (fig. 25). In the period of the maximum development of phytoplankton, still under the ice cover (March-

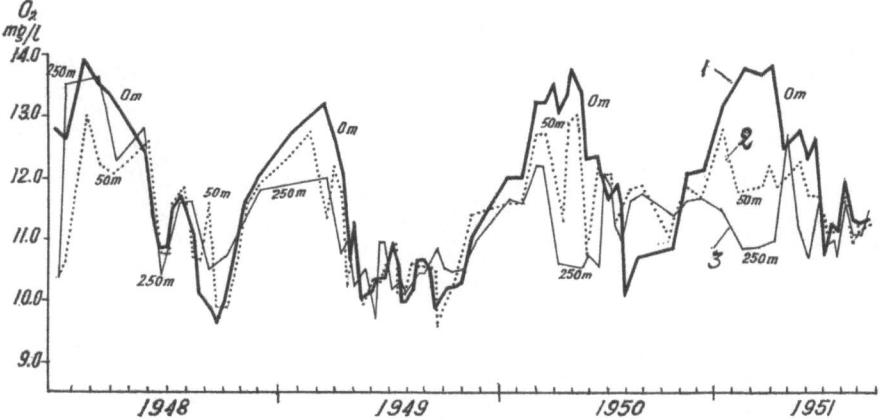

Fig. 25. Seasonal changes in oxygen content in the open waters of Baikal (the Bolshiye Koty) in 1948—1951; 1 - on the surface; 2 - at 50 m; 3 - at 250 m. After VOTINTSEV, 1961.

April), an oversaturation in oxygen is observed. After the ice break-up the oxygen content decreases towards the onset of homothermy and becomes more or less evened up throughout the upper zone, nearing 98—100% of saturation. In August, during the aestival development of phytoplankton, the O_2 content in the photosynthesis zone grows again and in autumn evens up in the whole of the upper zone. The amplitude of seasonal fluctuations in O_2 content in the surface layers (0—25 m) averages 3.67—3.52 mg/l. With increase in depth it gradually decreases, and no seasonal changes can be detected deeper than 250—300 m.

During the peak development of phytoplankton the photosynthesis zone is subjected to diurnal changes in O_2 content (fig. 26). For example, in the spring of 1953 the amplitude of these diurnal fluctuations in the surface layer reached 2.79 mg/l. At daytime

52

oxygen accumulates in this layer usually to 105—110% of saturation, dropping to 85% at night. Diurnal fluctuations are clearly manifest at a depth of 5 m but weak already at 10—25 m. On clear

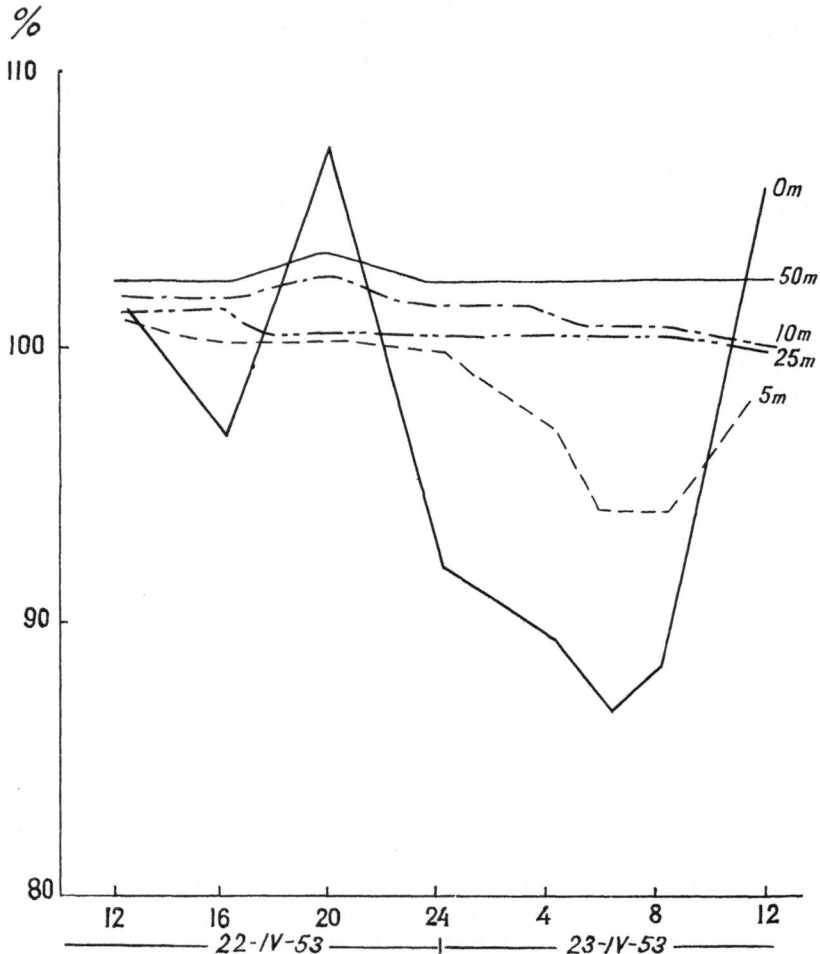

Fig. 26. Diurnal changes in oxygen content (per cent of saturation) in the open waters in Baikal (the Bolshiye Koty) on April 22—23, 1953. After VOTINTSEV, 1952.

calm days the oxygen content in the littoral can reach 150—175% of saturation.

The total amount of O_2 consumed annually on account of biochemical oxidation processes in the upper zone of the lake (0—250 m) was, according to observations by K. K. VOTINTSEV, 1,926 g/m² in

1951, 2,081 g/m² in 1952 and 2,919 g/m² in 1953, a year very rich in phytoplankton. Since biological consumption of oxygen takes place throughout the water mass down to the bottom, the total amount of O_2 expended on the oxidation of organic matter in the mass of water of the lake must be several times higher than its supply by photosynthesis. Consequently it is the atmosphere that is chiefly responsible for the annual supply of O_2 to the waters of Baikal (VOTINTSEV, 1961).

The behaviour of free CO_2 in the open waters of the lake is opposite to that of O_2. With increase in O_2 the CO_2 content often falls to the point of complete disappearance. Accordingly, the CO_2 content in the photosynthesis zone varies with seasons. Between 1948 and 1950 the mean annual content of CO_2 in the superficial layers was 1.11 mg/l in 1948, 1.26 mg/l in 1949 and 0.92 mg/l in 1950, the mean figure for the three years being 1.11 mg/l. With increase in depth the CO_2 content grows to 2.0 mg/l at 250 m and 5—6 mg/l at 1,000—1,500 m. The highest CO_2content in the upper layers is observed in January—February (up to 2.0—2.5 mg/l), dropping to 1.0—1.2 mg/l towards April and to 0.5—0.3 mg/l in phytoplankton-rich years. After the ice break-up the CO_2 content in the upper layers rises to 1.5 mg/l and drops sharply again in summer, in the period of the aestival development of phytoplankton. During water bloom and a sharp decrease in the free CO_2 content monocarbonates appear in the upper layers, reaching 0.5—1.0 mg/l, and in the littoral 6—8 mg/l (VOTINTSEV, 1961).

The deep waters of Baikal always contain carbonic acid the presence of which, as VOTINTSEV points out, results in an intense dissolution of carbonates. This explains the extreme poverty of the bottom sediments of Baikal in carbonates.

Table X.

Average Chemical Composition of Some Types of Soil in South Baikal according to 1948—1950 Analyses, in % in Air-dried Samples (after VOTINTSEV, 1961)

	Coarse aleurites	Diatom oozes
Number of samples	4	14
Depth of sampling in m	270—300	370—1,390
SiO₂	73.18	68.31
Fe₂O₃	4.36	6.72
Al₂O₃	12.96	13.19
CaO	1.57	1.78
MgO	0.44	0.83
P	0.088	0.096
Mn	0.047	0.103
N	0.17	0.19
Loss on ignition	4.95	3.56

The pH changes accordingly with changes in the oxygen content. In strong oversaturation in O_2 the pH shifts into the alkaline region towards 8.4—8.6; during a deficit in O_2 and increase in CO_2 the pH nears 7. In the course of the year the pH varies between 7.6 and 8.6 on the surface and 7.1 and 7.2 at a depth of 1,000—1,200 m (VOTINTSEV, 1961).

H_2S is not present in the open waters of the lake, since conditions for its formation are lacking even in the bottom layers of the deepest sections of the lake.

The chemical composition of bottom sediments in the deep-water regions of Baikal is shown in table X.

The sediments are for their most part formed by suspended substances brought by rivers and settling on the bottom. The products of the wear of the shores are evidently of secondary importance. According to KNYAZEVA (1954), all affluents of Baikal deliver into it more than 2.6 million tons of suspended material annually. K. K. VOTINTSEV considers that this supply of suspended substances by rivers should be of the order of 3.7 million tons, with the Selenga accounting for 2.9 million tons. These suspensions are admixed with substances of organic origin: diatom valves, siliceous spicules of sponges, remnants of the chitin of crustaceans, etc. In the deep-water regions these admixtures, diatom valves in particular, strongly influence the composition of the soil. According to data

Table XI.

Chemical Composition of Bottom Waters and Deep Mud Solutions in mg/l (after VOTINTSEV, 1961).

	Bottom waters	Interstitial waters
HCO_3	71.32	73.53
N in ammonia	0.157	0.563
N in nitrates	0.057	0.077
P in phosphates	0.065	0.209
Si	3.52	15.14
Fe	0.054	0.125
Oxidizability	6.85	18.73
pH	7.26	7.21

supplied by KNYAZEVA (1954), the diatom oozes of Baikal contain an average of 20.1% of authigenous SiO_2 and some samples yield up to 61.49%. All these products, being slowly dissolved in the waters saturating the soil, sharply increase the mineralisation of the interstitial water (table XI).

The interstitial waters have strongly increased contents of P,

Fe, N, Si, Al, Fe and Mg, and ammonium nitrogen is also found in them, whereas the nitrate N content is but slightly higher than in the water. The oxidizability of the interstitial waters is 4 to 6 times higher than that of the superficial waters. The content of biogenous elements in the interstitial waters increases from coarse-grained soils to silts.

The increased mineralisation of the soil water has a marked influence on the mineral content of the bottom layer of water approximately one metre thick. It can be supposed therefore that little exchange of waters of the layer lying directly above the bottom in the deep regions of Baikal with the upper layers takes place.

II

FAUNA AND FLORA

1. Systematic Composition of the Fauna

This review of the Baikalian fauna presents all animals in systematic groups, providing qualitative and quantitative characteristics of each group and the main data on the history of its study, its distribution in Baikal, comparative significance, connection with related groups from the waters of other countries, and, if possible, its origin.

Protozoa

Rhizopoda. No bottom rhizopods have been found in open Baikal, but gulfs and sors* contain rhizopod species common in Holarctic bodies of water, among them *Difflugia pyriformis* PERTY, *D. acuminata* EHRB., *D. lemani* BLANC., as well as *Arcella vulgaris* EHRB. and *Centropyxis aculeata* EHRB. (ROSSOLIMO, 1923).

A foraminifer test from the family Rotaliidae or Anomaliidae has been discovered in Tertiary deposits on the coast of the lake in the area of the River Polovinka.

Adjacent waters are inhabited by the heliozoans *Actinosphaerium* sp., which have not been found in Baikal proper.

Flagellata. Not a single form of colourless flagellates has so far been found in the open waters of the lake. The green flagellates are known to be represented by about 25 species, which are dealt with in the chapter on the flora.

Sporozoa. In 1910 B. A. SVARCHEVSKY found the gregarine *Lankestheria* sp., an intestinal parasite of the turbellarian *Sorocelis* sp., and gave a detailed description of its life cycle (SVARCHEVSKY, 1910).

V. N. TSVETKOV (1928) described two Baikalian species of gregarines, *Gregarina acanthogammari* Zw. and *Gregarina baicalensis* Zw. (fig. 27), found in the intestinal tract of Baikalian gammarids.

Baikalian Cottidae are often found to be affected by abscesses and tumours on the body and viscera, particularly in the area of the eyes. Some of these abscesses and tumours, especially those on the eyes, are caused by sporozoans. Recently two new species of these sporozoans were described, *Myxobolus taliewi* DOG. and *M. spatulatus* DOG., which are parasitic on miller's-thumbs of the Cottidae family (DOGEL and others, 1949; DOGEL & BOGOLEPOVA,

---

* "Sor" is the local name for a shallow gulf (firth) situated near a river mouth and separated from Baikal by sandy spits.

1957). Species of the genus *Myxobolus* are also said to parasitize the ide (kidneys and gills). The same authors have established the presence of two *Myxidium* sporozoans, *M. lieberkühni* BÜTSCHLI in

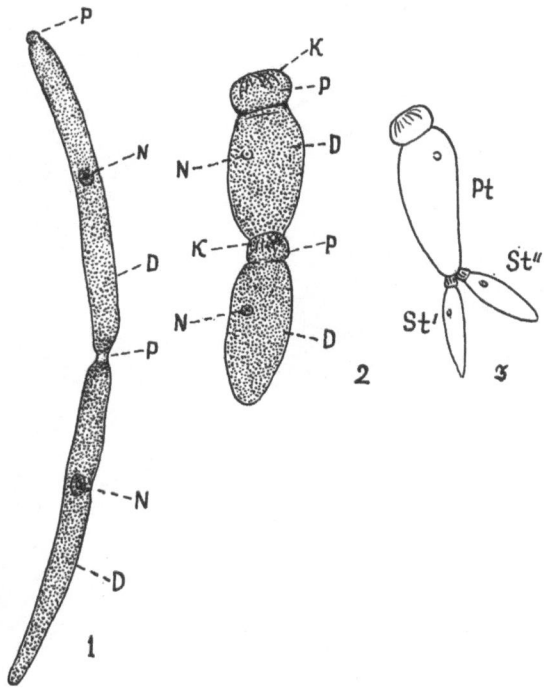

Fig. 27. Gregarines. 1. - two specimens of *Gregarina acanthogammari* ZWETK., body length 0.5—0.6 mm; 2 - two spicimens of *Gr. baicalensis* ZWETK., body length 0.35—0.40 mm; 3 - a colony of three specimens of the same species. P - protomerite; D - deutomerite; K - radial furrows; N - nucleus; St. - satellites. After TSVETKOV, 1928.

the pike and *M. perniciosus* DOG. in the gall bladder of *Comephorus*, and two *Henneguya* species, *H. baicalensis* DOG. in the urinary bladder of *Asprocottus* and *Henneguya* sp. in the ide *(Leuciscus idus)*. *Henneguya* sp. parasitizes the gwyniads, forming numerous comparatively large whitish sporocysts on their viscera.

Four of the eight species of sporozoans found in Baikalian fish are endemic. Two of them, *Myxidium perniciosus* and *Henneguya baicalensis*, are considered by DOGEL to be palaeoendemics brought to Baical from the sea by Cottidae species. Sporozoans related to these species are parasitic on Cottoidei species abiding in the Arctic Ocean.

Infusoria. Infusorians have been studied much better than other classes of Baikalian protozoans. Thanks to GAYEVSKAYA, SVAR-

58

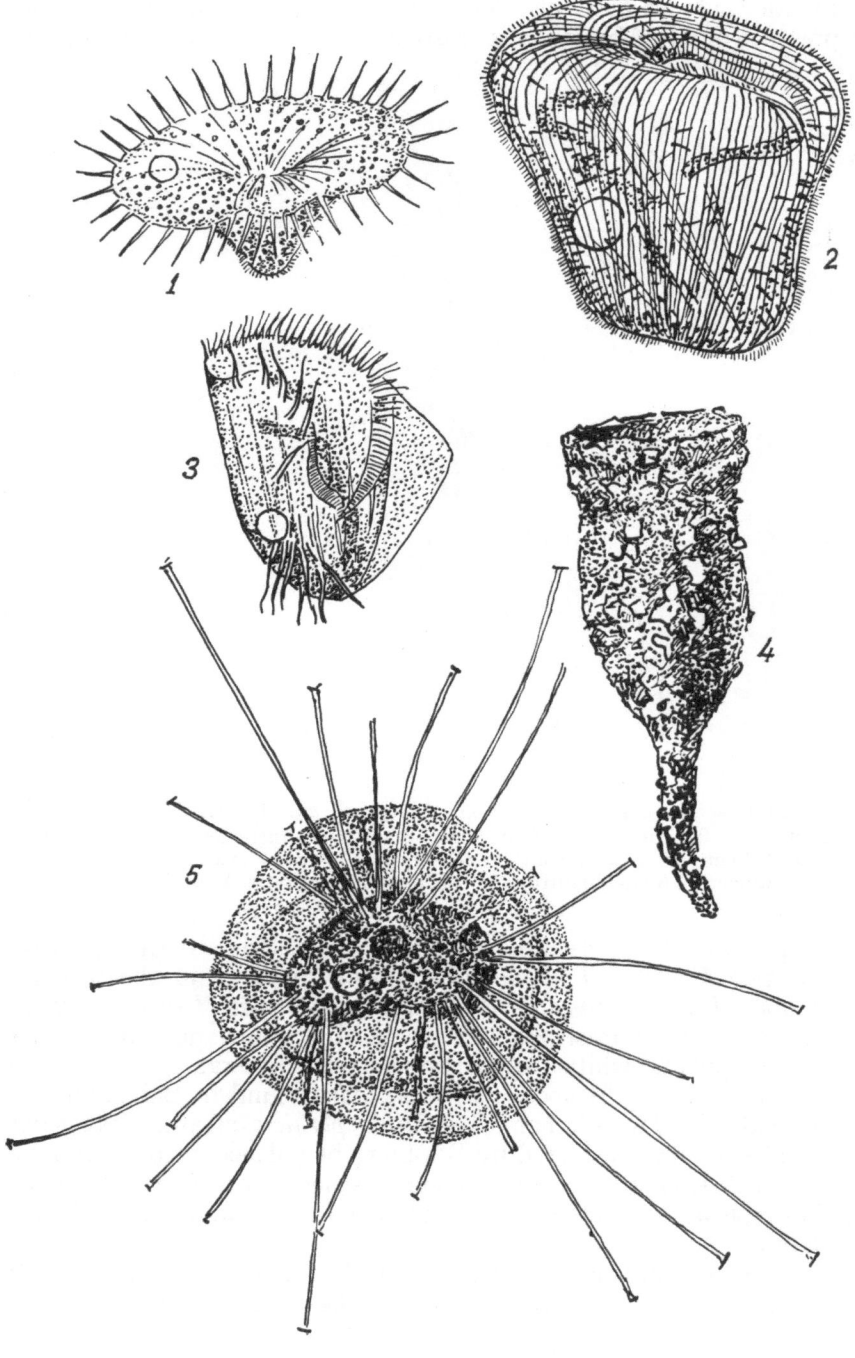

Fig. 28. Infusorians of Baikal's open waters: 1 - *Liliimorpha viridis* GAJEW., body length 0.11 mm; 2 - *Marituja pelagica* GAJEW., body length 0.14 mm; 3 - *Euplotes harpa baicailensis* GAJEW.; 4 - a test of *Coxliella* sp.; 5 - *Mucophrya pelagica* GAJEW., body length 0.11 mm. After GAYEVSKAYA, 1933.

CHEVSKY, ROSSOLIMO and KHEISIN, Baikal is now known to have about 300 species of Ciliata and Suctoria free living in the water or as commensals on gammarids. According to N. S. GAYEVSKAYA

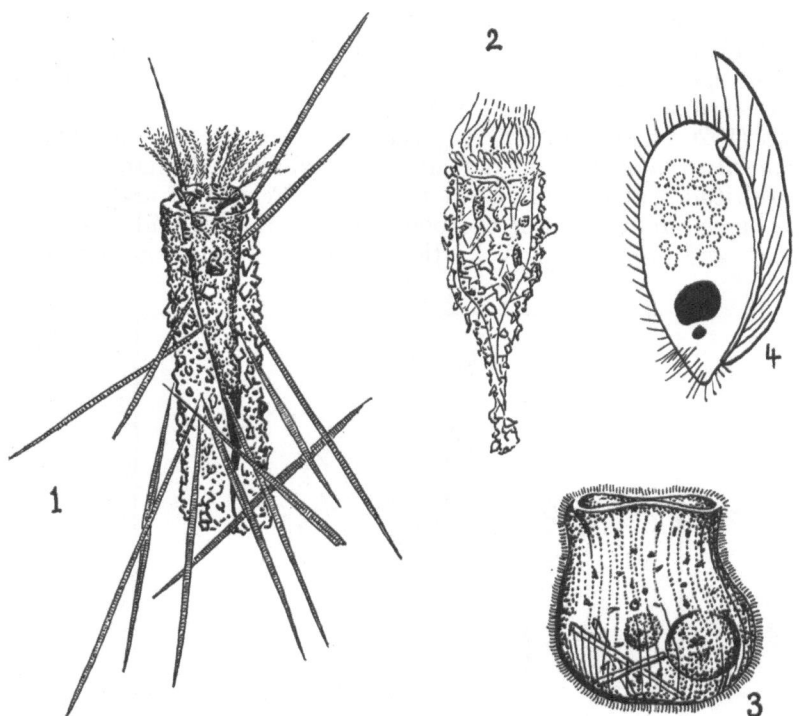

Fig. 29. Infusorians of Baikal's open waters: 1 - *Tintinnidium fluviatile* f. *cylindrica* GAJEW., body length 0.125 mm; 2 - *Tintinnopsis davidoffi* var. *cylindrica* f. *minima* GAJEW.; 3 - *Spathidiosus bursa* GAJEW., body length 0.095—0.110 mm; 4 - *Tiarella baicalensis* CHEJS., body length 0.05—0.07 mm. 1—3 after GAYEVSKAYA, 1933, 4 after KHEISIN, 1931.

(1929), 1933), Baikal is populated by 77 species of Holotricha, 49 species of Spirotricha, 44 species of Peritricha, 1 species of Chonotricha and 21 species of Suctoria — 192 species in all, from which 42 have been described as new forms and 10 new genera and 3 new families have been established. B. A. SVARCHEVSKY (1928, 1928—30) described an additional 86 new forms of infusorians parasitic on gammarids, although the validity of most of these species is doubtful. Besides, Y. M. KHEISIN (1930a, b, 1931, 1932) and L. L. ROSSOLIMO (1926) described more than 20 species of infusorians parasitic on various Baikalian invertebrates. Infusorians of the genus *Ichthyophthyrius* parasitize the integument and eyes of miller's-thumbs and other Baikalian fish. Most of the species of

60

free-living infusorians found in Baikal are said by N. S. GAYEVS-
KAYA to live in well-warmed sheltered shallows, bays, gulfs, sors,
near river mouths, etc. They are either common in fresh waters of
the Palaearctic in general, or have very close relatives in them.
But a considerable number of species also live in open Baikal, its
coastal belt and at various depths in the pelagic zone. In the coastal
belt of open Baikal infusorians are abundant in algal growths and
in areas more or less protected against turbulence. L. L. ROSSO-
LIMO (1926) described *Ophryoglena pyriformis* ROSS. and *Ophryoglena intestinalis* ROSS. as parasites of Baikalian turbellarians.

Multitudes of infusorians of the orders Peritricha and Suctoria
live on different parts of the bodies of gammarids and sometimes
worms and other animals. Particularly numerous among them are

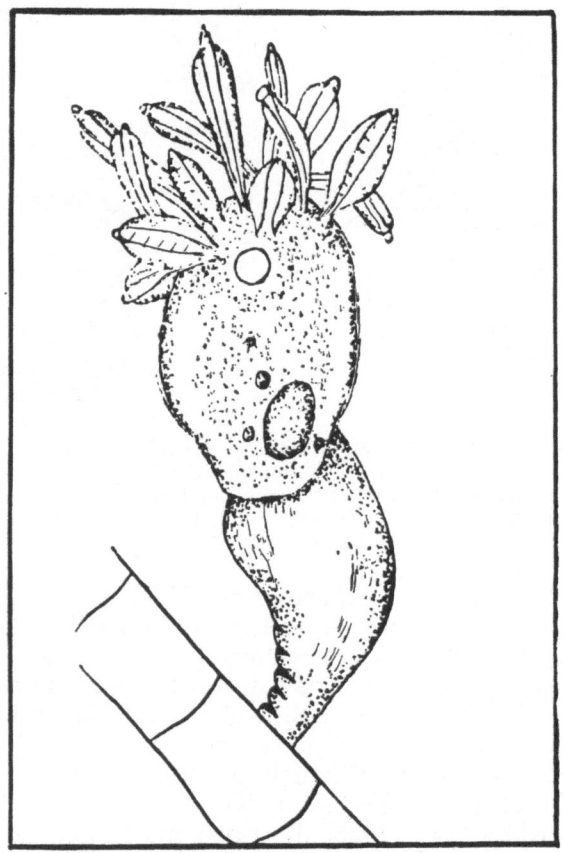

Fig. 30. Infusorian: *Dactylophrya collini* GAJEW., body length 0.04—0.05 mm.
After GAYEVSKAYA, 1933.

species of the genera *Vorticella, Vaginicola, Cothurnia, Spirochona, Dendrocometes* and *Dactylophrya collini* GAJEW. (fig. 30).

About 30 species of infusorians can be regarded as particularly characteristic of the pelagic zone of open Baikal, where they live at different depths. Many of them proved so peculiar that even new genera and families had to be established for them. The predominant part in the pelagic zone of open Baikal is played by *Liliimorpha viridis* GAJEW. (fig. 28,1) of the endemic family Liliimorphidae, and, of other families, *Longitricha flava* GAJEW. (endemic genus and species), *Marituja pelagica* GAJEW. (endemic genus and species, fig. 28,2), *Mucophrya pelagica* GAJEW. (endemic genus and species, fig. 28,5), *Prorodon morula* GAJEW. endemic genus and species, *Sulcigera comosa* GAJEW. the endemic family Sulcigeriidae, then *Tintinnidium fluviatile* f. *cylindrica* GAJEW. (fig. 29,1), *Tintinnopsis davidoffi* var. *cylindrica* DADAY (fig. 29,2), *Spathidiosus bursa* GAJEW. (endemic genus and species fig. 29,3), and *Prorodon garganellae* GAJEW. These species appear in early spring, under the ice. In summer the open waters of Baikal contain large quantities of *Vorticella* sp., which settle on planktonic algae, then *Didinium nasutum* O.F.M., *Bursella* sp., *Amphileptus* sp., and others.

In the opinion of GAYEVSKAYA (1932, 1933), close to 90% of the 192 species of infusorians known in Baikal are purely fresh-water or mixed fresh and salt-water species, while about 20 are marine salt-water species (species of *Coxliella*, fig. 28,4; Tintinnoidea, etc.). Therefore she finds it possible that the sea participated in forming the fauna of Baikal. It should be noted, however, that most of what GAYEVSKAYA considers to be marine species live only in Baikal's shallows, gulfs, sors and near river mouths. Some species do live in the open parts of the lake, but chiefly as parasites of deep-dwelling gammarids. Since the infusorian fauna of Siberian fresh waters has not been studied sufficiently, it is doubtful whether these species are immigrants from the sea.

It is to be particularly emphasised that the most widespread endemic species of infusorians from the pelagic zone of open Baikal seem to have no close relatives either among fresh-water or marine forms.

Among the parasitic and commensal infusorians of Baikal, 11 species were described by ROSSOLIMO (1926): 2 species of the genus *Anoplophrya*, 4 species of the genus *Mesnilella*, 2 species of the genus *Radiophrya* (a new genus), 1 species of the genus *Lada* (all of them from oligochaetes), and 2 species of the genus *Ophryoglena* (from turbellarians). Besides, about 15 species of infusorians were described by Y. M. KHEISIN (1930, 1931, 1932) from oligochaetes and mollusks. Of particular interest among them are infusorians of the Ancistridae and Boveridae families, found by him in the mantle cavity of Baikalian mollusks. In the Ancistridae, the new genera and species *Ancistrina ovata* CHEJS. and *Ancistrella choanomphali* CHEJS. were described, the former parasitizing the mantle cavity of the mollusks *Benedictia* and *Choanomphalus* and the latter *Choanomphalus*. In the Boveridae the new genus and species *Tiarella baicalensis* CHEJS. was described, living in the mantle cavity of *Benedictia* (fig. 29,4).

Y. M. KHEISIN is inclined to regard all of them as marine forms, since their relatives live only in marine organisms (mollusks and holothurians). But this argument is not conclusive, for species of infusorians very closely related to these forms live in the mantle cavity of such widespread fresh-water mollusks as *Planorbis*, *Unio*, *Anodonta* and *Sphaerium*.

DOGEL et al. (1949, 1957) pointed to the presence of the infusorians *Trichodina domerguei baicalensis* DOGEL and *Glossatella* sp. on the gills of miller's-thumbs (Cottoidei).

In recent years DE PUYTORAC (1959) undertook a comparative study of the infusorians parasitizing the oligochaetes of the Ohrid and Baikal lakes. In his opinion, those of them which are parasitic on oligochaetes of the Lumbriculidae family from both lakes are closely related and present relics of a very remote past which have retained some primitive characters and changed autochthonously parallel with the evolution of the Lumbriculidae in the two lakes. The genus *Radiophrya* established in Baikal is also representated in Lake Ohrid. Both lakes are inhabited by species of the genus *Mesnilella*, parasites of the Lumbriculidae. The genus *Juxtaradiophrya* from Ohrid is also close to these genera. All of them seem to be of a common origin. STANKOVIČ (1959) regards this group of infusorians parasitizing oligochaetes as a relic fauna which has survived in Ohrid and Baikal since Tertiary time. The genus *Ochridanus* from Ohrid is close to the Baikalian genus *Anthoniella* (?). The predominance of species of the subfamily Radiophrynae in both lakes was also stressed.

Spongia

The sponges of Baikal have been studied by W. DYBOWSKY (1882), B. SUKACHOV (1895), B. SVARCHEVSKY (1901, 1923a, 1925), N. ANNANDLE (1913), P. REZVOI (1936), M. KOZHOV (1930) and S. MARTINSON (1947). They are represented in the lake by several species belonging to three genera. Although differing markedly one from another, they all constitute one group singled out as an independent family, Lubomirskiidae, belonging to the order Cornacuspongidae. The Lubomirskiidae differ from the Spongillidae in that their skeleton is stronger and forms a more regular lattice and that the ends of their spicules are more deeply embedded in spongin. As distinct from the Spongillidae, the Baikalian sponges do not develop internal buds (gemmules) for asexual reproduction, but they can multiply by means of the so-called sorites evidently presenting parthenogenetic eggs from which free-swimming ciliate larvae (planules) develop.

Now for the most characteristic features of the composition of separate species.

The genus *Swartschewskia*. These sponges resemble minute graceful whitish papillae, incrustations or caps 1 to 3 cm in diametre (fig. 31, 33, B), often with a single osculum at the apex. The external tissue (fig. 35,1) is thin, hard and brittle, the internal tissue (inside the papilla or cap), soft and loose. The spicules (fig. 34,A) are thick and short, straight or slightly curved, with blunt rounded ends.

The genus is represented by the sole species *S. papyracea* Dyb., which lives on stones in the littoral and sublittoral zones.

The genus *Lubomirskia* is characterised by well defined longitudinal and transversal tufts of spinulated spicules sharpened at both ends (fig. 34,C, 35,2,3,5). The genus comprises three species: the Baikalian bark-like bright-green sponge *L. baicalensis* Pallas (fig. 32, 32a), the bark-like *L. fusifera* Souk. with loose soft tissue, and *L. abietina* Swartsch. (fig. 35,2), also a barklike sponge with a feeble skeleton.

Fig. 31. Baikalian sponge, *Swartschewskia papyracea* Dyb., with the osculum in the centre. Diametre of sponge 4—6 cm. Original.

64

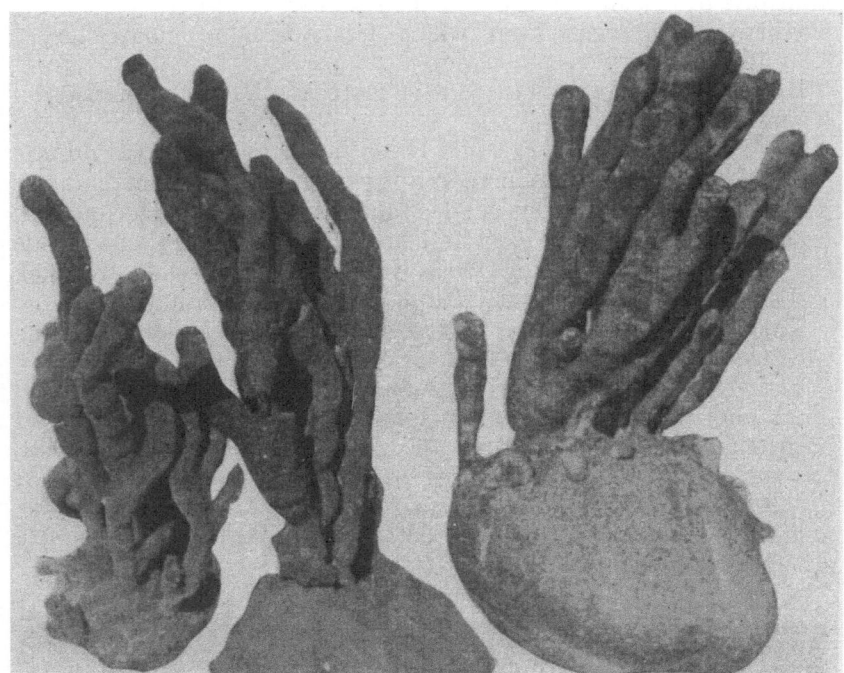

Fig. 32. *Lubomirskia baicalensis* PALLAS from a depth of 2—3 m, height of branches 40 cm. Original.

Fig. 32a. *Lubomirskia baicalensis* PALLAS from a depth of 5—10 m, height of branches up to 70 cm. Original.

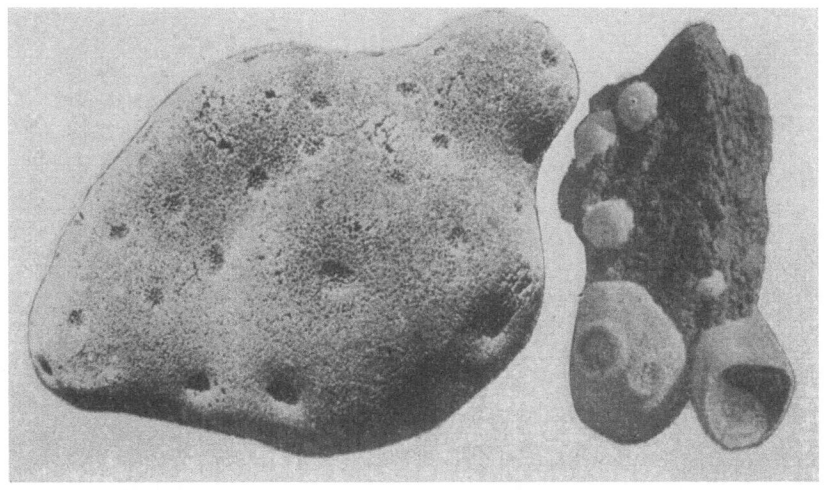

A   B

Fig. 33. A - *Baicalospongia bacillifera* DYB., diametre 8—12 cm. B - *Swartschewskia papyracea* DYB. at the top, *Baicalospongia bacillifera*. Below. Original.

L. baicalensis is particularly numerous. It lives on hard soil beginning with 1—1.5 m depths, where it is represented by a bark-like form which does not develop branches and covers bottom

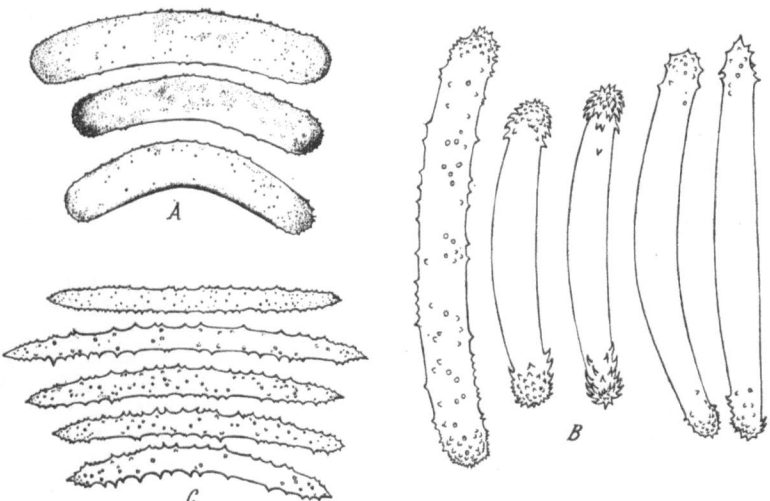

Fig. 34. Skeletal spicules (macroscleres) Lubomirskiidae. A - *Swartschewskia papyracea*, length of spicules 112—150 μ; B - *Baicalospongia bacillifera*, length of spicules 170—363 μ; C - *Lubomirskia baicalensis*, length of spicules 145—233 μ. After REZVOJ, 1936.

66

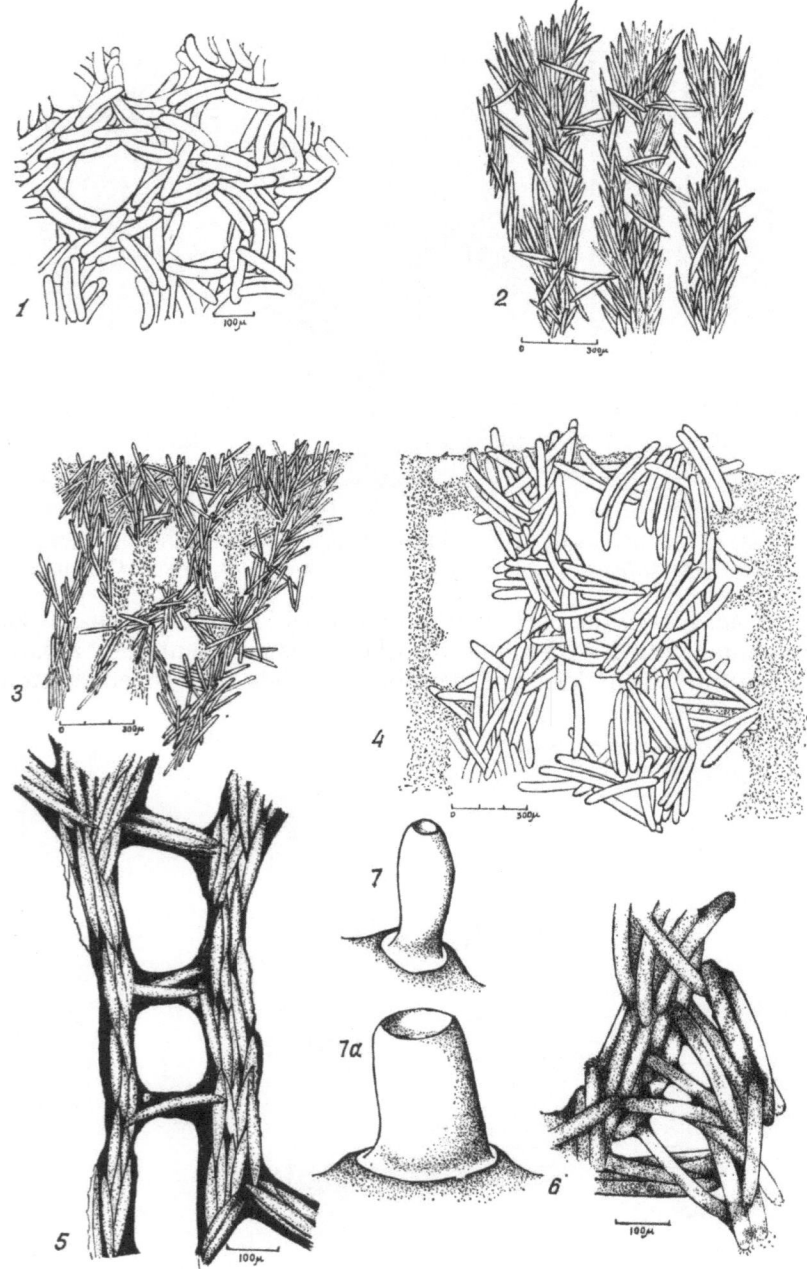

Fig. 35. Skeletons of Lubomirskiidae. 1 - *Swartschewskia papyracea* from the outside; 2 - skeleton of *Lubomirskia abietina;* 3 - skeleton of *Lubomirskia baicalensis;* 4 - skeleton of *Baicalospongia bacillifera;* 5 - structure of skeleton of *Lubomirskia baicalensis,* spongin covering spicules is shown with a point shading, compact masses of spongin are black; 6 - structure of skeleton of *Baicalospongia bacillifera.* After REZVOJ, 1936. 7, 7a - oscular tube of Lubomirskiidae, length of tubes are about 2 mm. After MARTINSON, 1947.

stones and flagstone in a bright-green carpet. It begins to form branches at 3—4 m depths and particularly in the belt of depths of 5—12 metres, with several thick cylindrical bright-green branches 60 to 70 cm long, often connected by bridges or fully intergrown, extending from the broad bark-like base attached to a stone or other hard substratum. Whole "forests" of these bright-green colonies can been seen on rocky ground in the littoral zone of Baikal. The sponge owes its colour to the microscopic alga *Zoochlorella* which lives symbiotically in its tissues. Its branches and bark-like base provide shelter to the emerald-green gammarid *Spinacanthus parasiticus* DYB. and many other minute animals, as well as oligochaetes, turbellarians, etc. Branching specimens of the Baikalian sponge often spread to depths of 25—30 m. It is found throughout Baikal and occurs, as a trailing bark-like form, in the upper section of the Angara.

The sponges of the genus *Baicalospongia* do not develop branches. They form massive crusts, columns and other growths, of pale-green colour and solid, hard consistency, attached to stones and other underwater objects (fig. 33). The skeletal spicules are pointed at both ends, spiny and rather irregularly spaced in the main tufts. This genus comprises two species: *B. bacillifera* DYB. (fig. 34b, 35,5,6) and *B. intermedia* DYB., which occur along the coastal belt of Baikal at depths from 4 to 500 m and more.

The Baikalian sponges prefer rocky grounds of open Baikal and are not to be observed in the waters of gulfs and bays, where they are replaced by widespread Siberian species from the genera *Spongilla* and *Ephydatia*. Life in the unusual conditions of these waters makes these species differ markedly from their relatives in the character of the coating of gemmules. This, and in particular the deformity and incomplete development of the siliceous needles covering the gemmules or their entire absence, prompted B. A. SVARCHEVSKY (1901) to describe them as independent species. But in fact they are merely varieties of widespread *Spongilla* and *Ephydatia* species (KOZHOV, 1947).

The Baikalian sponges are highly important consumers of microplankton, especially bacteria, and silicon, from which their skeletons are constructed (VOTINTSEV, 1948a).

Direct consumption of the sponges themselves by different animals is evidently insignificant, although there have been reports that sponges in Baikal can be used as food by sturgeons.

Prior to 1927 all the representatives of the family Lubomirskiidae were known only from Baikal; the statement by W. DYBOWSKY (1886a), repeated by many authors, that Lubomirskiidae species occur in the Bering Sea (the Komandorskiye Islands), is a result of a misunderstanding. Of great interest is the discovery by P. D. REZVOI (1927) of a new species of the genus *Baicalospongia* in the large but shallow lake Dzhegetai in the Western Sayan region.

The presence of sorites and some other traits make the Baikalian sponges somewhat akin to the marine siliceous-horny sponges of the family Haploskleridae. Some authors point out that Lubomirskiidae sponges are remotely reminiscent of the Caspian sponge *Metschni-kowia* and the Ohrid sponge *Ochridaspongia rotunda* ARNDT (ARNDT, 1937). The latter, in W. ARNDT's opinion, resembles in some res-pects the Baikalian genus *Baicalospongia* and also *Corticospongilla barroisi* TOPS. from Lake Tiberias in Palestine, *Pachydictym* from Lake Posso in Celebes and partly the genus *Nudospongia* from Lake Tiberias and Lake Tali-fu in South China. But this likeness is evi-dently due only to convergence.

W. ARNDT (1937) regards the whole of the above-mentioned group of sponges as Tertiary relicts. In the structure of the skeleton the *Ochridaspongia* resemble the Lubomirskiidae, but the latter are devoid of the subdermal spaces which exist in the *Ochridaspongia*. Like the Lubomirskiidae sponges, the *Ochridaspongia* do not develop gemmules and their embryos, formed sexually, are to be found throughout the year. S. STANKOVIČ (1960) considers the *Ochrida-spongia* to be a Tertiary relict having no close relation to the modern forms of *Spongia*.

In the opinion of P. D. REZVOI (1936), the Baikalian Lubomirskii-dae can be regarded as marine immigrants independent of the Spongillidae but evidently dating from a later period. Some authors (MAKUSHOK, 1925) assume that the Lubomirskiidae may have evolved from the Spongillidae in Baikal itself. Relying on the results of serological analysis, D. N. T. TALIYEV (1940) considers that the Lubomirskiidae are closer to the Caspian genus *Metschnikowia* than to the fresh-water family Spongillidae.

A final elucidation of the degree of kinship of the Baikalian sponges with marine and modern fresh-water sponges calls for a more detailed comparative investigation, especially in the history of development. No studies have yet been made of the sexual pro-cess, the formation of larval stages and the metamorphosis of the Baikalian sponges.

Fossil spicules of Baikalian sponges have been found in Tertiary deposits on the coast of Baikal (MARTINSON, 1936, 1938, 1948b) and in Tertiary (Miocene?) deposits in the Tunka trough close to South Baikal. Remnants of the Spongillidae have also been dis-covered there.

Coelenterata

The coelenterates are represented in Baikal only by species of the widespread fresh-water genus *Hydra* inhabiting gulfs and bays deeply indenting the coast. Among them B. A. SVARCHEVSKY (1923b) described a new Baikalian species, *Hydra (Pelmatohydra?)*

baicalensis SWARTSCH., which was found in *Potamogeton* growths in the shallow southern part of the Chivyrkui Gulf. This species has a slender stalky body reaching, when straightened out, 1—1.5 cm in length (without tentacles), It has 6 or 7 tentacles which are, as a rule, longer than the body.

Among algae in the littoral of open Baikal one can often come across a fairly large hydra resembling the above-mentioned *H. baicalensis*; its specific position has not yet been ascertained.

N. S. GAYEVSKAYA (1933) pointed to the presence in Baikal of *Hydra grisea* L. (= *N. vulgaris* PALL.)

Turbellaria

The Baikalian turbellarians include an unusual abundance of forms. The first data on them were published by GERSTFELDT (1859a) and GRUBE (1871—72, 1872, 1872—73), who described about a dozen species from the coastal belt of Baikal and the upper course of the Angara. A wealth of turbellarians was collected by the expedition of A. KOROTNEV in 1900 and 1901. They were studied by KOROTNEV himself (1912) and I. P. ZABUSOV (1901a, b, 1903, 1906, 1911).

The Baikalian Rhabdocoela became known as a result of the studies conducted by Academician N. V. NASONOV (1926, 1930, 1936) and I. A. RUBTSOV (1929). But despite the considerable amount of research work that has been performed, information about the turbellarians of Baikal remains incomplete and contradictory.

Among the Rhabdocoela, NASONOV discovered in Baikal 6 species and 11 forms belonging to the family Graffilidae and united in the new genus *Baicalellia*. All of them are minute worms (0.5 to 2 mm in length) with two black eyes, colourless or yellowish, pinky, brown, red or otherwise coloured. They live among algae in the coastal belt of open Baikal.

Characteristically, in the same genus *Baikalellia* NASONOV (1930) described one species from the brackish waters of the coast of the Peter the Great Gulf in the Sea of Japan; he also refers to this genus one species (*B. brevitulus* LUTHER) from the Gulf of Finland and Western Greenland. Besides, NASONOV established the presence in Baikal of *Macrostonum auriculatum* NASS. and *Polycystis angarensis* SIBIR. The latter form was described from the Angara near Irkutsk by O. A. SIBIRYAKOVA (1929).

In the summary given by BRESLAU (1933), the genus *Baicalellia* is designated to the family Provorticidae, whose representatives are known chiefly from seas, brackish waters and only partly from fresh waters.

The Rhabdocoela includes a very peculiar species, *Acrorhynchus baicalensis* RUBTZ., found at a depth of 80 m and described by I. A. RUBTSOV (1929).

70

It is a worm of a round-oval form, 2 to 3 mm long, with a small snout at the anterior end capable of protruding from a special capsule. When alive the worm is bright-red, with the exception of the brownish-yellow anterior end. The snout is of greyish-white colour. A closely related species, *A. fluvialis* SIB. from the Angara, was described by O. A. SIBIRYAKOVA (1929 fig. 36,C). The systematic position of *Acrorhynchus* is not quite clear. I. A. RUBTSOV designated it as belonging to the family Polycystidae, whereas Academician N. V. NASONOV (1930) considers that most probably it should be attributed to a special genus close to *Koinocystis* of the family Koinocystidae. Of the two species of the genus *Koinocystis* one is a marine and the other a fresh-water species.

The Rhabdocoela fauna of Baikal is certainly not restricted to the species mentioned above. Future studies will show that it is probably not poorer than in Lake Ohrid, which is known to have 25 species of this group, 44% of which are new (endemic) species

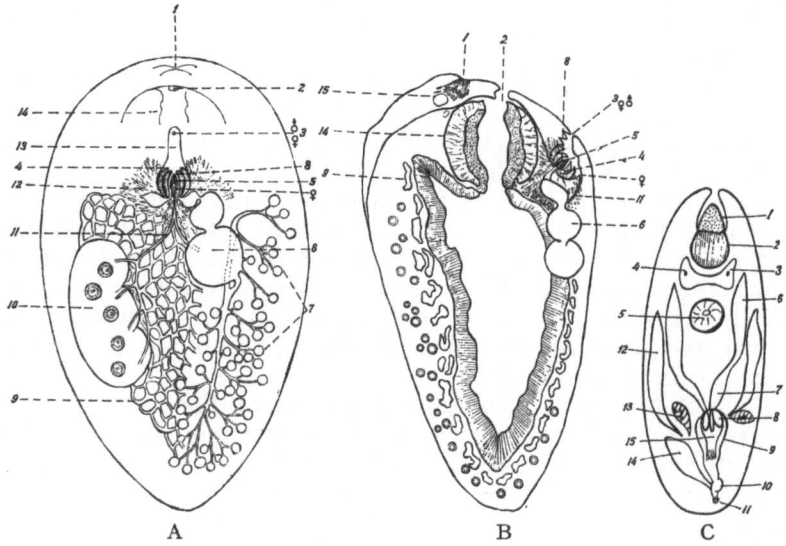

Fig. 36. Turbellarian from the order Alloecoela, *Baicalarctia gulo* FRIDM., body length up to 4 cm. The scheme of organisation. A - view from the ventral side; B - sagittal longitudinal section: 1 - frontal gland in the mouth area; 2 - mouth; 3 - genital orifice; 4 - penis; 5 - tongue of penis; 6 - vas deferens; 7 - testicles; 8 - receptaculum seminis; 9 - yolk sacs; 10 - ovaries; 11 - oviduct; 12 - glands; 13 - athrium genitale; 14 - pharynx; 15 - brain. After FRIDMAN, 1933. C - *Acrorhynchus fluvialis* SIB. The scheme of organisation: 1 - terminal cone of probosces; 2 - muscules of probosces; 3 – brain; 4 – eyes; 5 – pharynx; 6 – testicles; 7 - vesicula seminalis; 8 - ductus ejaculatorius; 9 - vesicula granulorum; 10 - athrium genitale; 11 - genital, foramen; 12 - yolk glands; 13 - ovaries; 14 - bursa copulathrix; 15 - chitinized penis. After SIBIRJAKOVA, 1929.

belonging to new genera (STANKOVIČ, 1960). No forms close to the Baikalian ones have been found there.

Academician NASONOV considered it possible to trace the origin of the Baikalian Rhabdocoela from marine forms. In doing so he referred to the heliotropism of the Baikalian species and the presence of representatives of the genus *Baicalellia* in sea and brackish waters (NASONOV, 1936). Without further and more detailed investigations the question of the origin of the Baikalian Rhabdocoela should be regarded as open.

Baikal has been found to contain an interesting form of the order Alloecoela, *Baicalarctia gulo* FRIDM. (fig. 36A, B), for which a new genus and even a new family, Baicalarctidae, had to be established. This worm possesses a very primitive organisation reminiscent of the order Holocoela, but at the same time it displays characteristics of more specialised groups, such as Rhabdocoela and Triclada (FRIDMAN, 1933). It lives in the near-bottom layers and on the bottom, at considerable depths, and feeds upon big and small oligochaetes.

Among the Tricladida, Baikal is known to have about 80 endemic species — a colossal number for a landlocked body of water. Regrettably, most of these species have been described very cursorily, and the systematic position of many of them is very vague.

All the triclads known in Baikal are distributed among 10 to 12 genera. Modern authors (BRESLAU, 1933) designate these genera to the families Planariidae (the genus *Planaria*, which is also represented outside Baikal) and Dendrocoelidae (the genera *Bdellocephala, Sorocelis*, etc.).

In the opinion of N. A. LIVANOV (1962), all Baikalian triclads belong to the family Dendrocoelidae, whereas the Planariidae and representatives of the genus *Planaria* are absent in Baikal. Owing to the predomination of the Dendrocoelidae and marked endemism there is a certain likeness between the triclad faunae of Baikal and Lake Ohrid. But the latter is much poorer in species than Baikal, and these species belong to other genera (STANKOVIČ, 1960). The genus *Sorocelis* is richly represented in Baikal by *S. nigrofasciata* GRUBE (fig. 37,1) and *S. hepatizon* GRUBE (fig. 37,2). For the species of the *Sorocelis guttata* group (fig. 39,1) LIVANOV established a new genus, *Baicalobia*. Also from the genus *Sorocelis* a new genus, *Armilla*, has been singled out, to which "*Sorocelis*" *pardalina* KOROTN. (fig. 39,2) has been referred.

Most of the Baikalian triclads display a vivid and varied coloration and often have complicated patterns on the upper surface of the body. Many of them are very large for fresh-water turbellarians. For instance, *Polycotylus* sp., when straightened out, is up to 30 cm long and 4 to 5 cm broad (fig. 38,a, b). Almost all of them are carnivorous.

The period of their propagation is greatly protracted. In winter reproductive processes disappear, the maximum of propagation evidently falling on May—September.

One of the species of the Baikalian genus *Archicotylus*, *A. viviparus* RUBTZ., was described by I. A. RUBTSOV (1928) from the upper section of the Angara. This viviparous species also lives in Baikal. Its embryos, numbering 3 to 18, develop in a special breeding pocket into which fertilised eggs are deposited.

Numerically, the triclads are very common in Baikal, attaining particular abundance on stones in depths of 1—20 m. Up to a hundred of these worms can often be found on one stone 20 to 30 cm in diametre. The species found most frequently on rocky bottom in the littoral of Baikal are *Bdellocephala angarensis* GERSTF. (fig. 37,3), *Baicalobia copulathrix* KOROTN., *Baicalobia guttata* GERSTF. (fig. 39,1), and on sandy soil, *Sorocelis nigrofasciata* GRUBE (fig. 37,1) and *Baicalobia variegata* KOROTN. (fig. 37,5). Somewhat deeper one can often find *Sorocelis hepatizon* GRUBE (fig. 37,2), *Armilla pardalina* KOROTN. (fig. 39,2) and *Thysanoplana papillosa* KOROTN. (fig. 37,4), *Sorocelis stringulata* KOROTN. (fig. 37,6), and *Graffiella lamellirostris* KOROTN. (fig. 37,7).

The Baikalian triclads are of great interest from the zoogeographical point of view. Their endemism is striking, with many species constituting a fairly distinct Baikalian group. They live exclusively in the open regions of Baikal, never entering the waters of shallow bays, gulfs, or sors.

Outside Baikal, they have so far been found only in the upper course of the Angara, which is inhabited by several species. V. N. GREZE (1954, 1956) pointed to the presence of turbellarians in the lower reaches of the Angara and in the Yenisei, where species of Baikalian origin live.

The genetic connection of the Baikalian turbellarians with congeneric groups has not been sufficiently elucidated, V. N. BE-KLEMISHEV (1923) pointed out that the Baikalian *Sorocelis* are a relict of the pre-Pliocene fauna. N. A. LIVANOV (1961) noted a highly peculiar progress of evolution of the Dendrocoelidae in Baikal, from the typical *Bdellocephala*, close to the species living in Kamchatka and Japan, to the very distinct unusually large Baikalian forms of the genus *Rimacephalus*.

Trematodes

Up to 20 species of trematodes parasitic on fish and other animals have been found in Baikal. Among the Monogena, 9 species are known. Parasitizing the gills of the Baikalian grayling is *Tetraonchus borealis* OLSS., which was originally described by N. M. VLASENKO (1928) as *Ankyrocotyle baicalense* WLAS. According to O. N. BAUER (1948a, b, c), *T. borealis* is widespread in Siberian rivers, parasitizing the gills of graylings.

Fig. 37. Turbellarians: 1 - *Sorocelis nigrofasciata* GRUBE, body length 1 cm; 2 - *Sorocelis hepatizon* GRUBE, length up to 6 cm; 3 - *Bdellocephala angarensis* GERSTF., length 4 cm; 4 - *Thysanoplana papillosa* KOROTN.; 5 - *Baicalobia variegata* KOROTN., length 2—3 cm; 6 *Sorocelis (?) stringulata* KOROTN.; length 0.7 cm; 7 - *Graffiella lamellirostris* KOROTN., length 0.5 cm. After KOROTNEV, 1912.

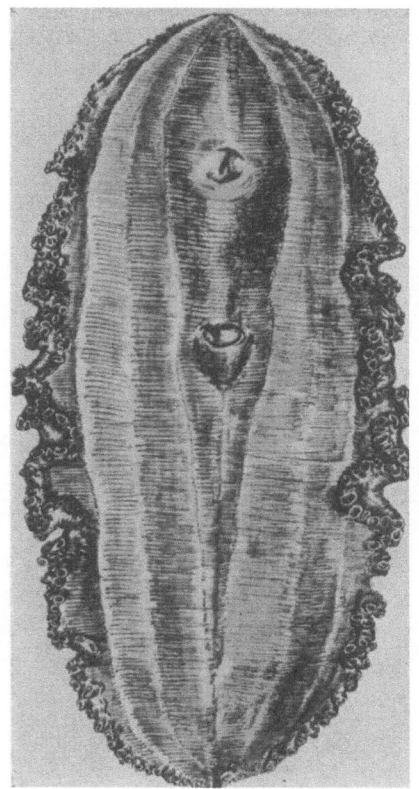

a

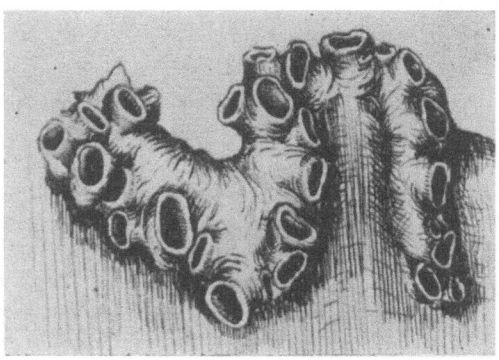

b

Fig. 38. *Polycotylus* sp., a - fixed specimen from the ventral side, body length when straightened out alive up to 30 cm, width up to 3 cm; b - magnified part of the lateral wall with suckers. After KOZHOV, 1947.

74

V. A. Dogel et al. (1949), I. I. Bogolepova (1950) and Dogel & Bogolepova (1957) have established the presence in Baikal of three species of the genus *Dactylogyrus*: *D. tuba* Linstow from the

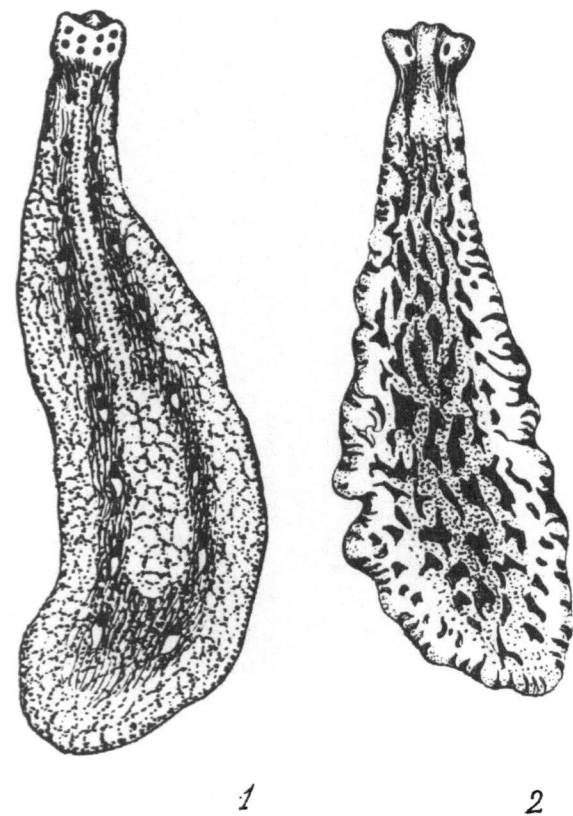

1 *2*

Fig. 39. 1 - *Baicalobia guttata* Gerstf.; 2 - *Armilla pardalina* Korotn., length up to 2.5 cm. After Korotnev, 1912.

ide, *D. leucisci* Zachw. from the dace and the endemic species *D. colonus* Bogol., parasitic on the fish of the Baikalian genus *Limnocottus* obtained from depths of 500—700 m. Three species of the genus *Gyrodactylus* have also been described: *G. baicalensis* Bogol. from the cottids *Limnocottus* and *Batrachocottus*, *G. bychows-kianus* Bogol. from *Cottocomephorus* and *G. comephori* Bogol. from *Comephorus*. Species close to *Dactylogyrus colonus* live on the Siberian Cyprinidae; species of the genus *Gyrodactylus* have so far stood apart.

In addition to the Monogena species mentioned here, we know

Diplozoon paradoxum NORDM. from the dace and *Diclibothrium armatum* LEUCK. from the sturgeon.

Among the Digena, species are known parasitic in the digestive tract of *Coregonus, Thymallus* and also *Cottocomephorus grewingki* DYB. and *Paracottus kneri* DYB. Among them mention can be made of *Crepidostomum baicalense* LAYMAN, which also lives in the intestines of graylings from the Yenisei. Another species, *Cr. auriculatum* WEDL. (= *Acrolichanus auriculatus* WEDL.), parasitizes the intestines of Baikalian sturgeons. It has also been found in the sturgeon and sterlet from Siberian rivers. The fluke *Allocreadium polymorphum* LOOS. lives in the intestines of the Baikalian Cottoidei; *A. isoporum* LOOS. in the intestines of the dace and ide; *Tetracotyle percae fluviatilis* LINST. and *Diplostomulum spathaceum* RUDOLPHI in Cyprinidae fish; *Bucephalus polymorphys* BAER. in pikes; and *Azygia robusta* ODHNER. in burbots.

Six of the 17 trematode species parasitizing Baikalian fish are endemics and live in endemic species of Cottoidei. V. A. DOGEL et al. (1949) considered three of these endemics belonging to the genus *Gyrodactylus* to be either marine relicts which invaded Baikal together with the progenitors of the modern Baikalian Cottoidei, i.e., palaeoendemics, or neoendemics, which have evolved in Baikal itself. The other three species, *Dactylogyrus colonus, Allocreadium polymorphum* and *Crepidostomum baicalense*, have originated, in DOGEL's opinion, in Baikal itself from widespread species in connection with the latter's attachment to the endemic species of Cottoidei.

Cestoda

At the present time 12 species of tapeworms are known, parasitizing Baikalian fish and people residing on the coast of Baikal. Special mention should be made of *Amphilina foliacea* G. WAG., a Baikalian and Yenisei sturgeon parasite, adult specimens of which live in the body cavity and not in the intestines. It is supposed that its intermediate hosts in the Yenisei are Baikalian gammarids, immigrants from Baikal: *Gmelinoides fasciatus* and *Eulimnogammarus viridis* (BAUER, 1948a).

In the Baikalian grayling, gwyniad and perch one can also often find *Cyathocephalus truncatus* PALL. LAYMAN (1933) writes that its intermediate hosts are gammarids. *C. truncatus* PALL. also parasitizes Coregonidae (BAUER, 1948a) in the Yenisei. *Proteocephalus longicollis*, a small tapeworm, lives in the intestines, body cavity and other organs of the grayling, gwyniad and omul.

The Baikalian Cottoidei and some Cyprinidae are infested with the ligula *(Ligula intestinalis* L.)

Mention should also be made of the following tapeworms:

Diphyllobothrium minus CHOLODK. Its plerocercoids live in the Baikalian omul, gwyniad and grayling. The final form has been found in residents of the trans-Baikal area. The pleurocercoids of the same tapeworm also infest *Coregonus peled* GMEL. and *Coregonus autumnalis* from the Yenisei and other Siberian rivers. For this reason *D. minus* cannot be regarded as a Baikalian endemic (BAUER, 1948a, b, c).

Diphyllobothrium strictum TALYSIN. It was found as a parasite of fishermen on the Olkhon Island in Baikal. Its pleurocercoids parasitize the Baikalian omul, gwyniad and grayling. They have also been found in the gwyniads and graylings of Yenisei and in fish from Lake Taimyr. V. A. DOGEL & I. I. BOGOLEPOVA (1959) also mention *Triaenophorus nodulosus* PALL. from the pike, dace and some Cottoidei species, then *Caryophyllaeoides fennica* SCHNEIDER from the dace and ide, *Proteocephalus torulosus* BATSCH. from the dace, and *Eubothrium* sp. from *Cottocomephorus*.

Not a single endemic has been found among the Baikalian tapeworms as yet. It is supposed (DOGEL et al., 1949) that the above-mentioned tapeworms *D. strictum* and *D. minus* penetrated into Baikal from the north, together with the omul. Other species of tapeworms are common in the Palaearctic.

Nematodes

The littoral of Baikal is populated by a considerable number of species of free nematodes most of which seem to be distinct Baikalian forms. But they have not yet been sufficiently studied.

The nematodes parasitic on the Baikalian fish and seal are better known. Eight species of these parasites are known from Baikal (LAYMAN, 1933; DOGEL et al., 1949, 1957; RYZHIKOV & SUDARIKOV, 1951, 1954). Among them, species of the genus *Cystidicola, C. impar* SCHN. and *C. skryabini* LAYM., deserve attention. *Cystidicola impar* parasitizes the swim-bladder of the Baikalian grayling, gwyniad, omul and dace and the Yenisei and Taimyr salmonids. O. N. BAUER (1948a) considers it to be an immigrant from the north. *Cystidicola skryabini* is a small nematode (0.4 to 0.9 mm) parasitizing the alimentary canal (the gullet, the stomach and the intestines) of the Baikalian grayling, sturgeon, *Cottocomephorus grewingki, Paracottus kneri* and *Procottus*. O. N. BAUER (1948a) refers this species to the genus *Capillospirura (= Ascarophis?)* and points out that *C. skryabini* parasitizes also the intestines of salmonids of the Angara section of the Yenisei.

A special place among the parasitic nematodes of Baikal is occupied by representatives of the endemic genera *Comephoronema* and *Cottocomephoronema* established by E. M. LAYMAN (1933), which parasitize the alimentary canal of *Comephorus, Cottocomephorus, Paracottus*, etc.

The species *Comephoronema werestschagini* LAYMAN has been

described, of the genus *Comephoronema*, and *C. problematica* LAYMAN of the genus *Cottocomephoronema*.

V. Y. SUDARIKOV & K. M. RYZHIKOV (1952) attribute the genus *Cottocomephoronema* to the new family Haplonematidae (from the superfamily Spirurata) established by them. In the same family they also list the genus *Haplonema* with the sole species *H. hamulatum* MOULT. from the burbot of the Great Lakes of North America, and the genus *Ichthyobronema*, the only species of which parasitizes the burbots of the northern rivers of the European part of the U.S.S.R. These authors point to the primitiveness of the entire group.

One species of the genus *Raphidascaria*, *R. acus* BLOCH., has been found in the alimentary canal of the perch and pike. This species also parasitizes many Yenisei fishes.

In the body cavity of *Cottocomephorus* species one can often find larvae of the ascarid *Contracoecum osculatum baicalensis* BOGOL. Sometimes they can also be found in *Batrachocottus baicalensis* DYB. Adult worms of this species have been discovered in the stomach and intestines of the Baikalian seal. It is supposed that *Cottocomephorus* is merely an intermediate host of this parasite, where its third-stage larva lives. The second stage evidently develops in the pelagic amphipod *Macrohectopus branickii* DYB. and the adult stage, in the seal (SUDARIKOV & RYZHIKOV, 1951). The length of the adult worm of this species from the seal is 13 to 24 mm, that of the larva from *Cottocomephorus*, 6.7 to 11 mm.

A. A. MOZGOVOI & K. M. RYZHIKOV (1950) pointed out that the nematode *Contracoecum osculatum baicalensis* parasitic on the seal is very close to the typical form *C. osculatum* RUD. parasitizing the seals of the northern seas, *Phoca hispida* in particular. In recent years this parasite was found in the omul and gwyniad.

V. A. DOGEL et al. (1949) pointed to the presence in the Baikalian dace of the nematode *Philometra rischba* SKRJABIN, which parasitizes the hypodermic tissue of this fish. A new species, *Capillaria baicalensis* RYZIK. & SUDAR., has been described by K. M. RYZHIKOV & V. Y. SUDARIKOV (1951) from the intestines of *Batrachocottus*.

Thus, today we know 8 nematodes parasitizing the Baikalian fish and seals. Three of them from the genera *Capillaria*, *Cottocomephoronema* and *Comephoronema* are endemic, the last two genera also being endemic. Besides there is an endemic subspecies, *Contracoecum osculatum baicalensis*. V. A. DOGEL considers that the genus *Comephoronema* has diverged from the widespread genus *Cystidicola*. He points also, as a peculiarity of the Baikalian nematode fauna, to the absence in it of many forms which are common in Siberian rivers.

Acanthocephala

These are very often found in the alimentary canal of many

78

Baikalian fish, *Batrachocottus baicalensis* being especially heavily infested with them. Species of *Procottus, Cottocomephorus*, the burbot, perch and others are also affected.

V. A. DOGEL et al. (1949, 1957) note two species of Acanthocephala from the fish of Baikal, *Echinorhynchus salmonis* MÜLLER and *E. clavula* DUJARDIN from the intestines of the Baikalian *Thymallus, Coregonus* and Cottoidei. Both species have an extensive distribution, the former in North Siberia and the latter throughout the U.S.S.R.

The Baikalian amphipod *Gmelinoides fasciatus* has been found to contain larvae of *Pseudoechinorhynchus lenok* ACHMEROV & DOM-BROWSKAJA-ACHMEROVA. The adult form of this acanthella has been described from *Brachymystax lenok* PALL. in the drainage of the River Amur (AKHMEROV & DOMBROVSKAJA-AKHMEROVA, 1941).

Rotatoria

Forty-eight species of rotifers are known to reside in Baikal today, distributed among 21 genera. The table printed below contains summarised data on the systematic composition of the Baikalian rotifers. The table has been drawn up on the basis of investigations conducted by G. YAKHONTOV (1904), N. VORONKOV (1925, 1927), V. YASHNOV (1922), L. ZENKEVICH (1922a), P. TIKHOMIROV (1927, 1929), V. YASNITSKY (1926, 1930, 1952) and K. GAIGALAS (1958).

Table XII.

Rotifers of Baikal.

Genus	Number of species	Distribution
Synchaeta	3	One of the species, *S. pachypoda* JASCHN., an endemic, lives in the open waters of Baikal. *S. pectinata* EHRB. and *S. stylata* WIERZ. live in shallows. The latter occurs also in open regions (in summer and sometimes in winter).
Polyarthra	2	*P. trigla* EHRB. and *P. euryptera* WIERZ., in the plankton of shallows and sors.
Trichocerca	5	Sors and shallows.
Diurella	1	Sors and shallows.
Dinocharis	2	Sors.
Euchlanis	3	The plankton of sors and shallows.
Lecane	2	Sors and shallows.
Monostyla	1	*M. lunaris* EHRB., in shallows.
Ascomorpha	1	*A. ecaudis* PERTY, a rare species, in shallows.
Brachionus	5	Ordinary lacustrine species, in shallows and sors; rare in open Baikal.
Schizocerca	1	*Sch. diversicornis* DADAY, in shallows and sors.
Keratella	2	*K. quadrata* MÜLL. and *K. cochlearis* GOSSE are widespread in the plankton of Baikal's open waters, where they comprise special Baikalian forms, typical forms of these species live in shallows.

Table XII (continued).

Genus	Number of species	Distribution
Notholca	10	*N. acuminata* EHRB. (=*N. striata* f. *acuminata*) abounds in open waters, comprising a special form there; so does *N. longispina* KELLIC., represented by the form *baicalensis* and *N. grandis* VOR. (=*N. striata acuminata* f. *grandis*). The species *N. olchonensis* TICHOM., *N. jasnitskii* TICHOM., *N. baicalensis* JASCHN., *N. lyrata* TICHOM. are regarded as endemic, but a re-investigation is required. A confirmation is needed of the presence in Baikal of *N. triarthroides* SCORIKOW described from the River Neva (SKORIKOV, 1903). The other *Notholca* species live in shallows (*N. striata* EHRB., *N. labsis* GOSSE).
Gastropus	1	*G. stylifer* IMHOF., in shallows and sors.
Ploesoma	2	*Pl. hudsoni* IMH., *Pl. truncatum* LOW., in shallows and sors.
Asplanchra	2	*A. herricki* GUERNE, *A. priodonta* GOSSE live in open waters in summer and sometimes in winter; more common in shallows.
Conochilus	2	*C. unicornis* ROUSS., common in shallows and sors; in summer occurs in open waters. *C. hippocrepis* SCHR. lives in sors.
Filinia	1	*F. longiseta* EHRB., in shallows. In open waters it comprises a special Baikalian form.
Collotheca	1	*C. mutabilis* BOLT., in shallows and open waters.
Albertia	1	The endemic species *A. voroncovi* ZENK., parasitizes oligochaetes.
Total	48	

Six or seven of these 48 species live in the open waters of Baikal. There is only one endemic among them, *Synchaeta pachypoda* (fig. 40); the rest are represented by Baikalian forms of widespread Palaearctic species, such as *Notholca acuminata* (= *N. striata* var. *acuminata*), *N. longispina*, *Keratella cochlearis*, *K. quadrata*, *Filinia longiseta*, *Notholca grandis*.

In summer and sometimes in the cold period one can see in open waters *Asplanchna priodonta*, *Asplanchna herricki*, *Synchaeta stylata* (sometimes also in spring), *Collotheca mutabilis*, *Polyarthra trigla* (rarely), *Conochilus unicornis*, and others. All other species indicated in the table are observed only in sors, shallow gulfs and pre-estuaries of rivers. The species permanently residing in open regions are distinguished by the absence of seasonal forms clearly marked in the dwellers of sors and shallows. They are also of a larger size when compared with typical forms living in shallows (YASNITSKY, 1926, 1952). Of particularly great size is *Synchaeta pachypoda* (600—750 μ in length). The Baikalian forms of *Keratella quadrata* and *K.*

80

cochlearis are characterised by the absence of well defined plates on the dorsal side of the lorica.

The species *Notholca jasnitskii* and *N. olchonensis* described by

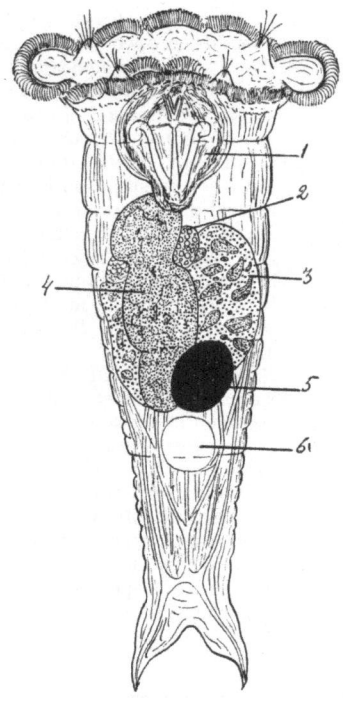

Fig. 40. Rotifer, *Synchaeta pachypoda* Jaschn. Body length 0.7 mm. 1 - pharynx; 2 - ovary; 3 - yolk gland; 4 - stomach; 5 - egg; 6 - urinary bladder. Original.

P. V. Tikhomirov (1927, 1929) and *N. triarthroides* have not been found in the open waters of Baikal.

N. L. Akatova (1949) pointed out in her work on the zooplankton of the Kolyma drainage (North-East Siberia) that the species *Synchaeta* abiding in the lower reaches of the Kolyma resembles, on the one hand, *Synchaeta johanseni* Harring from Alaska and, on the other, the Baikalian *Synchaeta pachypoda*.

In general, it should be noted that the complex of rotifer species from the open waters of Baikal is clearly of a northern habit. The majority of species live only in the oligotrophic lakes of Europe and Asia.

Keratella quadrata f. *baicalensis* is a species particularly abundant in the plankton of Baikal's open waters. It usually has two maxima in development. One maximum, comparatively weak, falls on the sub-ice period (March—April), the other in August—September. *K. cochlearis* f. *baicalensis* occurs the year round with the maxima in March—April and in autumn. *Notholca longispina* lives in Baikal also throughout the year, with the maxima in the sub-ice period and in autumn, the autumnal maximum being higher. *Filinia longiseta* (the Baikalian form) behaves likewise, reaching two maxima in development, in spring and in autumn. *Notholca acuminata* and various forms of *N. striata* appear, as a rule, only in the sub-ice period (January—April) and remain till July. A similar behaviour is displayed by *Synchaeta pachypoda*, which appears in large numbers in January and disappears in June. The maximum of its numerical density is reached in April—May. It feeds upon infusorians, young *Epischura* and *Cyclops* and also algae.

Polychaeta

The Baikalian fauna has a remarkable representative in the polychaete *Manayunkia baicalensis* (fig. 41) from the family Sabellidae, subfamily Sabellinae. It is a small worm up to 15 mm long, with a tuft of thread-like branchiae, two tentacle-shaped palps and two preoral appendages; the first and last (caudal) segments are not armed, the other 11 segments carry hair-like dorsal and hooked ventral setae. There are two eyelets on the first segment (NUSBAUM, 1901; ZENKEVICH, 1922b, 1925; TIMOFEYEV, 1928).

Manayunkia live in tubes constructed from silt and sand particles cemented with a chitinised substance. Vast accumulations of this worm can be observed on silty soils throughout the coastal belt of Baikal, often including bays and gulfs (the Proval Gulf, for instance). In gulfs and bays large numbers of *Manayunkia* can frequently be found on the stems and leaves of *Potamogeton*, to which they attach their tubes. They also enter the Angara and spread far down the Yenisei up to the Arctic coast.

Manayunkia can often be found in the littoral on stones, in the oscular openings of sponges and empty cases of caddis flies, among algae covering stones, etc. The form living on stones differs markedly from the typical silt form in that it is much shorter (2—3 mm in length) and lives in a short chitinised tube. The form which inhabits sands and constructs its tube from sand grains also somewhat differs from the typical silt form.

As distinct from its marine relatives, during its development from the egg *Manayunkia* does not form the free-swimming larva trochophore; the entire course of development takes place in the tube, where an incubation chamber is formed between the body of the mother and the wall of the tube; at both ends this chamber is bounded by cutaneous ridges growing between the 6th—7th and 8th—9th segments. The eggs are deposited in this chamber, and the development of the worm proceeds in it.

Polychaetes are sea dwellers, but some species live in brackish and fresh waters, particularly in those of India, Malaya and other tropical regions. There are also cave-dwelling species, such as

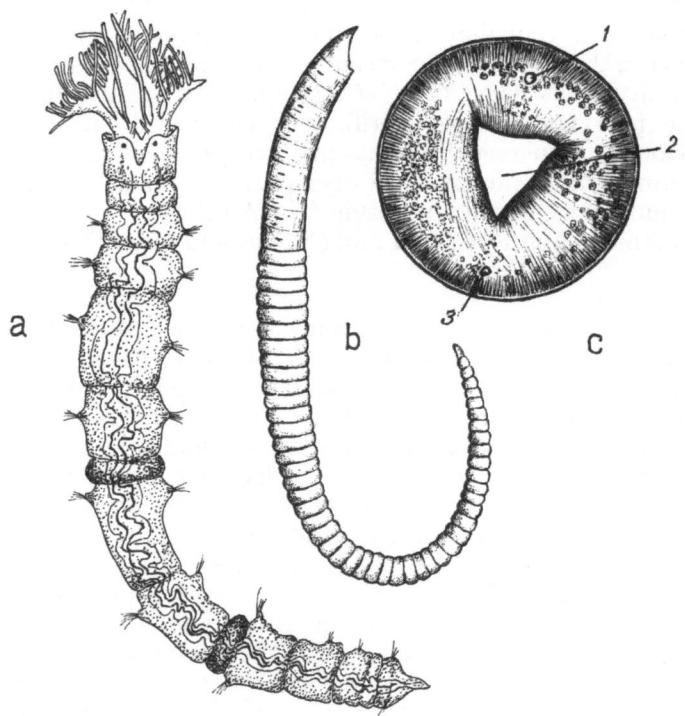

Fig. 41. a - Polychaete, *Manayunkia baicalensis* NUSB., removed from the tube, body length up to 15 mm. After KOZHOV, 1947. b, c - Oligochaete, *Agriodrilus vermivorus* MICHLSN, body length up to 45 mm; c - transverse section through the front part of the body; 1 - blood vessel, 2 - gullet, 3 - abdominal nerve. After MICHAELSEN, 1926.

Marifugia in the Balkans and *Troglochaetus* in Switzerland. Of especially wide fresh- and brackish-water distribution is the genus *Manayunkia*, of which up to 10 species are known. The genus is closely allied to the genus *Fabriciola*. Both of these genera are distinguished from the other Sabellinae by ordinary unplumed brachial tufts.

The *Manayunkia* species closest to the Baikalian species are *M. polaris* ZENK. (the tidal zone of the Murman coast), *M. aestuarina* BORN. (river mouths of the Irish Sea, the Channel, the North Sea, the Baltic Sea and the Gulf of Finland) and *M. caspia* ANNENK. (the Caspian Sea, the Black Sea, the mouths of rivers emptying into the Black Sea). The *Fabriciola* species include *F. atlantica* FREAD. (the Atlantic coast of Canada), *F. speciosa* LEIDY (rivers in North America near Philadelphia, the Great Lakes of North America), *F. pacifica* ANNENK. (the coast of the Komandorskiye Islands, the Pacific), *F. spongicola* SOUT. (the brackish lake Chilka in India), *F. baltica* FREAD and *F. blochmani* FRIDRICH (the Baltic Sea) (ZENKEVICH, 1935; ANNENKOVA, 1930).

Of great interest are the data on the Siberian distribution of forms close to or identical with the Baikalian *M. baicalensis*. Thus, *M. baicalensis hydani* SLASTN. was found in the brackish waters of the Gyda Gulf (the Arctic coast between the Yenisei Gulf and the Ob Bay) and in the fresh-water relict lakes of the Gyda drainage situated in the same region (SLASTNIKOV, 1940, 1941). In recent years *M. baicalensis* was found in some sections of the Yenisei and also in Lake Taimyr and the running-water lakes of the Lower Tunguska and Pyasina drainage. V. N. GREZE (1947, 1953, 1957) believes that in post-glacial times these basins communicated with the Yenisei and that, consequently, the Baikalian polychaete, as other Baikalian forms, penetrated into these basins from Baikal through the Angara and the ancient Yenisei (fig. 94).

Of still greater interest are the finds of *M. baicalensis* in the large running-water lakes of the Lena drainage (KOZHOV, 1942b; KOZHOV & TOMILOV, 1949). Large numbers of this *Manayunkia* were found in the lakes of the drainage areas of the Vitim and the Olekma, the Lena's important tributaries, situated in the areas of tectonic depressions of the Baikal system (Baunt, Busani, Oron-Vitim, large lakes of the Chara drainage, Davatchanda, Leprindo, and others (fig. 94). The *Manayunkia* of these lakes somewhat differs from the typical form.

There are several opinions on the origin of the genus *Manayunkia* and the ways of its migration:

1. A typical fresh or brackish water genus, in Tertiary time (Pliocene) *Manayunkia* had a broader distribution than now. It is an old inhabitant of fresh and continental brackish waters, from where it could migrate to the coastal regions of seas (L. S. BERG, 1949a).

2. The genus *Manayunkia* could be formed in Tertiary inland seas of Central Asia which gradually turned into more or less fresh-water bodies. From there it spread to the north, reached Baikal and from it penetrated into the Arctic Ocean (ZENKEVICH, 1935).

3. An old genus, formerly *Manayunkia* was widespread along the sea shores of the world ocean and was also represented in the Paratethys which preceded the Sarmatian Sea. *M. baicalensis* diverged from the common stock very long ago, prior to *M. caspia*'s penetration into Baikal together with the water masses successively con-

nected with the basins of marine relict origin (VERESHCHAGIN, 1940b).

L. A. ZENKEVICH's opinion appear to us to be more correct.

Oligochaeta

Although several important works are devoted to its, the oligochaete fauna of Baikal is still far from being fully known. Its main features became known thanks to the study by MICHAELSEN of a wealth of material obtained by the expeditions of KOROTNEV and VERESHCHAGIN (MICHAELSEN, 1901, 1905, 1926a, b, 1933, 1935; MICHAELSEN & VERESHCHAGIN, 1930). Over the last decades more material on oligochaetes has been collected and the list of species of this group has been gradually supplemented and specified (BUROV, 1931; IZOSIMOV, 1949, 1962; BUROV & KOZHOV, 1932).

At the present time Baikal is known to have approximately 62 species of oligochaetes, distributed among 23 genera (one of them endemic) and 5 families: Naididae, Enchytraeidae, Tubificidae, Lumbriculidae and Phreoryctidae.

The oligochaete species living in the open waters of Baikal, including the endemic emigrants to the Angara-Yenisei, are at least 90% endemic. In the abundance of oligochaetes and in their endemism Baikal can be compared with Lake Ohrid, which, according to STANKOVIČ (1960), has 27 species, 44.4% of which are endemic. In Baikal, oligochaetes live from the water's edge, where *Mesenchytraeus bungei* MICHLSN is very numerous, down to the greatest depths, which are inhabited by *Peloscolex inflatus* MICHLSN (to a depth of 1,200 m), *Phreoryctes ascaridoides* MICHLSN (to 1,300 m), *Rhynchelmis brachycephala* (to 900—1,650 m), *Lamprodrilus buthius* MICHLSN (to the maximum dephts), *L. inflatus* MICHLSN, etc. But the majority of species are concentrated at depths not exceeding 100—150 m, where the population density often stands at 300—1,000 and more specimens per 1 m² of the bottom (in some places 10,000 and more) weighing up to 20 g and in some cases 50—90 g/m². In the coastal belt of Baikal oligochaetes account for not less than 20—30% of the total mass of invertebrates. On silty and sandy soils at greater depths their share increases to 40—70%, and sometimes the population almost wholly consists of oligochaetes. Few of them are carnivores *(Agriodrilus vermivorus)*, most of them being detritus-eaters, and they themselves serve as food for other animals.

The family Lumbriculidae has the greatest number of representatives in Baikal. According to V. V. IZOSIMOV (1949, 1962), it comprises 32 species. Among the species of the genus *Lamprodrilus*, particular numerical density is attained at small depths by *L. nigrescens* MICHLSN.

Teleuscolex korotneffi MICHLSN is fairly common. It is a

big worm reaching 12 cm and more in length and 2—2.5 cm in diameter (in the front part of the body), with ordinary pointed setae. On the dorsal surface and on the sides of the front segments it has a characteristic pattern of almost circular black and dark-grey stripes.

The endemic genus *Agriodrilus* has only one, very peculiar species, *A. vermivorus* MICHLSN (fig. 41,b, c), which reaches 45 mm in length and 1.5—2 mm in diametre in the front part of the body.

In this species, the perivisceral cavity in the front part of the body (the 2nd—11th somites) is almost completely reduced and filled with interjoined muscles forming the wall of the gullet. Owing to this a transverse section of the gullet of the worm has an outline of an equilateral or isosceles triangle, one of the angles of which points to the ventral cord (fig. 41c, 1).

The Baikalian oligochaetes have an interesting representative in *Rhynchelmis brachycephala* MICHLSN, one of the biggest and deepest-dwelling species. It reaches 20 cm in length and has a diametre of 4 to 5 cm. The body of a fixed worm has a characteristic tetrahedral form. It occurs from 14 m to the greatest depths.

The Tubificidae is another family plentifully represented in Baikal. All the Baikalian species of this family belong to four widespread genera: *Clitellio, Rhyacodrilus, Limnodrilus* and *Peloscolex*. There are especially many species in the genus *Limnodrilus*. Many species of this genus were separated as *Lycodrilus*, a genus which was considered to be a Baikalian endemic. But later on, as a result of the study of more extensive material collected in Baikal and other bodies of water in Europe and Asia, these two genera were united into one. Among the *Limnodrilus* species, the more characteristic Baikalian form is *Limnodrilus dybowskii* GRUBE (= *Lycodrilus dybowskii* GRUBE). *Clitellio multispinus* MICHLSN can be mentioned as another representative of the family Tubificidae abundantly represented in Baikal.

In 1905 MICHAELSEN counted 5 oligochaete genera endemic to Baikal, but some of these genera turned out to be represented in other countries, and today only one genus of the Lumbriculidae family, *Agriodrilus* and one genus of the Naididae are considered to be endemic.

The Lumbriculidae is a typical fresh-water family of the northern (up to Alaska, Karelia, Norway and the Novosibirskiye Islands) and temperate latitudes. More than half of all the known species of this family (33 out of 55) live in Baikal.

The second genus of this family, *Styloscolex* (4 species), was regarded as a Baikalian endemic, but subsequently a new species belonging to it, *St. japonicus* JAMAGUCHI, was described from the creeks of Japan (H. JAMAGUCHI, 1937).

The third genus, *Teleuscolex* (2 species), which was thought to be endemic for Baikal, is included by S. HRABE (1929—1931) as a subgenus in the genus *Lamprodrilus*. The species *Lamprodrilus (Teleuscolex) pygmaeus* MICHLSN, which belongs to it and which was described from Baikal, has also been found in Lake Ohrid, where it is fairly numerous and comprises two forms. One of them, *intermedia*, combines traits of *Lamprodrilus* and *Teleuscolex*. STAN-KOVIČ (1960) refers *L. pygmaeus* to the group of Tertiary relicts with a rather discontinuous present distribution.

The other genera of the family Lumbriculidae populating Baikal include the genus *Lamprodrilus* with 13 species; in addition, there is *L. tolli* MICHLSN, which has since long been known from the middle reaches of the River Yana and from the Novosibirskiye Islands (Lyakhov Island). MICHAELSEN considered it to be an immigrant from Baikal, but not long ago representatives of this genus were found in Europe: *L. michaelseni* LAST. (in Macedonia near Monastir, in Lake Ohrid and also in lakes of the European part of the U.S.S.R.), *L. mraženi* HRABE in Czechoslovakia (near Prague and in bogs near Brno in Moravia), and *L. isoporus* MICHLSN, which was found in Ladoga, Onega and other lakes of Karelia. It is also said to live in Lake Pskov.

The genus *Bythonomus* has only one species in Baikal, *Bythonomus asiaticus* MICHLSN. Other representatives of this genus are known from Europe and Japan. According to S. HRABE (1929), *Bythonomus asiaticus* belongs to the genus *Stylodrilus*.

The genus *Rhynchelmis*, in addition to two species in Baikal, has widespread representatives in Europe *(R. limosella)*, including the Balkans (*R. komareki* HRABE), and in Japan (*R. orientalis* MICHLSN and others).

Recently Baikalian species of the family Lumbriculidae have been discovered in the Angara and the Yenisei. Among them, *Lamprodrilus satyriscus* f. *ditheca* MICHLSN has been found on sand at a depth of 6 m in the Yenisei Gulf, *L. wagneri* MICHLSN on sand in the Yenisei delta, and *Bythonomus opisthoannulatus* ISOSSIM. in the Yenisei delta. *Bythonomus mirus* CZEKANOWSKAJA, described from the lower reaches of the Yenisei, may also be of Baikalian origin (CHEKANOVSKAYA, 1956).

The Enchytraeidae, partially a terrestrial and partially an aquatic family, are spread mostly in the northern hemisphere. Baikal is inhabited by four species of this family, distributed among three genera. The genus *Propappus* is of special interest. Two species of this genus are known: *P. volki* MICHLSN, which lives in Baikal, the Elbe, Volga, Dnieper and other rivers of Europa and in Lake Ladoga, and *P. glandulosus* MICHLSN, an exclusively Baikalian inhabitant. Recently the presence of both these species has been established in the Yenisei, downstream to the outflow of the Angara. MICHAELSEN believes the genus *Propappus* to be the progenitor of the entire family Enchytraeidae. The genus *Mesenchytraeus*, which is very richly represented in Europe and North America, has only one species in Baikal, *M. bungei* MICHLSN, which leads an amphibious life and is usually observed at the edge of water. The genus *Enchytraeoides* is widespread in Europe and North America.

The family Tubificidae also lives mainly in the northern hemisphere. In Baikal it is represented by four genera and 13 species. One of the genera, *Rhyacodrilus*, has one species in Baikal, *R. coccineus* VEJD., represented by a special variety. Outside Baikal this species has an extensive European distribution. To this species, but as a variety, MICHAELSEN refers a form described from the lakes of New Zealand. The genus *Limnodrilus* is represented in Baikal by 7 endemic species. Prior to 1926 five of them were united in a special endemic genus, *Lycodrilus*. This genus is very common in Europe. Representatives of it are also known from North America. The genus *Peloscolex* has three species in Baikal, two of which are endemic. The third one, *P. velutinus* GRUBE, is also known from the lakes of West Europe. The other non-Baikalian genera of the *Peloscolex* family reside in the lakes and rivers of Europe and South Siberia, the River Amur and North America. The genus *Clitellio* is represented in Baikal by two endemic species. The third species of this genus, *Cl. arenarius*, is known to inhabit the littoral of European seas (Germany, Denmark, Britain, Iceland, Sweden, the Murman coast of the U.S.S.R., etc.).

The following species of the family Tubificidae have been found in the Yenisei: *Rhyacodrilus korotnevi* MICHLSN (the lower course of the Yenisei, the Angara), *Peloscolex* sp., which is close to *P. inflatus* MICHLSN (the middle reaches of the Yenisei), *Lycodrilus schizochaetus* MICHLSN (the Yenisei and the Angara), *Lycodrilus dybowskii* GRUBE (the Yenisei).

The Phreoryctidae is a purely fresh-water family residing in the temperate zone of the northern hemisphere. In Baikal it is represented by the endemic genus *Phreoryctes ascaridoides* MICHLSN, which lives at depths of up to 1,300 metres.

The family Naididae is widespread in fresh waters. In Baikal it has 9 genera and 20 species, most of them common in the northern hemisphere. In Baikal they live mostly in sors and pre-estuarine regions of big affluents. But five species seem to be peculiar to Baikal alone and are described by N. L. SOKOLSKAYA in litt. as new ones. A new genus with a sole species has also been established. All these five species live in the more or less open littoral of Baikal and its gulfs.

MICHAELSEN (1901), a noted authority on the Baikalian oligochaetes, stresses that *Lamprodrilus*, *Teleuscolex* and *Propappus* are primary genera for their families. In his opinion, among the oligochaetes of Baikal, along with a few phylogenetically young forms, there have been preserved (and in a considerable amount

88

at that) archaic forms which have retained the original peculiarities of the Lumbriculidae. For this reason the oligochaete fauna is to be regarded as being of a fairly old geological age.

V. V. IZOSIMOV (1949, 1962), who has revised the Lumbriculidae fauna of Baikal on the basis of new ample material, holds that MICHAELSEN's point of view cannot be accepted. The entire Lumbriculidae family, represented outside Baikal in alpine bodies of water, creeks, subterranean waters and other such places, was formed in the Tertiary period in the mountain belt stretching from the Alps and the Balkans eastward to Baikal and even Sakhalin and North Japan. By that (Tertiary) period such genera represented in Baikal as *Lamprodrilus*, *Bythonomus*, *Styloscolex*, *Lumbriculus* and *Rhynchelmis* had already been formed. Later some of these genera acquired a discontinuous distribution, which was probably due, in some measure, to the cooling connected with the glacial epochs.

In the opinion of V. V. IZOSIMOV, the oligochaete fauna of Baikal is both relict, having been formed long ago (in the Tertiary), and endemic, since there have evolved in Baikal a number of new species with peculiar traits of adaptation to life in this great lake.

Hirudinea

In the waters of sheltered gulfs and bays of Baikal one can come across leech species of wide Siberian distribution: *Protoclepsis tesselata* O.F.M., *Hemiclepsis marginata* O.F.M., *Piscicola geometra* L., *Glossiphonia heteroclita* L., *G. complanata* L., *Herpobdella octoculata* L., *Helobdella stagnalis* L., *Haemopis sanguisiga* L. (LIVANOV, 1902; PLOTNIKOV, 1906). There is no doubt that the gulfs of Baikal can be inhabited by other general Siberian species of leeches. But besides them Baikal has several peculiar species from the families Piscicolidae and Glossiphonidae.

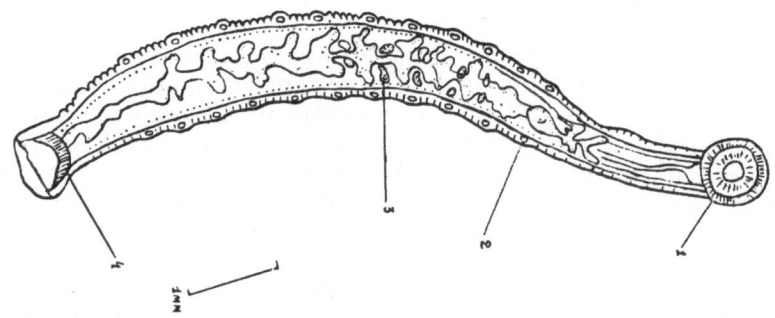

Fig. 42. Leech, *Codonobdella truncata* GRUBE, taken from a *Cottinella* cottid caught at a depth of 1,000 m. 1 - anterior sucker; 2 - lateral vesicles; 3 - testicles; 4 - posterior sucker. After DOGEL & BOGOLEPOVA, 1957.

Among the Piscicolidae, E. GRUBE (1871, 1872) very briefly described four species: *Piscicola multistriata* from Baikalian sturgeons, *P. torquata*, *P. conspersa* (from the Angara) and *Codonobdella truncata*, which is parasitic on Baikalian species of Gammaridae and Cottoidei.

In recent years Y. I. LUKIN & V. M. EPSTEIN (1959, 1960a, b; EPSTEIN, 1959) have conducted a revision of the Baikalian species of the Piscicolidae family and established the presence in Baikal of only two endemic species from those described by GRUBE: *Trachelobdella (Piscicola) torquata* (fig. 43) and *Codonobdella truncata* (fig. 42). The former parasitizes gammarids and fish from Baikal's littoral. It is a minute leech, up to 11 mm long and 3 mm broad, with small suckers and a highly variable body coloration with predomination of brown pigment in patches or throughout the body. *Codonobdella truncata* is parasitic on Gammaridae and Cottoidei living at depths of up to 1,100 m.

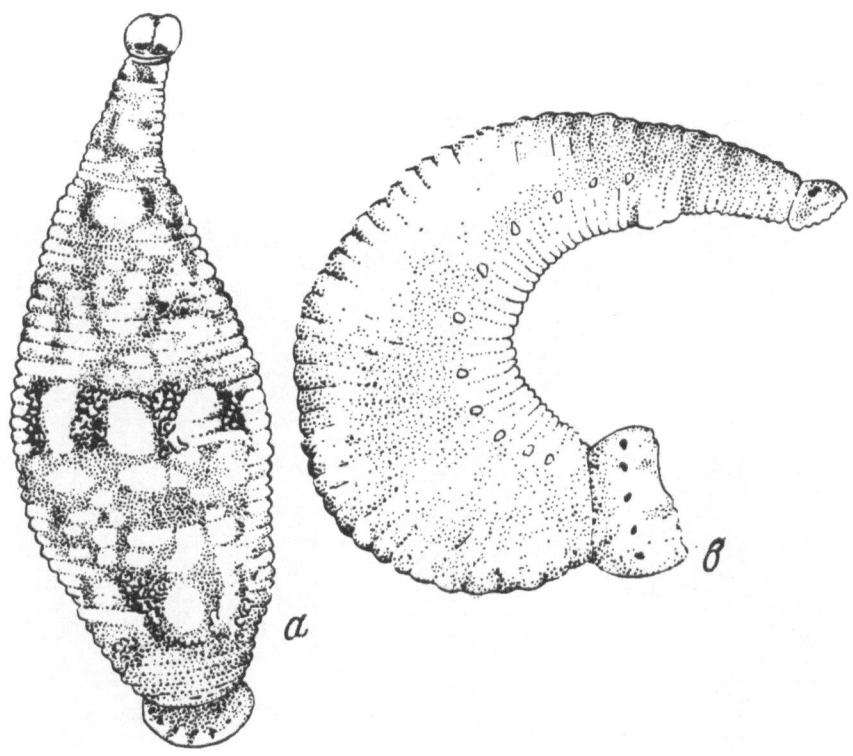

Fig. 43. Leech, *Trachelobdella torquata* GRUBE: a - dorsal surface; b - side view; body length up to 11 mm. After EPSTEIN, 1959.

In the opinion of Y. I. LUKIN & V. M. EPSTEIN (1959) the Glossiphonidae family is represented in the open waters of Baikal only
by species of the new subfamily Toricinae established by them,
which includes *Torix baicalensis* described by G. G. SHCHEGOLEV
(1922) and two species of the new and endemic genus *Baicaloclepsis*:
B. echinulata, briefly described by GRUBE (1871, 1872) as *Clepsina
echinulata*, and the new species *B. grubei* LUKIN & EPSTEIN. *Torix
baicalensis* (fig. 44,2,3; 45) described by SHCHEGOLEV was excluded
by these authors from the genus *Torix*, and a new endemic genus,
Paratorix LUKIN & EPSTEIN, is established for this species.

All the Baikalian Toricinae species are of a comparatively large
size (up to 40 mm), with a broad and moderately flat body densely
papillated on the dorsal surface. In the *Baicaloclepsis* species the
papillae are of various sizes, running along the body in parallel
rows. The ventral surface is either smooth or covered only with
small papillae (fig. 44). There are two pairs of eyes evincing a trend
to reduction. A full somite of a leech consists of three uneven rings,
the first being the shortest, the second very long and the third
short. The stomach has six pairs of processes (fig. 45).

The genus *Torix* with the species *T. mirus* BLANCH. was established by BLANCHARD in 1893 after a study of specimens collected in
Lake Kao Bang (South Asia) from the fresh-water mollusk *Melania*.
The genus *Torix* is mainly distinguished by the presence in the
somite of two rings of uneven size, the bigger ring on the ventral

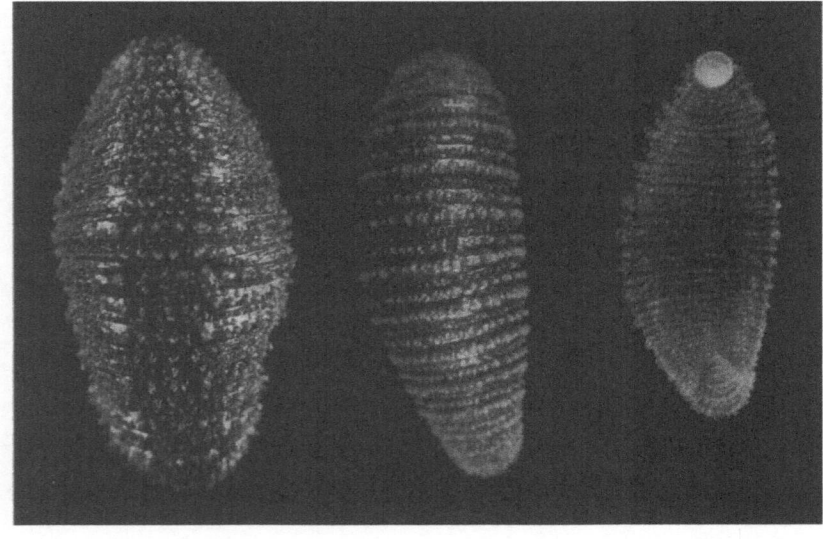

1   2   3

Fig. 44. Leeches: 1 - *Baicaloclepsis echinulata* GRUBE, body length up 40 mm;
2—3 - *Paratorix baicalensis* from the dorsal and ventral sides. Original.

surface being divided by a furrow into two rings of a second order. The other species of the same genus, *T. cotylifer*, is parasitic on the turtle *Tryonix*. There are no traces of papillae on the Chinese *Torix*.

LUKIN & EPSTEIN (1960a, b) came to the conclusion that the endemic Baikalian Glossiphonidae of the subfamily Toricinae are connected by their origin with the small natural group of leeches belonging to the genera *Torix* (China), *Oligobdella* (Japan, Korea, North America) and *Oligoclepsis* (Japan) and are very far removed

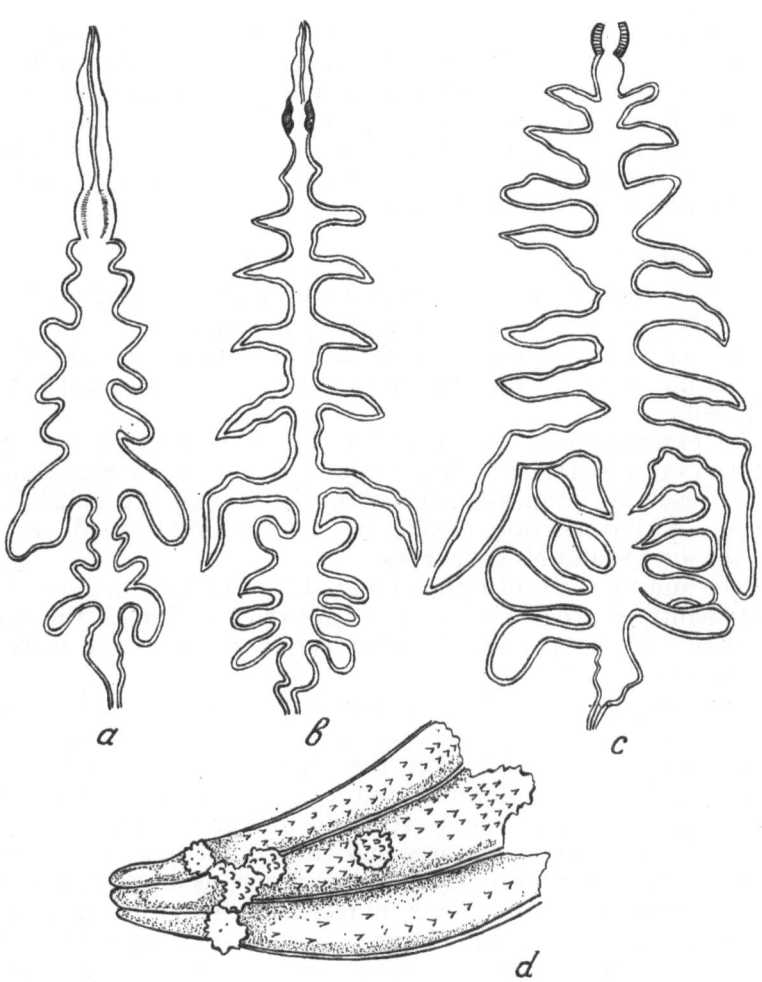

Fig. 45. a,b,c, - diagrams of the digestive tracts of leeches: a - *Baicaloclepsis grubei;* b - *B. echinulata;* c - *Paratorix baicalensis;* d - total somit of *B. grubei.* After LUKIN & EPSTEIN, 1960.

from the Glossiphonidae currently widespread in Europe and North Asia. Among the Baikalian and non-Baikalian Toricinae the genus *Baicaloclepsis* is more primitive than the other above-mentioned genera of this subfamily. These authors consider that the Toricinae appeared and spread in East Asia and North America in the Miocene period, when the two continents were still closely connected.

It can be supposed that the Baikalian Toricinae species *Baicaloclepsis echinulata* GRUBE, *B. grubei* LUK. & EPST. and *Paratorix baicalensis*, being very closely related, have evolved in Baikal from a common progenitor.

Thus, today, Baikal is known to have 8 to 10 Hirudinea species widespread in the Palaearctic, living in sors and shallow gulfs, and 5 to 6 endemic species from the families Piscicolidae and Glossiphonidae with endemic genera from the latter, *Baicaloclepsis* and *Paratorix*. Endemic Baikalian leeches do not occur in sors, shallow bays and tributaries of Baikal. *Trachelobdella torquata* only is found in the upper part of the Angara.

Entomostraca

No representatives of Phyllopoda are known in Baikal. At any rate, they are not to be found in its open waters.

Among the Calanoida copepods, a highly important part is played in the life of the Baikalian plankton by *Epischura baicalensis* SARS. (fig. 46).

Its extensive distribution in Baikal was noted long ago by A. KOROTNEV (1901), but a fuller information about it was obtained as a result of studies conducted by B. YASNITSKY (1928, 1930, 1934), A. ZAKHVATKIN (1932), B. GARBER (1941, 1946), as well as the author and his colleagues.

The adult females of *Epischura* are 1.3 to 1.6 mm long, the males are slightly shorter. The antennae are thickly plumed, the caudal filaments are of equal length, being long in the males and shortened in the females.

Epischura live in Baikal the year round. In autumn the crustaceans descend to the deep layers to propagate. In January, after the formation of the ice cover, the young nauplii rise to the upper layers and gradually accumulate there, and the maximum of their numerical density is usually achieved in March—April. In May—June this generation matures and the crustaceans descend again (to 100—200 m and deeper) for propagation, as a result of which a new, usually higher maximum numerical density of young nauplii is observed in June-July. On rare occasions a female is found with an egg sac attached to the genital segment and containing several dozen eggs. As a rule, however, the eggs float freely in the mass of water.

The total mass of the crustaceans in the open waters of Baikal reaches its maximum in August—September, when most of the specimens of the aestival generation turn into older copepodit stages. This period is the main feeding time for the plankton-eating fish of Baikal. During this period the bulk of the crustaceans live in the upper 50-metre layer of water.

Epischura is the chief consumer of planktonic algae; it can even filter bacteria (O. KOZHOVA, 1953, 1956a). The crustacean itself constitutes the staple food of such plankton-eating fish as the omul, *Cottocomephorus, Comephorus* and the young of other fish. The Baikalian pelagic amphipod *Macrohectopus branickii* DYB. also feeds to a considerable degree upon *Epischura* (VILISOVA, 1951).

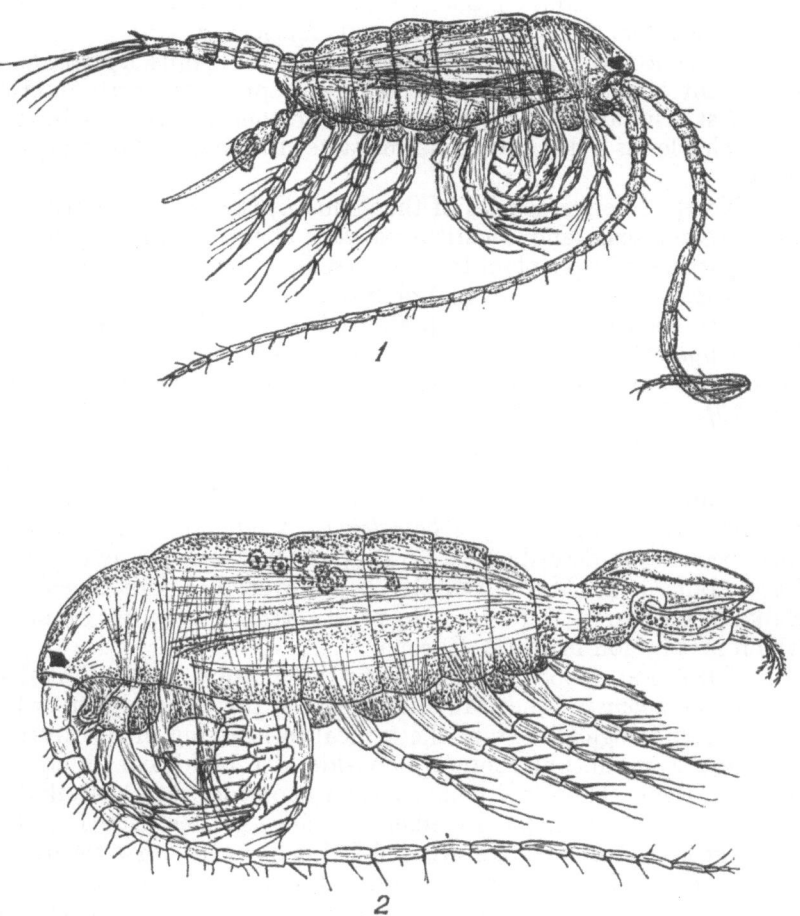

Fig. 46. *Epischura baicalensis* SARS. 1 - male; 2 - adult female with spermatophore, body length up to 1.6 mm. After SARS, 1900.

In sors and well-sheltered shallow gulfs *Epischura* occurs individually, and only in the cold period of the year.

Epischura is of a considerable zoogeographical interest (ZENKE-VICH, 1922b; SMIRNOV, 1929). In the Holarctic region species of this genus are to be found only in the following widely separated places: Lake Baikal (*Epischura baicalensis* SARS), Lake Khanka in the Far East (*E. chankensis* RYLOV*)*, Lake Udil, which communicates with the River Amur through a gullet *(E. udulensis* BORUZ.*)*, Lake Kronotskoye in Kamchatka, which has an area of about 200 square kilometres and is more than 60 m deep (*Epischura baicalensis* SARS). Besides, four or five species of this genus are known in North America (Alaska, California, Newfoundland, North Carolina, the Great Lakes). The Palaearctic species comprise the subgenus *Epischurella* and the American species the subgenus *Epischura*.

Y. V. BORUTSKY (1947a) considers that the genus *Epischura* is a remnant of a group which was once widespread in North America and East Asia and after which separate colonies have remained in Asia. This lends a discontinuous character to the distribution of the genus.

In North America species of the genus *Epischura* are distributed on a greater scale and constitute a very characteristic component of lacustrine pelagic plankton, superseding the Palaearctic genus *Heterocope* which is absent in North America.

The genus *Diaptomus* is represented in the gulfs of Baikal by *D. graciloides* LILLIEB. In summer small numbers can be found in the open regions. The bays and gulfs are also populated by *Heterocope appendiculata* SARS. The common Europe-Siberian *Diaptomus denticornis* WIERZ. and *D. pachypoditus* RYL. live in sors and gulfs.

The order Cyclopoida has a considerable number of forms in the open waters of Baikal. A fairly high density in the plankton is attained by *Cyclops kolensis baicalensis* WASIL. (fig. 47). Initially this species was described by G. L. VASILYEVA (1950) as *C. baicalensis*, but G. F. MAZEPOVA (1957), who studied this species in Baikal, considered it to be a form of *Cyclops kolensis* LILLIEB., which is common in North Europa.

K. LINDBERG (1955) confirmed the designation of the Baikalian cyclops as *kolensis*. The Baikalian form of this species differs but little, morphologically, from the typical form living in the lakes of Southern Sweden, but there are considerable biological differences between them. Therefore it is correct to consider it as a Baikalian form of *Cyclops kolensis*. In Siberia *C. kolensis* has been found in the large lake Munduiskoye situated 70 km downstream of the mouth of the Lower Tunguska on the ancient terrace of the Yenisei, together with the Baikalian bryozoan *Hislopia placoides* KOROTN. and the relict amphipod *Pallasea quadrispinosa* SARS (GREZE, 1953). It also lives in the lakes of the Angara drainage (VASILYEVA, 1956;

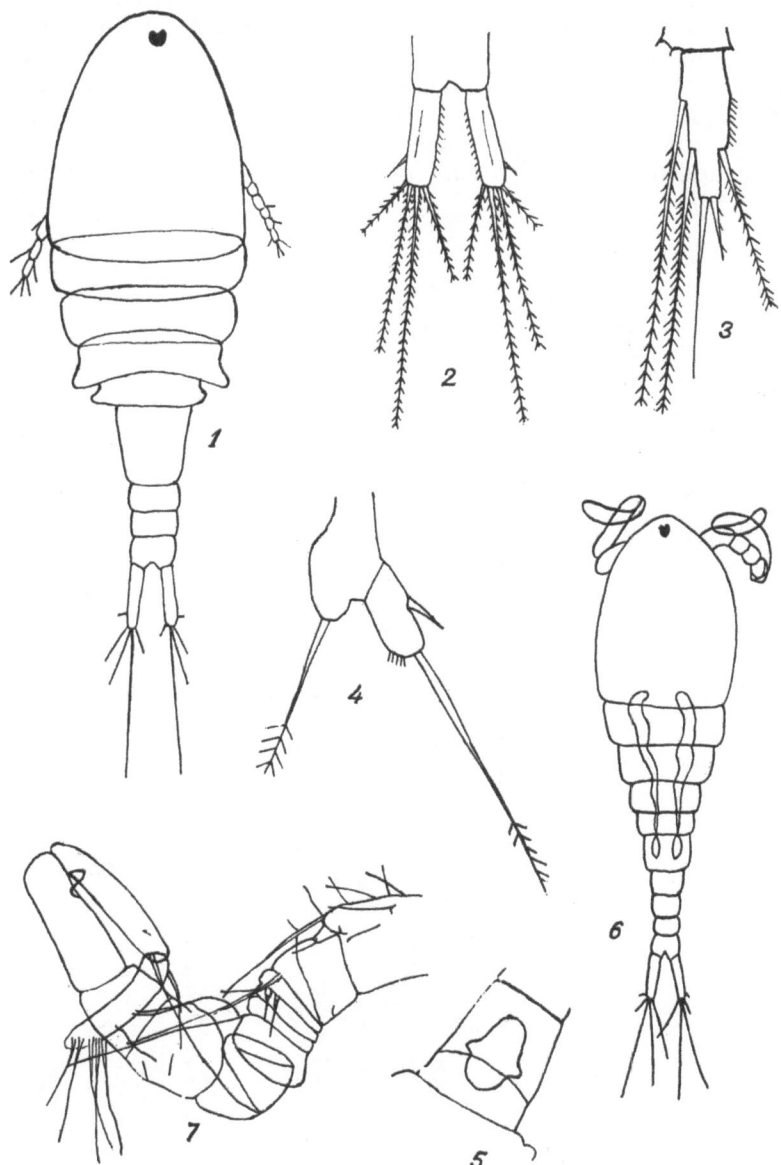

Fig. 47. *Cyclops kolensis* LILLIEB. from Baikal. 1 - female (general view), body length 1.2 mm; 2 - furcal ramus; 3 - distal segment of the exopodite IV; 4 - a limb of the Vth pair; 5 - genital segment with receptaculum; 6 - male (general view); 7 - antenna of the male. After MAZEPOVA, 1960.

96

VILISOVA, 1954) and in the Urals (ULOMSKY, 1957). In Baikal, *C. kolensis* is particularly numerous in extensive shallows and big gulfs and bays. But in well sheltered shallow bays, gulfs and sors it occurs only in the cold period of the year, being rare there in summer. In zooplankton-rich years with a warm summer the cyclops spread widely throughout Baikal, consuming young *Epischura*.

In addition to *C. kolensis baicalensis*, the plankton of the gulf

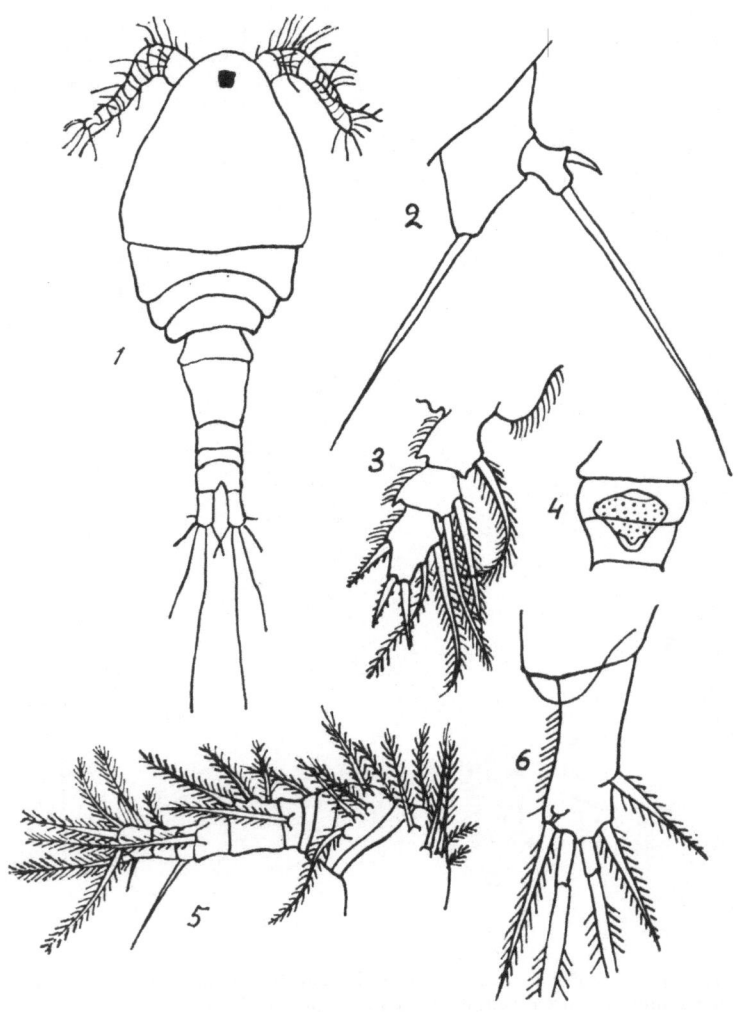

Fig. 48. *Acanthocyclops profundus* MASEP., female, body length 1.4—1.5 mm. 1 - general view; 2 - a 5th pair limb; 3 - endopodite of a 4th pair limb; 4 - receptaculum; 5 - front antenna; 6 - furcal ramus. After MAZEPOVA, 1950.

contains *Acanthocyclops viridis* JUR. The sors are populated by *Cyclops strenuus* FISCH., *C. vicinus* ULJAN., *Mesocyclops leucarti* CLAUS, *Eucyclops serrulatus* FISCH., *Acanthocyclops viridis*, and others.

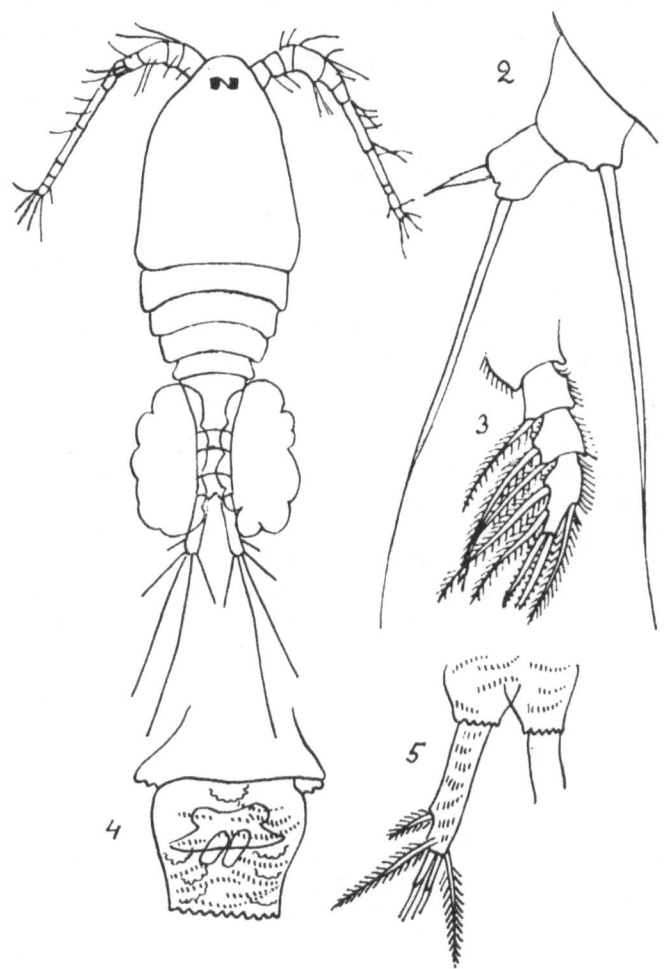

Fig. 49. *Acanthocyclops rupestris* MASEP., female, body length 3 mm. 1 - general view; 2 - a 5th pair limb; 3 - endopodite of a 4th pair limb; 4 - genital segment with receptaculum; 5 - furcal ramus. After MAZEPOVA, 1950.

Baikal has been found to contain a highly peculiar fauna of bottom-dwelling cyclops. In recent years G. F. MAZEPOVA (1950, 1952a, 1955) has described a series of species found in the benthos of open Baikal: *Acanthocyclops profundus* MASEP. (fig. 48), *A.*

arenosus MASEP., *A. rupestris* MASEP. (fig. 49), *A. intermedius* MASEP., *A. signifer* MASEP., *Orthocyclops bergianus* MASEP. *Acantho-cyclops notabilis* MASEP. has been described from the gulfs and open sections of Baikal. The rocky bottom of the open littoral is densely populated by *Eucyclops serrulatus* FISCH., which is represented by a special form in Baikal, *baicalocorrepus* MASEP., and furnishes, in the opinion of G. F. MAZEPOVA (1955), an example of the invasion of the open waters of Baikal by the modern Siberian fauna. In all, about 25 cyclops species are known in Baikal today, 72% of which are endemic. According to MAZEPOVA (1957), the bottom-dwelling cyclops occur in Baikal down to the extreme depths, but they are especially numerous in the zone of depths between 1—2 and 40—50 metres.

Some characteristics make many of the bottom-dwelling species akin to the group of species living in subterranean waters in West and South-East Europe and North America (species of *Acantho-cyclops*, *Orthocyclops bergianus*, etc.). Evidently these are descendants of a widespread subterranean Tertiary fauna which has survived in the conditions of Baikal and produced new species there.

Thirteen species of parasitic Copepoda are known in Baikal (MESYATSEV, 1926, KORYAKOV, 1951a, b, 1952, 1954; MARKEVICH, 1956). They include one species of the genus *Ergasilus*: *E. sieboldi* NORDMANN (= *E. baicalensis* MARK.), parasitic on the pike and roach; three species of the genus *Achtheres*: *A. strigatus* MARK. on the omul, *A. percarum* NORDMANN on the perch, and *A. extensus* KESSL. on the gwyniad; three *Salmincola* species: *S. extumescens*

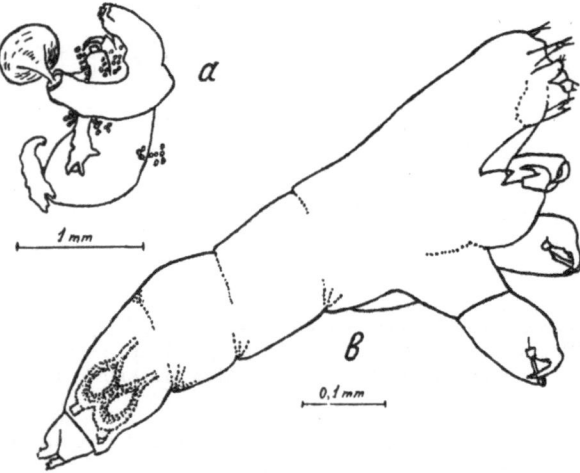

Fig. 50. *Salmincola cottidarum* MESSYATZEFF; a - female, b - male, body length 0.8 mm. After KORYAKOV, 1951.

GADD. on the omul and gwyniad, *S. cottidarum* MESSJATZEFF on Cottoidei species (fig. 50), and *S. thymalli baicalensis* MESSJATZEFF on the grayling and omul; one *Tracheliastes* species: *T. polycolpus*

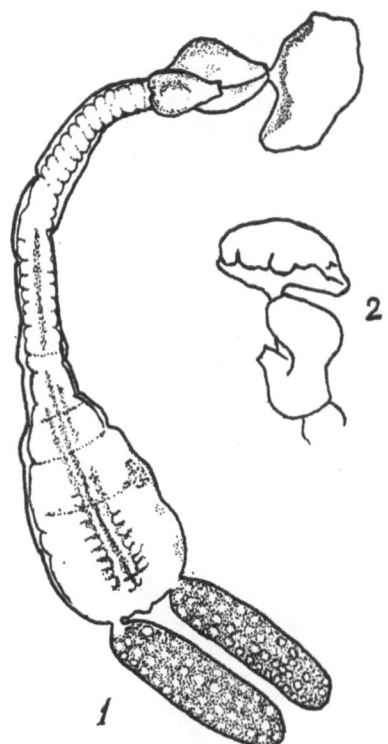

Fig. 51. *Coregonicola baicalensis* KORYAK., a parasite of *Abyssocottus*, body length 15—20 mm. 1 - general view; 2 - sticking apparatus. After KORYAKOV, 1951.

NORDMANN on the dace and roach. Besides, a new species of the Baikalian genus *Coregonicola* has been described, *C. baicalensis* KORYAK. (fig. 51), found by Y. A. KORYAKOV (1951a) in the oral cavity of the abyssal cottid *Abyssocottus bergianus* TAL.

V. N. GREZE (1951a) pointed to the presence of *Paraergasilus rylovi* MARK. in the plankton of Baikal's sors. This species was first described from the Caspian Sea, where it was also found in plankton. Parasitism is supposed to be facultative in the life of this species.

Recently Y. A. KORYAKOV (1954) has discovered in Baikal three other species of parasitic copepods: *Ergasilus briani* MARK. on the dace, roach and minnow, and two *Basanistes* species, *B. briani* MARK. and *B. woskoboinikowi* MARK., on the lenok and taimen.

Thus, seven out of the 13 species of parasitic copepods are found on general Siberian fish species usually living only in sors and gulfs (Cyprinidae and *Perca fluviatilis*); two species of the genus *Basanistes* parasitize fluvial fish entering the coastal zone of open Baikal (*Brachymystax lenok* PALL., *Hucho taimen* PALL.); five species are parasitic on fish permanently living in Baikal (two of them parasitize endemic Cottoidei species).

The species parasitizing Coregonidae (*Achtheres strigatus, A. extensus, Salmincola extumescens*) must have penetrated into Baikal from the north, together with the omul and other fish (DOGEL et al., 1949). Y. A. KORYAKOV (1954) considers that only one of the 13 species known in Baikal, *Paraergasilus rylovi*, is undoubtedly a relict of pre-glacial time. Two neoendemics of autochthonous origin, *Salmincola cottidarum* and *Coregonicola baicalensis*, are derivatives of widespread Siberian forms. The other 10 are common Palaearctic species; four of them can be observed also in the Amur drainage.

A very rich and peculiar Harpacticoida fauna has been found in Baikal. According to Y. V. BORUTZKY (1931, 1932, 1947b, c, 1948, 1949, 1952), it contains 43 species of this group, belonging to 9

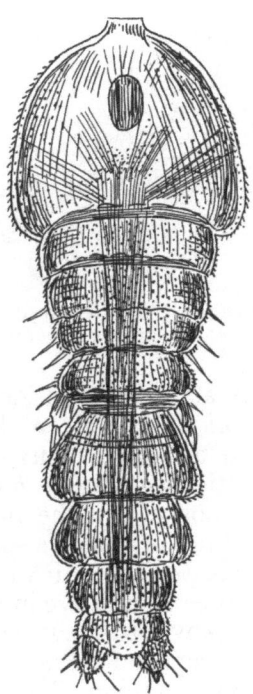

Fig. 52. Harpacticidae. *Canthocamptus (Baicalocamptus) werestschagini* BORUTZKY, female, body length 0.9 mm. After BORUTZKY, 1952.

genera. Thirty-eight of them are endemic. The genus *Canthocamptus* is represented by 5 species; 2 of them have been singled out by BORUTSKY as the endemic subgenus *Baicalocamptus* (fig. 52).

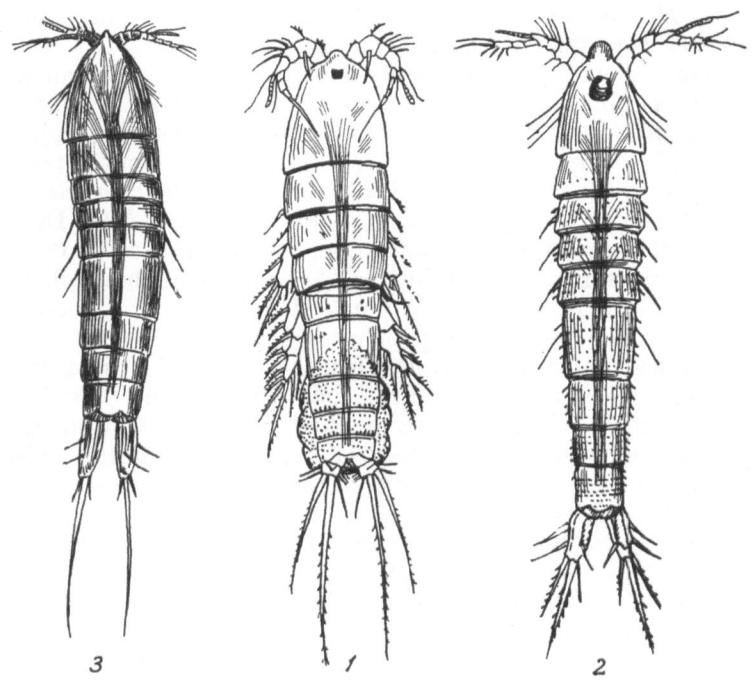

Fig. 53. Harpacticidae: 1 - *Echinocamptus (Limocamptus) hiemalis* var. *werestschagini* BORUTZKY, female, body length 0.6 mm; 2 - *Moraria (Baicalomoraria) baicalensis* BORUTZKY, female, body length 0.8 mm; 3 - *Morariopsis typica* BORUTZKY, female. After BORUTZKY, 1952.

One species is known from the genus *Paracamptus*; 10 endemic species from the genus *Bryocamptus*; four species from the genus *Echinocamptus*, (subgenus *Limocamptus*) among them, only *E. hiemalis* BORUTZKY is not endemic, but this species, too, is represented in Baikal by the endemic subspecies *E.h. werestschagini* BORUTZKY, fig. 53,1); two species from the genus *Attheyella*, of which *A. (Ryloviella) baicalensis* BORUTZKY is endemic; and 17 species from the genus *Moraria* (only one of them is not endemic). Fifteen species of this genus belong to the endemic subgenus *Baicalomoraria* (fig. 53,2). The genus *Morariopsis*, which comprises 2 species, is endemic to Baikal (fig. 53,3). The genus *Maraenobiotis* is represented by one non-endemic species, *M. insignipes* LILLIEB.

From the Harpacticidae family, the species very abundantly represented in the gulfs of Baikal is *Harpacticella inopinata* SARS (1908). Most of the genera of this family live in seas, but *Harpacticella* is a fresh-water genus and comprises only 4 species: *H. inopinata* in Baikal, *H. paradoxa* BREHM. in Lake Tali-fu in Yunnan (China), *H. lacustris* SEWELL in Lake Chilka in India, and *H. amurensis* BORUTZKY in the River Amur.

Endemic species live mostly in the open sections of Baikal. In its sors and gulfs, only five species have been found so far: *Harpacticella inopinata* SARS, *Moraria duthiei* SCOTT. (a widespread Palaearctic species), *Maraenobiotis insignipes* LILLIEB., which is peculiar to the Arctic zone of Siberia, *Attheyella (Bremiella) dogieli* RYL., which also occurs in China, and *Moraria schmeili* VAN DOUWE, which is widespread in the Palaearctic. The Harpacticoida fauna of the gulfs and sors of Baikal, just as of the lakes of the Baikal and trans-Baikal areas, is probably much richer than has been indicated above, but it has been poorly studied.

BORUTSKY points out that the Harpacticoida of Baikal are, on the whole, of a Palaearctic nature, their roots having been formed on the territory of the ancient Angarida. A confirmation of this proposition is seen by BORUTSKY (1952) in the affinities existing between the faunae of primitive forms of North America, East Asia and Baikal (*Bremiella, Moraria*, etc.).

It is a noteworthy fact that many Baikalian species are close to North-American ones. Such are, for instance, *Attheyella (Ryloviella) baicalensis, Bryocamptus (Pentacamptus) incertus* BORUZKY, and many others. A similarity between many Baikalian species and cave-dwelling species of West Europe is also stressed.

The Baikalian species *Harpacticella inopinata* was appraised by G. Y. VERESHCHAGIN (1940b) as a marine element in the fauna of Baikal, but Y. V. BORUTSKY points out that this species is a remnant of the upper-Tertiary fresh-water fauna of North Asia.

The Harpacticoida fauna in Baikal is very rich also quantitatively. Among algae and on stones in the coastal belt one can easily notice large numbers of these minute (0.5—1 mm) crustaceans which serve as food for many Baikalian invertebrates. They are especially numerous on the Baikalian sponge and stones amidst water plants in the coastal belt. It seems very likely that the crustaceans living on sponges feed at their account and constitute, in turn, the diet of the gammarid *Acanthogammarus parasiticus*, vast accumulations of which can be found on any sponge.

All the 33 ostracod species known in Baikal today (BRONSTEIN, 1930, 1939, 1947) are distributed between the families Cypridae (genera *Candona* and *Pseudocandona*) and Cytheridae (genus *Cytherissa*). In the former family, we know 14 species of the genus *Candona* and 8 species of the genus *Pseudocandona* (fig. 54). In the

Cytheridae family, 14 species of the genus *Cytherissa* are known (fig. 55).

BRONSTEIN (1947) points out that only three non-Baikalian

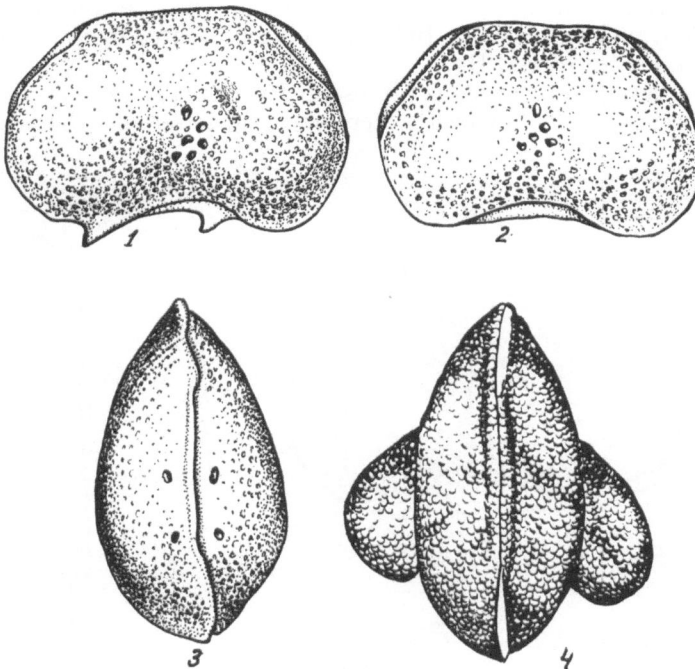

Fig. 54. *Pseudocandona bispinosa* BRONST.; 1—2 - side view of male and female test; 3 - male test from above, length 1.2 mm; 4 - *Pseudocandona gajewskaja* BRONST., male test from above, length 0.6 mm. After BRONSTEIN, 1947.

species of the genus *Pseudocandona* are known: *P. zschokkei* WOLF. from the subterranean waters of Switzerland, *P. insulpta* G. W. MÜLL., plentiful in small water bodies of Europe, and *P. elongata* HOIMES in lakes of Britain and in Lake Ohrid. It is possible to refer to this genus also two species from the cave waters of North America and one species from the cave waters of Yugoslavia. In some characteristics (trapeziform test, reticulate sculpture on the valves, absence or poor development of genital appendages, etc.) the Baikalian species stand somewhat apart from the other species and are characterised, on the whole, by characteristics of primitive organisation indicative of their more ancient origin.

The genus *Candona* is generally very rich in species. The Holarctic fauna includes up to 45 species, more than a third of which are Baikalians. More than 20 *Candona* species are known from Lake

Ohrid. The tests of some of them are trapeziform and have asymmetrical valves, thus being reminiscent of some *Candona* species from Baikal (STANKOVIČ, 1960). But this similarity is probably due to convergence. Among the Baikalian *Candona* species, only *C. inaequivalvis* G. O. SARS has been found outside Baikal, near Verkhoyansk on the Lena. In Baikal it is represented by the endemic subspecies *baicalensis* BR.

The family Cytheridae is represented principally in seas. Few of its species live in brackish waters and still fewer in fresh waters.

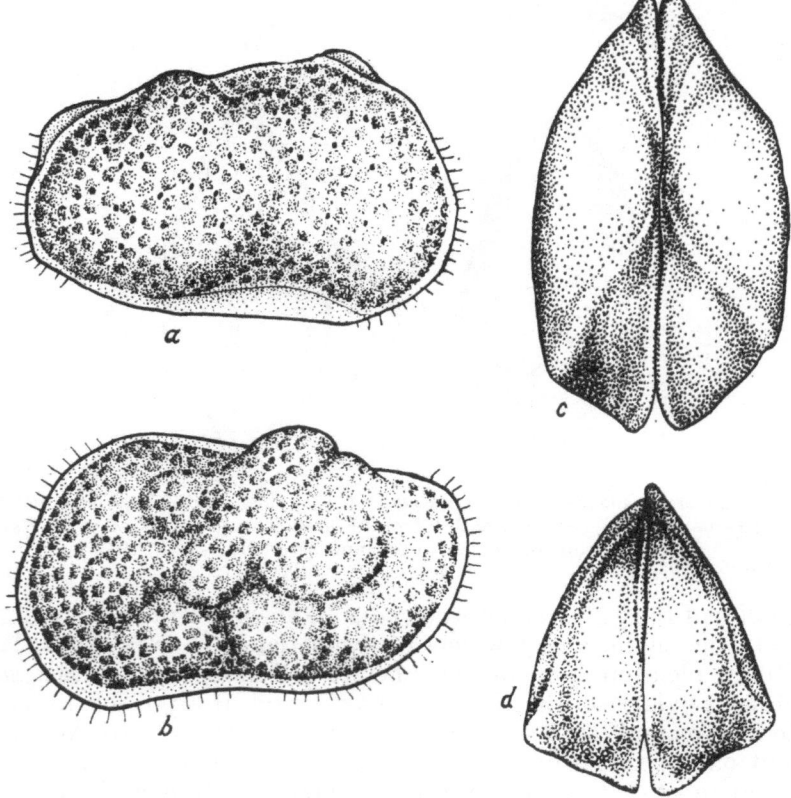

Fig. 55. a,b - Ostracoda *Cytherissa tuberculata* BRONST.; a - female; b - male, shell length 1.16 mm; c,d - *Cytherissa triangulata* BRONST., view from above and from the front, length 1.1 mm. After BRONSTEIN, 1947.

Cytherissa is a fresh-water genus, and all its 14 species are to be found in Baikal. Only one of them, *Cytherissa lacustris* SARS, has a wide distribution, being represented in Baikal by the endemic subspecies *baicalensis* BR. Thus, Baikal is a centre of intense specialisation in the genus *Cytherissa*. BRONSTEIN (1939), pointing out that representatives of the genus *Cytherissa* have been found in Tertiary deposits in the USSR, supposes that by its origin this genus is connected with brackish basins.

In this way, 31 out of the 33 ostracod species known in Baikal are endemic, while two are represented by endemic subspecies. It is a surprising fact that all these species belong to only three genera out of the 30-odd genera represented in the Holarctic. In the variety of the ostracod fauna Baikal is evidently inferior to Lake Ohrid, which has been found to contain 36 species distributed among 10 genera. The endemism of these species is also striking (66%). BRONSTEIN (1939, 1947) points out that in their characteristics the ostracods of Baikal are closely connected with ancient (most probably Tertiary) primitive fresh-water ostracods dating from pre-Pliocene time.

The numerical density of ostracods in Baikal is very high. On rocky and sandy bottom they comprise a sizeable part of the microbenthos. Among the species known to us, *Candona unguiculata* var. *baicalensis* BR. and *Cytherissa elongata* BR. are particularly numerous. More than 10,000 specimens of the former are sometimes counted per square metre on the stones of the littoral. Evidently ostracods play an important part in the diet of young fish and numerous Baikalian gammarids and mollusks.

There is a regrettable lack of more detailed data on the distribution of the various species of this group among the regions of Baikal. It is to be expected that studies in this direction will make it possible to separate purely Baikalian forms peculiar only to the open parts of Baikal from forms living in bays and gulfs. The latter, as can be supposed from an analogy with the other faunal groups, also live in other water bodies of the Baikal area.

The Cladocera of Baikal have not been studied sufficiently. In the plankton of its open waters one can sometimes come across *Daphnia longispina* O.F.M. and *Bosmina longirostris* O.F.M. Small amounts of them appear in summer and linger on till autumn. In some years individual specimens can be found also in the sub-ice period. Besides, *Chydorus sphaericus* O.F.M. at times occurs in open waters. But cladocerans are not typical of deep-water regions. They abound only in gulfs, shallow bays and in areas lying opposite to river mouths, where large numbers of them are often observed. The Cladocera species residing there include the above-mentioned *Daphnia longispina*, *Bosmina longirostris* and *Chydorus sphaericus*, as well as *Bosmina coregoni* BAIRD., *Daphnia galeata* SARS, *Sida crystallina* O.F.M., *Alona affinis* LEID., *Ceriodaphnia pulchella* SARS, *Leptodora kindti* FOCKE and *Daphnia cristata* SARS.

The sors and sheltered shallow parts of gulfs are evidently in-

106

habited by other Cladocera species. There are no specifically Baikalian forms of cladocerans in Baikal.

In pre-estuarine regions and in gulfs cladocerans comprise a considerable part of the crustacean plankton, often having a higher biomass as compared with copepods and serving as food for numerous young fish, especially amidst thick plant growths in the littoral.

Malacostraca

Anaspidacea. In recent years species of the genus *Bathynella* belonging to the Anaspidacea, the oldest Malacostraca order, and the family Bathynellidae have been found in Baikal at depths of from 20 to 1,440 metres.

A. Y. BAZIKALOVA (1954c) has described two species of Bathynellidae from Baikal, *Bathynella baicalensis* BAZ. and *B. magna* BAZ. (fig. 56). Both of them have a white semi-transparent wormlike drawn-out body 1.7 to 3.4 mm long, with short antennae and

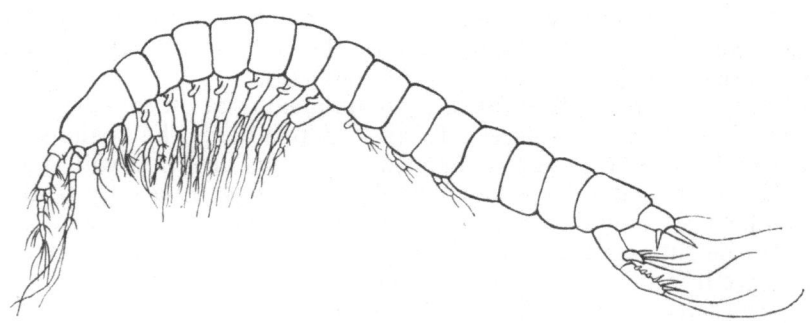

Fig. 56. *Bathynella magna* BAZIK., side view, body length 2.1—3.4 mm. After BAZIKALOVA, 1954.

no eyes. The Baikalian species *B. baicalensis*, according to A. Y. BAZIKALOVA, is related to *B. notans* VEJD. and *B. chappuisi* DELACHAX, especially to the latter, which lives in subterranean waters in South-East Europe (the Balkans) and West Germany. *B. magna* stands somewhat apart.

The family Bathynellidae consists of three genera: *Bathynella*, *Parabathynella* and *Thermobathynella*, with a sole species living in a hot spring in the Congo at a temperature of 55° C (CHAPPUIS, DELAMARE-DEBOUTTEVILLE, 1954). The genus *Parabathynella* is known from the subterranean waters of the Balkans, the Malay Peninsula and Japan. The genus *Bathynella* is represented by several species in West and South-East Europe and 6 species are known from Japan (UENO, 1954). Everywhere species of the genus *Bathynella* live in wells, cave waters, capillary ducts between sand grains, pebble and rock debris on the coasts of seas and lakes and in the valleys of rapidly flowing rivers. In Europe, *Bathynella* species dwell mostly in subsoil waters from where they penetrate into wells, cave waters, springs and other such places. In the opinion of CHAPPUIS, the European Bathynellidae are a relict of the fauna of the Sarmatian Sea which covered a great part of Central Europe. He thinks, however, that the genus *Bathynella* is a descendant of a more ancient fresh-water fauna, fossils of which are dated back to the Palaeozoic. The Japanese authors stress the common origin of the Japanese and European Bathynellidae. The Japanese species live in subterranean waters among Palaeozoic deposits at a temperature of more than 15° C.

According to BAZIKALOVA (1954), the Bathynellidae emigrated from the Tethys in Cretaceous time or in the Palaeogene: as a result of the gradual retreat of that sea and the freshening of lagoons and residual lakes the Bathynellidae turned into fresh-water forms and moved to subsoil waters. In her opinion, the Bathynellidae also lived in the region which is now covered by the waters of Baikal and found themselves in the depths of the lake as a result of the gradual subsidence of that region.

Isopoda. The isopods are represented in Baikal by a sole genus, *Asellus*, from the family Asellidae.

Baikal is known to be populated by 5 species of this fresh-water genus, which is altogether poor in species. Evidently they present a fairly uniform group well differing from the widespread European *A. aquaticus* L. and its closest relatives. Among the Baikalian species. *A. angarensis* GRUBE (fig. 57,b) and *A. baicalensis* GRUBE (fig. 57,a) are more numerous than *A. korotnewi* SEMENK., *A. minutus* SEMENK. and *A. dybowskii* SEMENK. (fig. 57,c). According to BIRSTEIN (1939, 1951), the Baikalian species can be grouped in two subgenera: *Baicaloasellus* and *Mesoasellus*. The latter has only one species, *A. dybowskii*; the other four belong to the former.

A. angarensis usually inhabits stones in the coastal belt and enters the Angara. *A. baicalensis* lives mostly on rocky patches of the open littoral and among algae. It also enters the Angara.

Very peculiar characteristics are displayed by *Asellus dybowskii* (fig. 57,c). It has a strongly flattened body reaching 1 cm in length, with unusually long lower antennae exceeding many times the length of the body, thin and long thoracic feet and very small, almost rudimentary eyes. It lives at 100—1,000 m and deeper.

G. B. GAVRILOV (1949) found that the *Asellus* species in Baikal feed upon detritus, bacteria and benthonic diatoms. The period of propagation is protracted and females with eggs occur the year round, but more frequently from January till June.

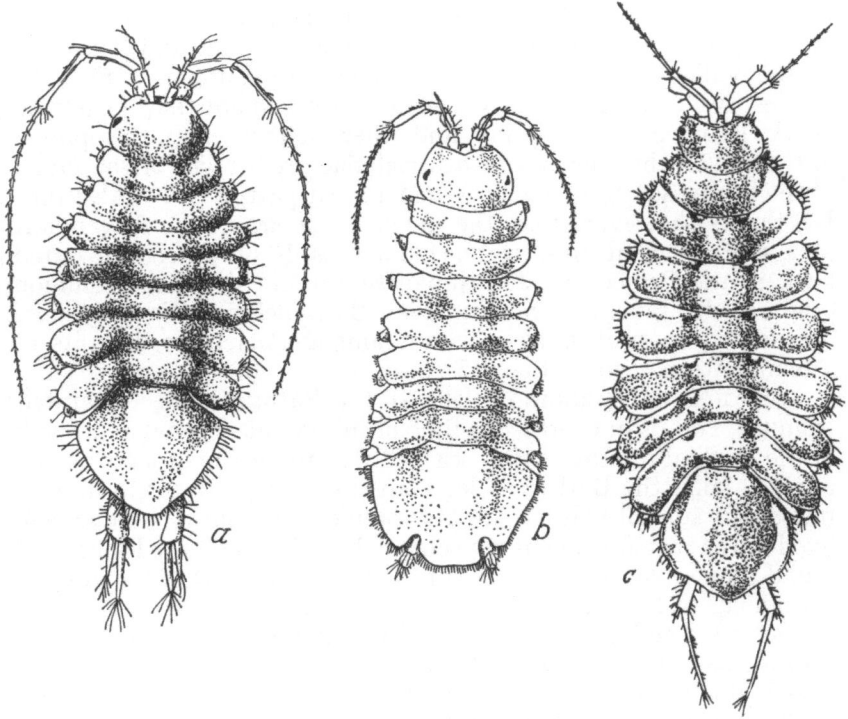

Fig. 57. a - *Asellus (Baicaloasellus) baicalensis* GRUBE; b - *A. (B.) angarensis* GRUBE, body length 5—6 mm; c - *A. (Mesoasellus) dybowskii* SEMENK. body length 10 mm, lower antennae (removed) are three times as long as the body. After BIRSTEIN, 1951.

None of the Baikalian *Asellus* species live in the waters of the Baikal area, with the exception of the Angara, the upper reaches of which they penetrate from Baikal.

In Siberia, Asellidae have been found only in the Ob-Irtysh drainage area (*Asellus latifera* BIRST.) and in the Tsipa-Tsipikan lakes of the drainage area of the Vitim, a tributary of the Lena (*Asellus epimeralis* BIRST.) *A. hilgendorfi* BOVAL. lives in the delta of the Lena and the drainage area of the Amur. All these species belong to the subgenus *Asellus*, which also has an extensive European distribution. One of the varieties of *Asellus hilgendorfi* (*A. h.martynovi* BIRST.) has been discovered in the middle reaches of the Lena (KARANTONIS, KIRRILLOV & MUKHOMEDIAROV, 1956).

According to Y. A. BIRSTEIN (1939, 1951), the subgenus *Meso-asellus*, represented in Baikal by the sole species *A. (M.) dybowskii*, is a relict of a very remote past. Besides Baikal, *Mesoasellus* species live in caves of Japan and California, In Y. A. BIRSTEIN's opinion, *Mesoasellus*, as more primitive, could be the initial form for the subgenus *Asellus*. As regards the subgenus *Baicaloasellus*, which has four species in Baikal, BIRSTEIN considers it possible that this subgenus has been formed in Baikal itself and has remote ancestors in forms close to *Mesoasellus*. He disagrees with V. V. ALPATOV (1923) and Y. N. SEMENKEVICH (1924) as to the existence of affinities between the Baikalian and North-American *Asellus*.

However, representatives of the subgenus *Baicaloasellus* have been found recently in caves of the State of Virginia in North America. In the opinion of BRESSON (1955), they are by their characters intermediary between the Baicalian species of *Baicalo-asellus* and the genus *Proasellus* living in West Europe.

Amphipoda. This order is represented in Baikal by the family Gammaridae, which has developed on an exceptionally large scale there.

The first few species of gammarids were discovered in Baikal and the Angara by PALLAS (1776) and GERSTFELDT (1859a), but the foundations of the knowledge of the Baikalian gammarids were laid by B. DYBOWSKY.

B. DYBOWSKY (1874, 1875, 1884) provisionally referred all the Baikalian gammarids known to him (97 species) to the genus *Gammarus*, establishing a separate genus, *Constantia*, only for *Macrohectopus branickii*. STEBBING (1899) distributed these species among 20 genera four of which (*Pallasea, Brandtia, Gammarus* and the above-mentioned *Constantia-Macrohectopus*) were already known at that time. V. P. GARYAYEV (1901) described a dozen new species and attempted to unite the Baikalian genera in three groups: Pachygammarini, Eugammarini and Acanthogammarini.

In his work published in 1915 V. K. SOVINSKY analysed the rich collections of A. A. KOROTNEV's expedition and supplemented the known species with several dozen new ones, establishing ten new genera.

After V. K. SOVINSKY the gammarids of Baikal and the Angara were studied by V. C. DOROGOSTAISKY (1917, 1922, 1930, 1936) and A. Y. BAZIKALOVA (1935, 1941, 1945, 1948b, 1951a, b, 1954a, 1957). BAZIKALOVA established several new genera and described more than 50 new species.

In recent years A. Y. BAZIKALOVA (1959) has described several more new gammarid species from the Maloye More and established a new genus, *Metapallasea*. Today the number of described species stands at 240 and that of genera at 34. Such a variety of gammarids is not observed in any other continental body of water in the world.

About a third of all species of gammarids and more than a third of all genera inhabiting the fresh and marine basins of the world have been described from Baikal.

It must be said, however, that despite the tremendous amount of work performed in the study of the systematic composition of gammarids, some genera, and in particular such polymorphic ones as *Eulimnogammarus* and *Micruropus*, call for a re-investigation, because a number of species should evidently be brought together in synonymy in view of the intraspecific variability peculiar to them. But even in this instance no essential changes in the systematics of the Baikalian gammarids can be expected*.

In all the tremendous variety of forms of Baikalian gammarids, the least specialised group not armed with cutaneous outgrowths in the form of spines, carinae, etc., is constituted by the genus *Eulimnogammarus*, which comprises up to 40 species (fig. 61,1). It has a related genus in the deep-dwelling *Abyssogammarus* (fig. 58,2). This central group is adjoined by the considerably specialised but unarmed genera *Ommatogammarus* (fig. 58,1), perhaps *Macropereiopus* and the well-defined genus *Odontogammarus* (fig. 61,4) A. Y. BAZIKALOVA also lists here the genera *Pachyschesis*, *Lobogammarus* and *Polyacanthisca* established by her.

The most common features of this group are: a more or less elongated, smooth, slender, laterally compressed body without or with only rudimentary cutaneous outgrowths, well developed rudder feet, long antennae, with a multiarticulate flagellum on the upper antennae, a normally developed caudal section, numerous spinules on the caudal and, not infrequently, ventral segments.

Species of this group inhabit a wide diversity of biotopes; they are very mobile and, as a rule, good swimmers. The size of the body (without appendages) ranges from 10 to 70 mm. Body coloration is most varied: blood-red, green, violet, brown, pink, etc., often with a very complicated pattern. All this motley group of genera can be called, after V. P. GARYAYEV, Eugammarini.

The next group unites forms characterised primarily by the presence of cutaneous outgrowths on body segments in the form of carinae, spines, denticles, knolls, ridges, etc., situated both on the sagittal dorsal line of all or some segments and sometimes on their sides. As the first group, in most cases the species of this group possess a fairly slender laterally compressed body, well developed antennae,

---

* Recently A. Y. BAZIKALOVA (in litt.) has revised the genus *Micruropus*, as a result of which some of the species were brought together in synonymy, some species were described as new ones, and a new genus, *Pseudamicruropus*, was singled out from the genus *Micruropus*. Thus, instead of 32 species of the old genus *Micruropus*, 30 species are considered to be real and four of them have been included into the genus *Pseudamicruropus*.

strong rudder feet (with few exceptions) and, as rule, a multi-articulate flagellum on the upper antennae.

This group of armed gammarids, which partly corresponds to

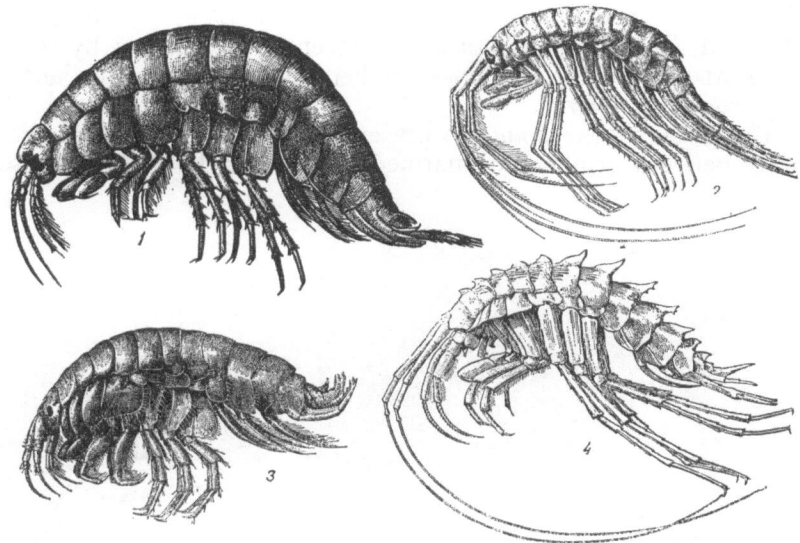

Fig. 58. Gammarids: 1 - *Ommatogammarus albinus* DYB., body length up to 25 mm; 2 - *Abyssogammarus sarmatus* DYB., body length up to 63 mm; 3 - *Crypturopus pachytus* DYB., body length up to 18 mm; 4 - *Garjajewia cabanisi* DYB., body length up to 80 mm. After BAZIKALOVA, 1945.

V. P. GARYAYEV's Acanthogammarini, includes several genera, many representatives of which are the largest in size of the entire gammarid fauna of Baikal. Among them are species of the genera *Acanthogammarus* (fig. 61,5, 7), *Garjajewia* (fig. 58,4) and *Paragarjajewia*. They are evidently closely related to the peculiar genus *Boeckaxellia* (= *Axelboeckia*, fig. 61,6) and also the genus *Brandtia* (the range of *Brandtia lata* forms, fig. 59,1), distinguished by a broad truncated caudal section. Also highly peculiar are species of the genus *Spinacanthus* (after A. Y. BAZIKALOVA, a subgenus of the genus *Brandtia*) living on sponges. They are tenacious spined crustaceans with a vivid mottled coloration.

The genus *Hyalellopsis* (fig. 61,2) is still more peculiar. Its species, which lead a fossorial life, have a broad compact body with a strongly truncated caudal section and slightly outlined knolls, ridges and other cutaneous projections. This genus is closely allied to the genera *Gammarosphaera* and evidently *Carinurus* and *Coniurus*.

A special branch of armed gammarids is comprised by the genera

Pallasea (fig. 59,2), *Parapallasea* (fig. 60,2), *Metapallasea* and the genera *Hakonboeckia* and *Ceratogammarus* adjoining them.

According to A. Y. BAZIKALOVA (1945), the genus *Pallasea* has a close relative in the genus *Poekilogammarus* (fig. 61,3) from which the pelagic species *Macrohectopus branickii* DYB. (fig. 60,1) has diverged. The genus *Metapallasea*, recently established by A. Y. BAZIKALOVA (1959), combines, in her opinion, traits of *Pallasea* and *Poekilogammarus*.

The genus *Carinogammarus* presents, as it were, an intermediate stage between armed and unarmed gammarids. It is characterised

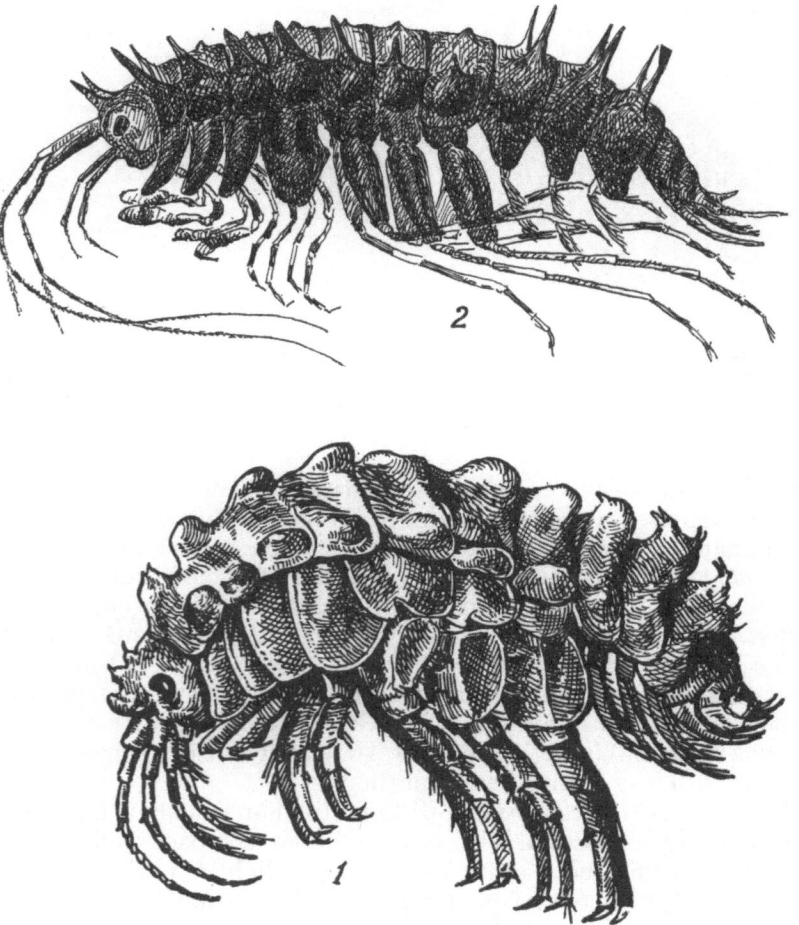

Fig. 59. 1 - *Brandtia lata* DYB., body length up to 18 mm. After BAZIKALOVA. 2 - *Pallasea bicornis* DOROG. After DOROGOSTAISKY, 1930.

by the presence of small carinae or knoll-like protuberances along the sagittal line of the dorsal surface of segments.

A group quite apart is comprised by the genera *Crypturopus,* .

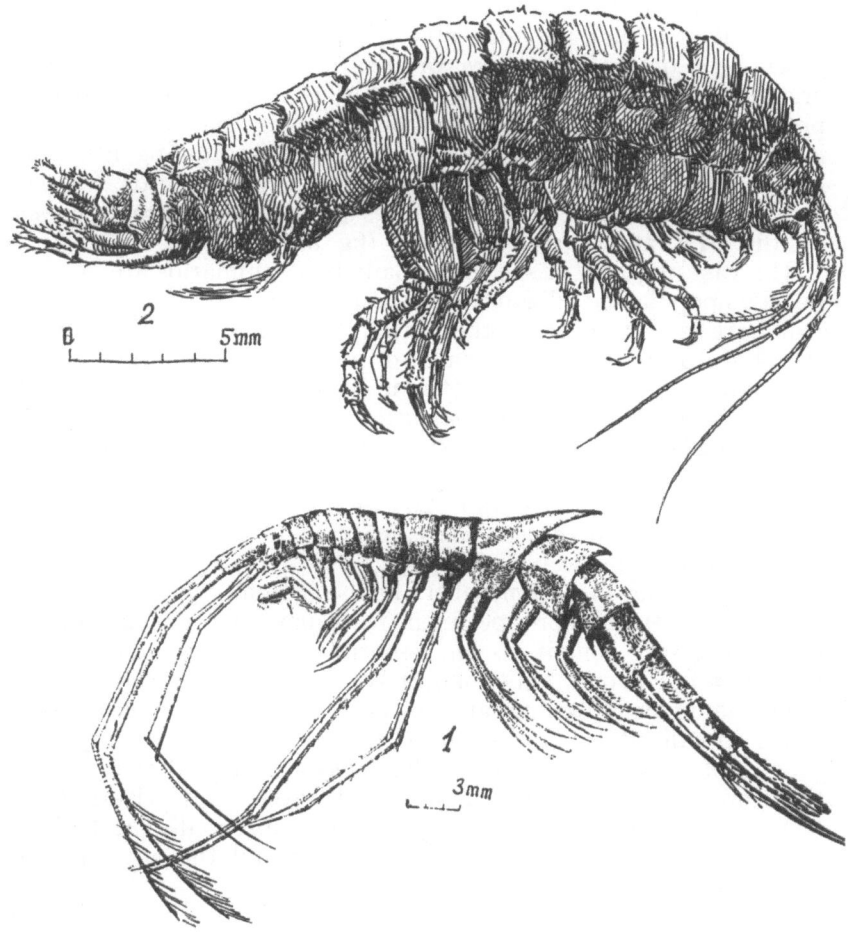

Fig. 60. 1 - pelagic gammarid *Macrohectopus branickii* DYB., body length 25—35 mm; 2 - *Parapallasea puzylli* DYB. body length up to 30—40 mm. 1 - after BAZIKA-LOVA, 1945; 2 - original.

(fig. 58,3) *Micruropus* and *Homocerisca*, characterised by a truncated, relatively broad and smooth body devoid of any projections, usually with short antennae and a uniarticulate flagellum on the upper antennae. The numerous species of these genera are predominantly burrowing forms. The genera *Gmelinoides* and *Baicalogammarus* also stand somewhat apart.

In the opinion of A. Y. BAZIKALOVA (1945), all this variety of Baikalian amphipods can be grouped around a few central genera, namely, 1. *Homocerisca-Crypturopus*, 2. *Acanthogammarus-Garjaje-wia*, 3. *Pallasea*, 4. *Eulimnogammarus*.

The Baikalian gammarids are also very abundant. They populate highly diverse biotopes and play an important part in biological processes. Some of them are weed-eaters or feed upon decaying organic remains, bodies of big animals, etc. Many are carnivores, preying upon small invertebrates or attacking fish caught in nets; it happens sometimes that, instead of an omul or grayling, fishermen haul up a skin sack literally packed with big gammarids. Particularly notable in this respect are species of the genera *Acanthogammarus* (fig. 61,7) and *Ommatogammarus* (fig. 58,1).

The vast majority of species of Baikalian gammarids are bottom-dwellers populating all zones of Baikal right up to extreme depths. Some are less attached to the bottom and rapidly move about in the above-bottom layers in search of prey; for this reason they can be characterised as above-bottom, benthonectic species (for instance, species of *Ommatogammarus* and *Abyssogammarus*). The mysidiform amphipod *Macrohectopus branickii* DYB. (fig. 60,1) leads an exclusively pelagic life, living in the mass of water and migrating between the deep layers (150—200 m and deeper), where it stays during the day, to the surface layers, where it preys upon small planktonic organisms at night.

Concentrating in vast numbers in the littoral, where they often number 30,000 and more specimens per square metre of the bottom, the gammarids constitute an essential part of the diet of the Baikalian fish population (BAZIKALOVA & VILISOVA, 1959).

The gammarids dwelling at great depths have a number of interesting peculiarities. For instance, the eyes of littoral forms are usually black, whereas those of deep-dwellers are red or pink; furthermore, in typically abyssal species the eyes are deprived of pigment and a deformity in their position is observed. The body of deep-dwelling forms is usually pale-yellow and the antennae are very long.

Most of the gammarid species propagate almost throughout the year and the propagation period is very long both in the abyssal and littoral species. Still, in many species one observes periods of particularly vigorous multiplication, predominantly in spring, and an attenuated propagation in autumn and the first half of winter.

In their habitat, the majority of the Baikalian gammarid species and genera are restricted to Baikal. Up to 50 species live in the upper section of the Angara. About 20 of them penetrate into the middle and lower reaches of the Angara and also into the Yenisei, along which they spread right up to its mouth and the Yenisei Gulf. They include, for instance, *Gmelinoides fasciatus* STEBB., *Eulimno-*

gammarus viridis DYB., *E. verrucosus* GERSTF., *E. cyaneus* DYB., *E. lividus* DYB., *Micruropus wahli* DYB., *M. glaber* DYB., *M. litoralis* DYB., *M. vortex* DYB., *Pallasea cancelloides* GERSTF., *P. kessleri* DYB., *Hyalellopsis* sp. *Brandita lata* DYB., and some others. The Baikalian species *Gmelinoides fasciatus* has been observed by P. L. PIROZHNIKOV (1937a, 1941) in the big running-water lake Nalimye and other lakes of the Yenisei system and in the system of the River Pyasina. The River Upper Taimyra is inhabited by *Eulimnogammarus verrucosus* and the Arctic lake Taimyr, by *E. viridis*. It is a noteworthy fact that in these lakes Baikalian immigrants live side by side with such relictary forms as *Pallasea quadrispinosa* SARS and *Mysis oculata relicta* LOVEN, while in Lake Taimyr they occur together with *Pontoporeya affinis* LINDSTR. and *Gammaracanthus loricatus* SABINE (GREZE, 1953).

It should be noted that in Baikal all the above enumerated species which spread up to the lower reaches of the Yenisei, live exclusively in the littoral and are more capable of enduring non-Baikalian conditions than other species. For example, *Gmelinoides fasciatus* and *Micruropus wahli* are often observed in Baikal's sors and coastal lakes and also in the lower reaches of big affluents. Besides, the sors and lower reaches of rivers are visited by such species as *Eulimnogammarus verrucosus*, *Micruropus possolskii* Sow., *M. talitroides* DYB. and some others, which develop local forms there.

The ability to survive of some species of Baikalian gammarids in such places as the Yenisei Gulf and Arctic relictary lakes, where they comprise an important component of benthos, enabled V. N. GREZE (1951b) to express a supposition regarding the possibility of their acclimatisation in other rivers and lakes of the Arctic seaboard of Siberia.

According to modern data, outside Baikal (discounting the Angara and the Yenisei) the Baikalian gammarid genera are represented only by *Pallasea*, which, besides the Baikalian species, also includes *Pallasea quadrispinosa* SARS and *Pallasea laevis* EKM. *Pallasea quadrispinosa* is widespread in the big lakes of the coastal regions of the Baltic and White seas and also in the Arctic relict lakes of the Siberian coast (Lake Taimyr and others). *Pallasea laevis* is known only from the fresh waters of the coast of the Novaya Zemlya. Unlike the other relict amphipods *(Pontoporeya, Gammaracanthus)*, *Pallasea* species prefer fresh waters.

The Finnish scientist S. SEGERSTRÅLE (1956, 1957, 1958) considers the *Pallasea* species living in the lakes of North-West Europe to be Siberians which migrated to Europe after the Great Glaciation. It can be supposed that the genus *Pallasea* was generally formed in Siberia.

The inclusion of a number of species from the Balkans, the Caspian Sea, Kamchatka, etc., into the genus *Carinogammarus* has proved to be incorrect (BAZIKALOVA, 1945). Likewise, it was a mistake to unite the species *Boeckia* from the Caspian Sea and *Axelboeckia (= Boeckaxellia)* from Baikal into one genus. There is no confirmation of the presence in Baikal of species of the genus *Gammaracanthus* which is common in the relict lakes and brackish waters of North Eurasia.

According to A. Y. BAZIKALOVA, the genus *Eulimnogammarus*, which is noted for the greatest specific diversity, is closest to the genus *Gammarus* abounding in fresh, brackish and marine basins, and the subgenus *Marinogammarus*, from which it differs only by the absence of spines on the segments of the metasoma and mesosoma. It also has some likeness to the subgenus *Gammarus*. MARTYNOV (1935b) found an affinity between *Eulimnogammarus* and the genus *Ostiogammarus* represented in the fresh and brackish waters of the Balkan Peninsula.

Among the group of armed gammarids, the Baikalian genus *Acanthogammarus* and the genera adjoining it have a certain relation to the relict species *Gammaracanthus loricatus* SABINE. A. Y. BAZIKALOVA (1940, 1945) and MARTYNOV (1935b) point to a similarity between the Caspian species *Boeckia spinosa* and the Baikalian genus *Hyalellopsis*. SCHELLENBERG (1937a, b) considers that the Baikalian *Hyalellopsis* is related to the subterranean genera of the group *Grangonyx*.

The genus *Gmelinoides* is related by authors to *Gammarus kusceri* KARAM. from the Balkans and the genera *Gmelina* and *Pontogammarus* from the Caspian Sea.

While studying the problem of the origin of the Baikalian gammarids A. Y. BAZIKALOVA, Y. A. BIRSTEIN & D. N. TALIYEV (1946a, b) measured the osmotic pressure of cavital liquid in some species and found it to vary within a very wide range, with species of each systematic group having their own amplitude of fluctuations in osmotic pressure. The highest internal osmotic pressure was found in species of *Eulimnogammarus*. *Odontogammarus* and *Micruropus*, which, in the opinion of these authors, corresponds to the closer relation of the first two genera to marine ancestors *(Marinogammarus)*. The lowest values of osmotic pressure were found in *Garjajewia* and *Spinacanthus*. The authors believe that these genera have evolved in Baikal itself and are therefore more distant from their marine ancestors.

A very low osmotic pressure was recorded in the ordinary lacustrine species *Gammarus (= Rivulogammarus) lacustris*, which does not live in Baikal. The authors write in this connection that *G. lacustris* is farther removed from its initial marine progenitors than all Baikalian gammarids.

Analysing the suppositions, contained in the literature, regarding

Fig. 61. 1 - *Eulimnogammarus czerskii* DYB., body length up to 133 30 mm; 2 - *Hyalellopsis costata* Sow., length up to 10 mm; 3 - *Poekilogammarus cyanurus* DO-ROG., length 10 mm; 4 - *Odontogammarus calcaratus improvisus* DOROG., length up to 35 mm; 5 - *Acanthogammarus flavus* GARJAJEV, length up to 20 mm; 6 - *Axel-boeckia carpenteri* DYB., length up to 30 mm; 7 - *Acanthogammarus maximus* GARJAJEW., length up to 70 mm. After DOROGOSTAISKY. In litt.

the kinship of the Baikalian gammarids with gammarids from other bodies of water, we come to the conclusion that, with the exception of the genus *Pallasea*, no unquestionable and close relation is traced between the Baikalian genera and the genera represented in the Caspian Sea, the Balkans, the seas of the northern hemisphere and other basins, both fresh-water and brackish. The existing similar characteristics can evidently be well explained by convergence and are of no serious phylogenetic importance. Evidently the Baikalian range of genera has formed independantly of the Pontocaspian and Balkan centres of specialisation and only their most ancient roots could be common.

D. N. TALIYEV (1955) maintained that the Baikalian and Caspian gammarids had their original birthplace in the northern seas, from which they penetrated first into the Pontocaspian basin and from it, via running waters, into Baikal. A. MARTYNOV (1935b) expressed the supposition that Baikal and the Caspian Sea had received some related or common forms from the Sarmatian basin and that the latter acquired the main nucleus of this marine fauna in the Palaeogene from the more ancient body of water which lay in place of the Sarmatian basin. These ancestors made their way into Baikal through running waters and lakes in which, he believes, Siberia and Turkestan abounded in Tertiary time.

In the opinion of BAZIKALOVA (1945), the ancestors of the Baikalian gammarids originated from some unknown body of water, a derivative of the gradually freshening sea with a rich amphipod fauna which included ancestors of the Baikalian and Caspian genera and also the genus *Gammarus*.

Arachnoidea

In the waters of gulfs deeply indenting the coast of Baikal and also in sors one can observe aquatic spiders (Argyroneta) identical with Euro-Siberian species. They are not to be found in the open parts of gulfs, to say nothing of open Baikal.

Siberian forms of mites of the Hydrachnella group are often found in the sors of Baikal and in the mouths of its affluents, but they have not been studied systematically. I. I. SOKOLOV (1930) lists 91 species for Siberia and only two of them from the coast of Baikal: *Thermacarus thermobius* SOKOL. from the hot spring Khakusi on the north-eastern coast, and *Pionacercus leuckarti* PIERS from Baikal (near Maritui).

Open Baikal is inhabited by very small (1—1.5 mm) mites described by I. I. SOKOLOV (1944—45) first as *Baicalacarus vermiformis* n. gen. n. sp. (fig. 62) from the family Protziidae (Hydrachnellae) and later (1948) referred by him to the group Trombidida, genus *Stygothrombium*, subgenus *Cerberothrombium*. The group Trombidida is represented mostly by terrestrial forms, and only the genus

118

Stygothrombium includes several fresh-water species found in sub-terranean waters (wells) in Yugoslavia. I. I. SOKOLOV considered the genus *Stygothrombium* to be a Tertiary relict whose modern distribution is widely disconnected.

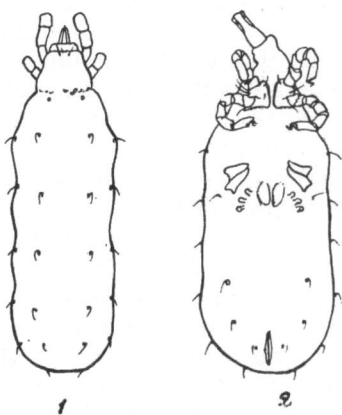

Fig. 62. Water mite *Stygothrombium vermiformis* SOKOL. 1 - female, dorsal view; 2 - ventral view, body length up to 1.5 mm. After SOKOLOV, 1952.

In Baikal this species lives in the littoral on stones among algae in empty leech and mollusk casts which the mites probably gnaw through, eating away their contents.

Not long ago three species belonging to the genus *Parasaldanel-lonix* of the order Halacarae were found at a depth of 34—40 m in the area of the Olkhonskiye Vorota Strait of Baikal (SOKOLOV, 1952). One of them, *P. baicalensis* SOKOL., is new to science; the other two are believed to be varieties of the species *P. parisculatus* WALT. inhabiting lakes of Switzerland, Germany, Scotland and the Kola Peninsula, and *P. typhlops* WIETS. The latter inhabits sub-terranean waters in Yugoslavia and Hungary. Evidently, the species of the genus *Parasaldanellonix*, just as those of *Stygothrombium*, present relicts of the Tertiary fauna of Eurasia which has survived only in a few places.

It can be supposed that the Hydracarina fauna of open Baikal is not limited to the species mentioned above.

Tardigrada. These curious poorly mobile semi-transparent microscopic animals can be often found on sandy soil in the open sections of Baikal's coastal belt. They also frequently occur on stones in empty leech sacks. The systematical composition of the Baika-lian tardigrades has not been studied.

Insecta

Plecoptera. In the waters of some gulfs one can now and then come across aquatic stages of stone flies which are apparently identical with ordinary Siberian species but have not been studied systematically. The small dull-black larva of a stone fly whose specific position has not been ascertained occurs in the open littoral of the region adjoining the outfall of the Angara. In the same region June is marked by the mass outcrop of imagines of the stone fly *Arcynopterix dichroa* KLP. Larval stages of this fly can also be observed in Baikal, but only in the immediate neighbourhood of the outflow of the Angara. They are very numerous in the Angara.

Trichoptera. Thanks to the studies conducted by A. V. MARTYNOV (1909, 1910—14) and I. M. LEVANIDOVA (1948), up to 50 imaginal stages of caddis flies are known on the coast of Baikal. But most of their larvae live in coastal lakes and sors and are not connected with Baikal's open waters. They include the Palaearctic species of the families Phryganeidae, Sericostomatidae, Molannidae, Leptoceridae, Limnophilidae, and others.

Few of them, in the larval stage, have been found in relatively open bays and gulfs, where, according to I. M. LEVANIDOVA (1948), 21 species from the above-mentioned families have been discovered. Especially frequent are *Phryganea rotunda* ULM., *Molanna palpata* Mc. L., species of *Leptocerus*, and some others.

These species do not enter the open parts of Baikal.

In the open regions, 14 species of caddis flies have been described belonging to the family Limnophilidae and subfamily Apataniinae (MARTYNOV, 1909, 1910—1914, 1924; BEBUTOVA, 1941). The family Limnophilidae is divided into two subfamilies: Limnophilinae and Apataniinae. All Baikalian species inhabiting open waters belong to the latter. The subfamily Apataniinae, in turn (after MARTYNOV), comprises two groups: Apatanini and Baicalinini. Of late SCHMID (1953—54), revising the caddis fly systematics of the subfamily Apatanini, established a separate tribe, Thamastini, abolishing the group Baicalinini, all the genera and species of which were included into the Thamastini tribe. In this review, however, we apply MARTYNOV's system.

The tribe Apatanini has the genus *Archapatania* (*Apatania* of SCHMID), represented in Baikal by two species: *A. baicalensis* MART. and *A. nigrostriata* MART. The other genera and species belong to the tribe Baicalinini.

The genus *Archapatania* is widespread in Central and East Asia and North America. The Baikalian species of this genus prefer gulfs, bays and shallows and do not occur, as a rule, in the open littoral.

Among the Baicalinini, only one species (*Radema infernalis* HAG.) is known also from the River Lena and creeks flowing into Baikal.

Thus we can regard this group as having been wholly formed in Baikal.

In A. MARTYNOV's opinion, the genera *Thamastes* and *Radema* (Baicalinini) originate from primitive *Apatania*-like forms in Baikal itself, and the Baikalian species of the genus *Archapatania* can be regarded as a remnant of these primitive forms. The most characteristic features of the whole of the Baicalinini group, according to MARTYNOV, include: strong sexual dimorphism, a big onion-shaped head, reduced wings in some species and primitive organisation of genital appendages. The larvae are characterised by the reduction of the branchial apparatus, with one pair of branchiae being preserved as inconspicuous appendages only in *Baicalina foliata*. The Baikalian species are also distinguished by a very long period of larval life (2 to 3 years) and a comparatively large size.

In the open coastal belt the Baikalian caddis flies occupy a place of no small importance among other animals. In April, still before ice break-up, large black pupae of some caddis fly species from the genus *Radema* emerge from the water and squeeze out into the surface through cracks and thaw pores in the ice. After casting the pupal skin they stretch out their tightly folded wings, run about on the ice and try to fly towards the shores. In broken spring ice one can find dead frozen-in pupae of these insects which failed to reach the surface. After break-up the surface of nearshore waters teems in vast amounts of minute (4 to 6 mm) dark-grey or almost entirely black insects, imagines of caddis flies predominantly of *Thamastes dipterus* HAG. (fig. 63,2) and *Baicalina reducta* MART. (fig. 63,1), swimming about briskly in groups or individually. Both species have only the front pair of wings; the back wings are reduced to small rudiments. The back and middle pairs of tarsi are modified for swimming; the body and wings are thickly haired, due to which the insects hold well on the surface and do not attempt to fly. The mating and deposition of eggs also take place in the water, although considerable numbers of the insects also accumulate on stones at the edge of water. Their larvae live on the rocky bottom mostly in the 2—4 m zone of depths.

At the beginning of June and later, till July inclusive, the mass metamorphosis of other species takes place. Among them, *Baicalina bellicosa* MART. is especially numerous. The big bright-yellow larvae of this species (12—17 mm) live on the rocky bottom in the littoral. On calm days in July, preferably in the morning, one can observe an interesting sight. Numerous pupae of *B. bellicosa* and other caddis fly species rise from the bottom to the surface, float about for some time, then change to the imaginal stage, try to fly up but alight and rapidly "swim" shoreward. The larvae of some species crawl along the bottom towards the shore and pupate at the edge of the water, crawling upon the stones to emerge.

The females deposit eggs in the form of small rounded jelly-like lumps. Some species lay these casts at the edge of the water and often even on the shore, putting them into water-filled pits between stones. The casts get fairly strongly glued to the substratum, such

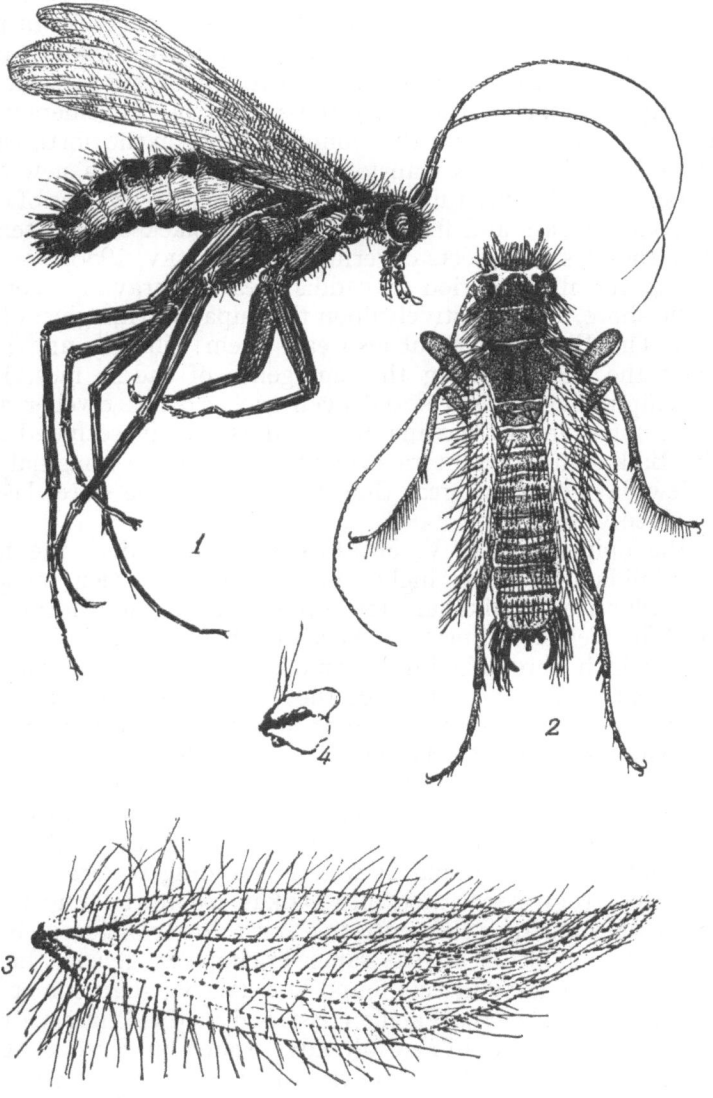

Fig. 63. Caddis-flies of the Baicalinini group: 1 - *Baicalina reducta* Mart., body length up to 5—7 mm; 2 - *Thamastes dipterus* Hag., body length up to 5—7 mm; 3 - a right front wing of the same species; 4 - rudiment of a back wing (3 and 4 strongly magnified). After Golyshkina, in litt.

as stones and other hard objects. Towards the summer and especially the autumn, thanks to the gradual rise in the level of the lake, the casts find themselves far from the shore.

Tremendous amounts of caddis flies accumulate at the edge of the water and on coastal rocks in June, which is the period of the maximum emergence. Thousands of insects can often be counted on one square metre of the shore. They swarm on rocks in places protected from wind in a layer sometimes up to 10 cm thick, rolling down in heaps to the foot of the rock at the slightest breeze or when somebody approaches them. In the period of the emergence of caddis flies from the water the mountain taiga on the north-eastern coast of Baikal becomes transformed, as if donning a new attire. The branches of larches turn absolutely black with swarms of caddis flies. Young cedars and firs look like fluffy black conical tents, so thick is the mass of insects covering them (GUSEV, 1956).

During the flight period of caddis flies the graylings converge near the shore, feeding actively upon the pupae and imagines of these insects. The Baikalian omul also eats them willingly and lingers on near the shores during the emergence of caddis flies. Forest birds, chipmunks, squirrels and even such a taiga dweller as the bear are not averse to feasting upon caddis flies in secluded spots.

The Baikalian Trichoptera evidently live in the imaginal stage for a few days only, whereas their larval life in the water lasts for up to three years.

In the opinion of A. V. MARTYNOV (1929, 1935a) the family Limnophilidae, which is highly varied and very rich in species and to which the Baikalian Trichoptera belong, was formed and evolved in Tertiary times in North-East Asia and North-West America which were linked in its time by the so-called Bering land. In the course of a large part of the Tertiary period, MARTYNOV thought, the climate of that region was temperate and partially quite cool. A strong impetus to the spread of this heat-loving fauna and in particular its migration far to the south (up to the Himalayas) and south-west (the Altai, Turkestan, Iran, Europe) was provided by the Miocene cooling. The ancient extinct ancestors of the Baicalinini group could live in running as well as standing waters; a certain part of this group penetrated into Baikal, where it found itself in isolation and so further differentiated. In the lower Oligocene Europe was already inhabited by the same Trichoptera genera that live there today. The fossil insects found in Tertiary (Oligocene or lower Miocene) deposits on the Pacific coast and in the Primorye region show that the Limnophilidae already existed in Europe at that time. MARTYNOV considers that the divergence of the Baicalinini group from the stock common with the Limnophilidae could begin at least in early Miocene time and probably even in an earlier epoch and that, since the larvae of these genera could live only in Baikal, it follows that in the Miocene or at the end of the Oligocene Baikal was a fresh-water lake.

We can assume that in their entire diversity the Baikalian caddis flies have originated from two or three Limnophilidae species which penetrated into ancient Baikal. New species and genera have evolved from them already in the lake itself. The peculiar conditions of life have made a sharp imprint on the life cycle and organisation of both the larval and imaginal stages of this group in Baikal. Particularly conspicuous among these specific features is the loss of the ability to fly, by the imaginal stages which could happen only in a vast and deep body of water where winged insects driven by wind far from the shore were bound to perish and could not have offspring.

The other peculiarity of the Baikalian caddis flies is the prolonged period of larval life resulting from the low temperature of the water.

It should be noted that the Baikalian caddis flies do not enter the Angara and that, conversely, the Palaearctic species living in the Angara do not appear in Baikal even in the vicinity of the outflow.

Diptera. Among Diptera, a significant part in the biocoenoses of Baikal is played by Chironomidae, but the systematics of this group has not been studied in detail yet. They were studied by A. A. LINEVICH (1948, 1957), A. A. CHERNOVSKY (1949) and I. M. LEVANIDOVA (1948). In gulfs deeply indenting the coast one can observe large numbers of larval stages of the same chironomids that inhabit abundantly ordinary bodies of water of Siberia and Europe. Dark viscous silt in these gulfs is inhabited by *Chironomus gr. semireductus* LENZ, which is characteristic of eutrophic and semi-eutrophic lakes; larvae of the groups *Endochironomus, Cryptochironomus* and many other also occur there. I. M. LEVANIDOVA (1948) offers a list of ten forms of chironomid larvae found by her in the gulfs of Baikal and five forms found in sors. A. A. LINEVICH (1948, 1957) counts in Baikal 60 chironomid species distributed among 20 genera; only 22 of them belonging to 10 genera live in the open regions of the lake. Eleven species from the latter group are endemic. Most of them belong to cold-loving representatives of the genera *Orthocladius, Diamesa, Micropsectra, Sergentia* and *Tanytarsus*. In the genus *Sergentia*, five species have been described constituting the endemic subgenus *Baicalosergentia*. They live on soft soils in all zones of the lake. The widespread endemic species are *Sergentia l. koschowi* LINEWITSCH (fig. 64) and *Diamesa baicalensis* TSHERN. (fig. 65). The *Sergentia koschowi* larvae are big (12 to 20 mm) and bright-red coloured with rudimentary eyes; they live on silty or sandy soils at a depth of up to 200 and more metres. The adults deposit eggs on the surface of the water, from where the eggs descend to the bottom layers. Stones in the coastal belt of open

124

regions are densely populated by tiny brown-green larvae of *Ortho-cladius l. setosus* TSHERN., *O. gregarius* LINEWITSCH and the above mentioned *Diamesa l. baicalensis*.

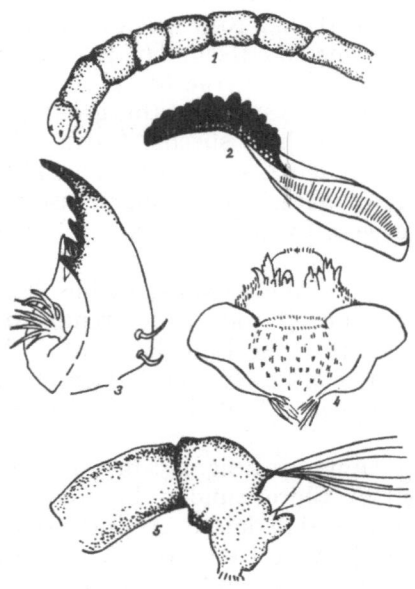

Fig. 64. Chironomid *Sergentia (Baicalosergentia) koschowi* LINEWITSCH. 1 - general view of the front part of a larva; 2 - submentum; 3 - mandible; 4 - hypopharynx; 5 - posterior part of the body. After LINEVICH, 1948.

Many littoral species of chironomids deposit casts at the edge of the water. *Sergentia* species deposit them on the surface of the water. Outside Baikal, species of chironomids related to or identical with Baikalian species live predominantly in mountain lakes and rivers of the Palaearctic. Two species of the genus *Sergentia* are known to live outside Baikal: *Sergentia longiventris* KIEFF. from European mountain lakes, big lakes in the drainage area of the Vitim, a tributary of the Lena, and other mountain lakes of the Baikal area, and *S. coracinum* ZETT. from the lakes of North Europe and Siberia.

Two species of *Sergentia* from the subgenus *Baicalosergentia* (*S. koschowi* and *S. baicalensis*) have been discovered in Frolikha, a deep mountain lake 7 km from Baikal. Evidently they have been brought there by the winds constantly blowing from Baikal. Some Euro-Siberian rheophile species (*Syndiamesa orientalis* TSHERN., *Diamesa longimanus* KIEFF, and some others) have been found in Baikal only opposite to the mouths of its mountain affluents.

Other Insecta. Larval stages of general Siberian species of Ephemeroptera occur in the depths of some gulfs and bays, but even

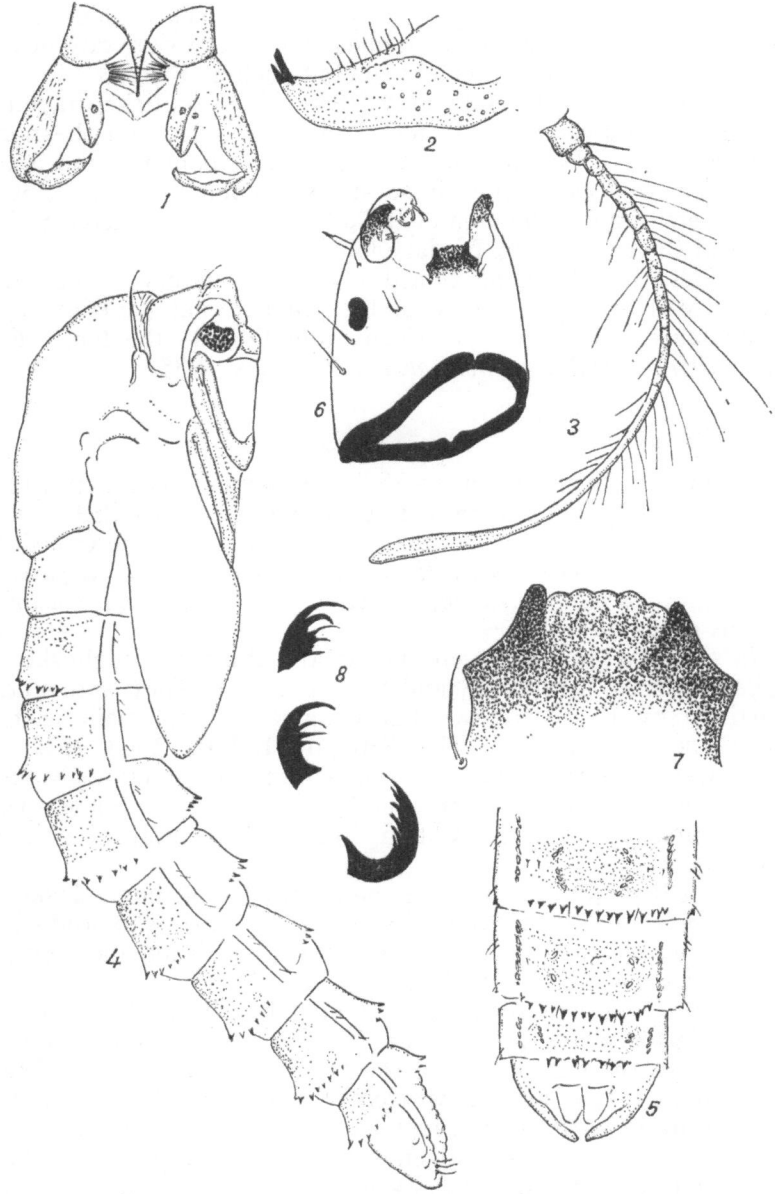

Fig. 65. Chironomid *Diamesa baicalensis* Tshern. 1 - hypopygium ♂; 2 - gonostyle of hypopygium ♂, distal part; 3 - antenna ♂; 4 - a pupa ♂, side view; 5 - the 6th to 9th tergites of the abdomen of the pupa ♀; 6 - the head of the pupa from the ventral side; 7 - submentum of the pupa; 8 - claws of the anterior pseudopods of the larva. After Linevich, in litt.

126

there they are very rare. They are not to be found in open Baikal. The same is true for Hemiptera (Hydrometra, Nepa), Odonata and Coleoptera, which are absent in open Baikal and very rare in gulfs and bays, but occur in sors and shallow sheltered bays.

Not long ago B. F. BELYSHEV (1956) discovered the dragon fly *Orthetrum albistylum* SELIS., found in a hot spring on the north-eastern coast of Baikal. The interesting fact about the find is that at present this species lives in sub-tropical countries (Japan, South Korea, etc.) and in the area of Baikal should be considered to be a relict of the heat-loving Tertiary fauna.

Besides the above-mentioned orders of insects whose larval stages lead an aquatic way of life, an external parasite of the Baikalian seal has been described from Baikal—the louse *Echinophthirius horridus* var. *baicalensis* Ass (Ass, 1935).

Mollusca

We find the first data on the Molluscan fauna of Baikal in GERST-FELD's work published more than 100 years ago (1859b). With R. MAACK's collections as the basis, GERSTFELD described the following five species: *Benedictia baicalensis, Baicalia angarensis, Pseudancylastrum sibiricum, Valvata baicalensis* and *Choanomphalus maacki*.

In the 'sixties ample material on the Baikalian mollusks was collected by B. DYBOWSKY and V. GODLEVSKY. They were studied partially by B. DYBOWSKY himself and in their main part by W. DYBOWSKY (1875, 1884b, 1886b, 1901, 1902, 1903, 1912). W. DYBOWSKY described 32 new species from the genera *Sphaerium, Pisidium, Benedictia, Baicalia* and *Choanomphalus*. Some species were cursorily studied by him from the anatomical point of view.

The next major step towards the knowledge of the mollusks was made by V. A. LINDHOLM (1909, 1924a, b, 1927), who studied the extensive collections of A. A. KOROTNEV's expedition and described 48 new species, and by A. STAROSTIN (1921).

After a thorough revision conducted by the author (KOZHOV, 1936b, 1945, 1950a, 1951) on the basis of new collections and the material of the expeditions of the U.S.S.R. Academy of Sciences, 84 mollusk species are known from Baikal today. They all can be divided into two groups: 1. representatives of the common general Siberian fauna, 2. the indigenous Baikalian fauna. The general Siberian (ordinary Palaearctic) genera or subgenera are represented only by 28 species; the other 56 species belong to indigenous Baikalian genera or subgenera, with the exception of one species of *Sphaerium* and two species of *Pisidium*.

Table XIII.

Mollusks of Lake Baikal
(After Кознov, 1936)

Name	Number of species		Note
	Baikalian endemic	General Siberian	
1	2	3	4
Prosobranchia			
Fam. Valvatidae			
Genus *Valvata*			
Subgenus *Megalovalvata*	4	—	Endemic to Baikal
Subgenus *Valvata*	—	1	
Subgenus *Cincinna*	—	1	
Fam. Hydrobiidae			
Subfam. Benedictiinae			
Genus *Kobeltocochlea*	2	—	Known only from Baikal and Lake Kosogol (Khubsugul)
Genus *Benedictia*	4	—	Endemic to Baikal
Subfam. Bithyniinae			
Genus *Bithynia*	—	1	
Fam. Baicaliidae			Endemic to Baikal
Genus *Baicalia*	32	—	
Genus *Liobaicalia*	1	—	
Total	43	3	
Pulmonata			
Fam. Physidae			
Genus *Aplexa*	—	1	
Genus *Physa*	—	1	
Fam. Limnaeidae			
Genus *Limnaea*	—	6	
Fam. Planorbidae			
Subfam. Choanomphalinae			Represented in Baikal and Lake Kosogol (Khubsugul)
Genus *Choanomphalus*	7	—	
Subfam. Planorbinae			
Genus *Planorbis*	—	7	
Fam. Ancylidae			
Genus *Acroloxus*	—	1	
Genus *Pseudancylastrum*	3	—	Endemic to Baikal
Total	10	16	
Order Bivalvia			
Fam. Unionidae			
Genus *Anodonta*	—	2	
Fam. Sphaeriidae			
Genus *Sphaerium*	1	2	
Genus *Pisidium*	2	5	
Total	3	9	
Grand total	56	28	

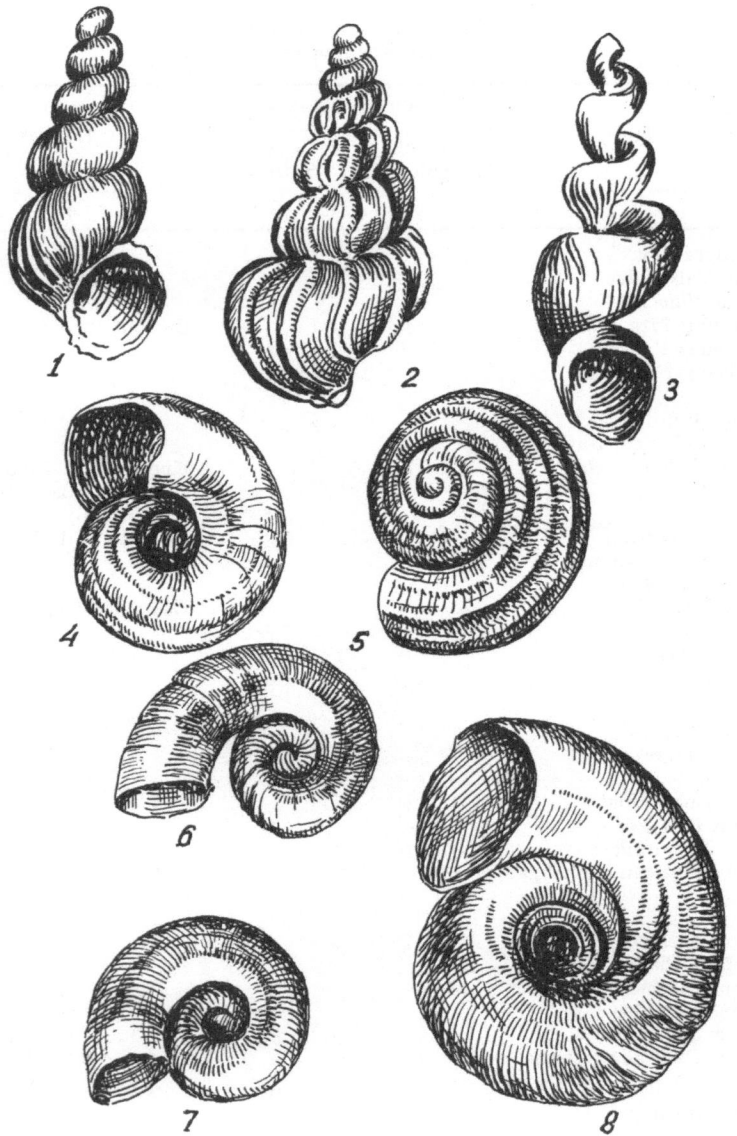

Fig. 66. Mollusks: 1 - *Baicalia korotnewi* LDH., shell height 15—19 mm; 2 - *Baicalia costata* DYB., shell height 9—10 mm; 3 - *Liobaicalia stiedae* DYB., shell height 10—11 mm; 4—5 - *Valvata (Megalovalvata) piligera nudicarinata* LDH., shell diameter 10—12 mm; 6—7 - *Valvata baicalensis* GERSTF., shell diameter up to 16 mm; 8 - *Choanomphalus maacki* GERSTF., shell diameter up to 11 mm. After KOZHOV, 1936.

The general Siberian group of species is alien to open Baikal and lives, as a rule, only in sors and depths of shallow bays and gulfs, whereas indigenous Baikalian species never occur in continental adjacent waters and sors and avoid the gulfs, disappearing as soon as environmental conditions begin to resemble those of ordinary lakes.

Mollusks occupy a very important place in the bottom community of Baikal. They are especially numerous in coastal shallows to a depth of 15—20 m (with the exception of the surf belt), attaining the highest numeric density on overgrown stones and also on sandy soil enriched with detritus. Beyond the 20—30 m depths the mollusk fauna grows markedly poorer, dropping to insignificant values on strongly silted soil. On silty soil at depths greater than 100—200 m it is conspicuously scant, and in the zone of great depths one can find only individual specimens of relatively abyssal forms, such as *Benedictia fragilis* DYB., *B. maxima* LDH. and *Valvata bathybia* LDH.

Being chiefly plant- or detritus-eaters, the Baikalian mollusks play a very important part in biological processes in the littoral zone of Baikal, and the places where mollusks live are also abundantly populated by other animals.

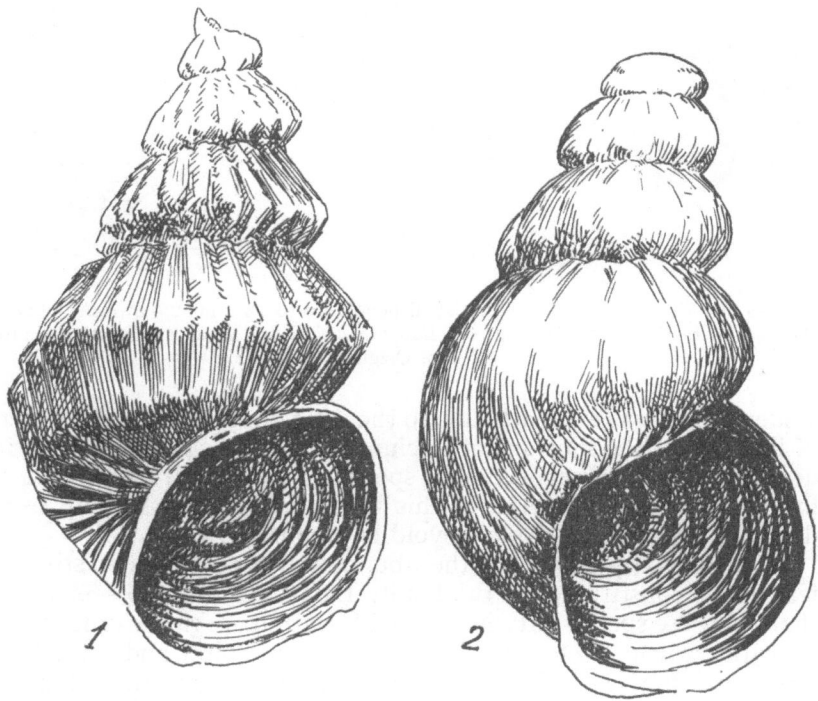

Fig. 67. 1 - *Baicalia variesculpta* LDH., shell height up to 8 mm; 2 - *B. herderiana* LDH., shell height 8 mm. Original.

130

A specific feature of the Baikalian mollusks is the extreme thinness of the walls of their shells; indeed, in some species *(Benedictia fragilis)* the last whorl almost entirely consists of superficial

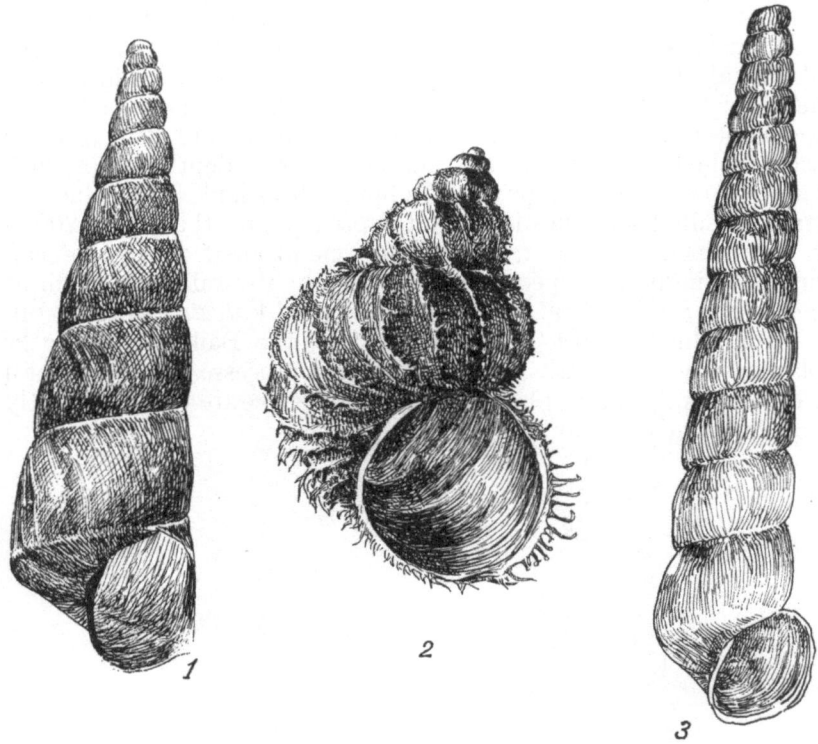

Fig. 68. 1 - *Baicalia carinata* Dyb., shell height up to 18 mm; 2 - *Baicalia ciliata* Dyb., shell height up to 10 mm; 3 - *Baicalia godlewskii* Dyb., shell height up to 20 mm. Original.

epidermis. This is evidently due to the low temperature of the water of Baikal and the shortage of calcium. The latter circumstance evidently prevents the evolution of specifically surf-belt forms with a firm reliable shell. In view of this the surf belt of Baikal up to a depth of 0.5 m is, as a rule, devoid of mollusks.

A group conspicuous for the abundance of species and specific features is constituted by the family Baicaliidae (fig. 66—69). The shells of these species vary in size and form and are richly ornamented with carinae, costae, knolls and very fine spiral and reticulate striation, often with filaments. Anatomically, they differ sharply from the families Micromelaniidae and Hydrobiidae. The nervous system of the latter has a metapodial commissure (fig. 69, 1, 2),

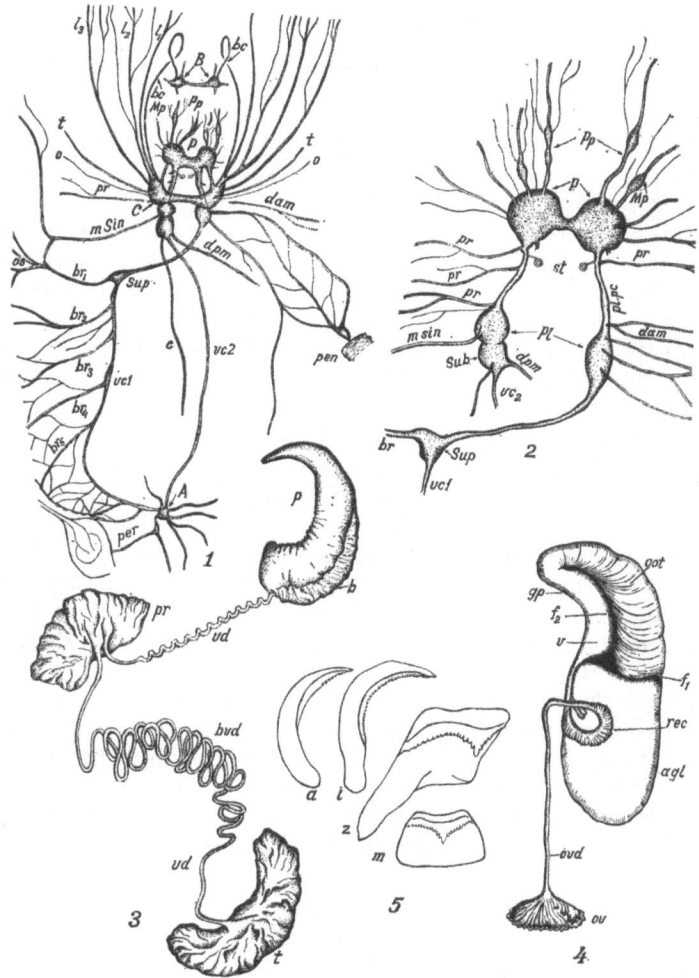

Fig. 69. 1—2 The scheme of nervous system of Baicaliidae: 1. *Baicalia florii*.
2. *B. oviformis* (pedal, pleural supraintestinal, subintestinal ganglion with its nerves)
A - abdominal ganglion; B - buccal ganglion; bc - buccal connective; br - branchial
nerves; C - cerebral ganglion; c - columellar nerve; dam - dextral anterior mantle
nerve; dpm - dextral posterior mantle nerve; L_{1-3} - labial nerves; Mp - metapodial
ganglion; m sin - sinestral mantle nerve; O - optic nerve; os - osphradium nerves;
P - pedal ganglions; Pl - pleural ganglions; pl.p.c. - pleuropedal connective; per -
pericardial nerve; pen - penis; Pp - propodial ganglion; pr - lateral nerves; $vc_1 vc_2$-
visceral connectives; Sub - subintestinal ganglion: Sup - supra-intestinal ganglion;
t - tentacle nerve; st - statocyst. 3 - Genitalia of the male *Baicalia florii:* t - testicle;
vd - vas deferens; bvd - ball of the vas deferens (it is rather elastic); pr - prostate;
p - penis; b - glandular swelling. 4 - Genitalia of the female *Baicalia oviformis:* oot -
ootype; gr - genital pore; v - vagina; f_1 - fissure dividing vagina from the ootype;
f_2 - fissure dividing ootype from the adventition gland; a.gl - adventition gland;
rec - receptaculum; ovd - oviduct; ov - ovary. 5 - The plate radula of *Baicalia florii:*
a - external-lateral; i - internal-lateral; z - intermediate; m - middle. After Kozhov, 1951.

132

whereas the Baicaliidae do not have it. The female genital apparatus of the Baicaliidae does not contain a distinct copulative sac (fig. 69,4) the presence of which is characteristic of the Hydrobiidae and Micromelaniidae. The outward similarity between shell structures of various representatives of these families from Lake Ohrid and the Caspian Sea, noted by many authors and especially by palae-ontologists, is certainly due to convergence and not to close re-lationship (KOZHOV, 1951).

Ranking second to it in the abundance of species is the subfamily Benedictiinae (fig. 70), represented by fairly large forms with a smooth rounded egg-shaped or round conoidal shell and a rather peculiar radula.

The Baikalian Valvatidae (subgenus *Megalovalvata*, fig. 66,4–7), giants compared to their relatives from other countries, are also represented by several species playing an important part in the biocoenoses of the littoral belt of Baikal.

Among the Pulmonata, an outstanding role in the life of the coastal belt is played by the genus *Choanomphalus* (family Planorbi-dae, fig. 66,8), represented by 7 species very densely populating littoral stones, sands, etc. Finally, among the Ancylidae we have 3 species from the endemic subgenus *Pseudancylastrum* (fig. 71), of special interest among which is the plical *P. kobelti* DYB. (fig. 71,1). Several species from the genera *Sphaerium* and *Pisidium* are also important.

Ties of relationship between the Baikalian species of mollusks and analogous modern groups from other countries as well as the fossil fauna present themselves in the following light. From Lake Biwa (Japan), *Choanomphalus japonicus* PRESTON was described (1916). By its funnellike umbilicus it strongly resembles the Baikal-ian species *Ch. maacki*, though the latter is several times bigger.

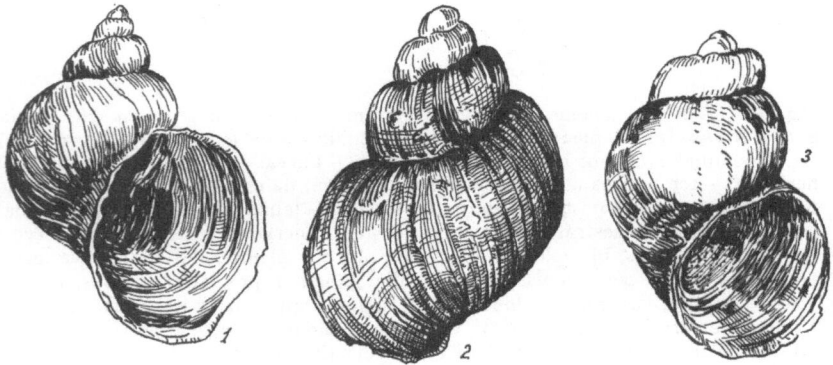

Fig. 70. 1—2 - *Benedictia fragilis* DYB., shell height up to 50 mm; 3 - *Benedictia baicalensis* GERSTF., shell height up to 22 mm. Original.

But nevertheless it is hardly a *Choanomphalus*, for its apex is very clearly submerged, which is never observed in *Choanomphalus* but is characteristic of *Gyraulus* species (LINDHOLM, 1927; KOZHOV, 1936).

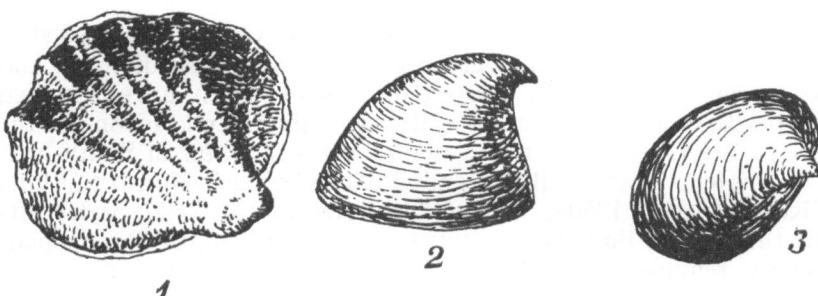

Fig. 71. *Pseudancylastrum kobelti* DYB., shell diameter 4 mm; 2—3 - *Pseudancylastrum sibiricum* GERSTF., shell diameter 8 mm. Original.

The literature (BERG, 1910) mentions, as relatives of the Baikalian *Choanomphalus*, some Planorbidae species, including *Planorbis paradoxus* STUR. from the Balkan Lake Ohrid and species of the genus *Carinogyraulus* from the same lake. In reality, however, they have no close relation to *Choanomphalus*. The same applies to the North-American planorbid genera *Pompholicodea* and *Carinifex* (KOZHOV, 1936b).

It is only in the big Mongolian Lake Khubsugul (Kosogol) lying at a distance of 200 km from Baikal that D. ANUDARIN found a new species belonging to *Choanomphalus, C. mongolicus,* which the author (KOZHOV, 1946) described.

The peculiar architecture of the shell of the Baikalian *Choanomphalus*, namely, the raised apex, the angularity of the last whorl or the presence of a carina around it, are undoubtedly of adaptive importance, for they lend rigidity to the shell. Apparently this character could be developed only in big bodies of water, in the coastal belt affected by the surf.

As a fossil, *Choanomphalus* has been found only in Tertiary deposits of the Baikal terraces (MARTINSON, 1956).

Recently Y. I. STAROBOGATOV (1958) has elaborated a new system of Planorbidae, under which *Choanomphalus* is merely one of the genera of the subfamily Planorbinae.

The entire Planorbidae family lives exclusively in fresh water. Fossil Planorbidae are known since the Mezosoic. Anatomically, species of the genus *Choanomphalus* are close to the genus *Gyraulus*, which is united by STAROBOGATOV with the genus *Anisus* and is widely represented in the fresh waters of the northern hemisphere.

It can be supposed that *Choanomphalus* separated from the stock common with *Anisus* back in Tertiary time and was formed in its present shape in Baikal and other big lakes of its system.

Close relatives of the Baikalian Ancylidae (subgenus *Pseudancylastrum*) have not been found so far either among fossil or modern faunae.

On the basis of anatomical investigations we have succeeded in proving the close affinity between the Baicalian Benedictiinae and the genus *Lithoglyphus*, representatives of which have survived in the Caspian Sea (*L. caspius* EICHW.), rivers and lagoons of the Black Sea (*L. naticoides* PF.), some bodies of water of South-East Europe (chiefly the Balkans) and also in some basins in North China (KOZHOV, 1945, 1950a). It should be noted that in Tertiary deposits on the coast of Baikal G. MARTINSON (1948a) found fossil gastropods highly reminiscent of the modern Chinese *Lithoglyphus* species. Benedictiinae also have some likeness to the North-American genus *Fluminicola*, but their true kinship can be established only after the study of their anatomy. *K. michnoi* LDH., certainly a representative of the Baikalian genus *Kobeltocochlea*, was found in Lake Khubsugul in Mongolia (LINDHOLM, 1929).

Distinguishable remains of representatives of Benedictiinae in a fossilised state have not been found yet, with the exception of some in Tertiary terraces on the south-eastern coast of Baikal (MARTINSON, 1956). The "*Benedictia*" species described by Y. RAMMELMEYER (1935, 1940) from lower Cretaceous deposits in the trans-Baikal area cannot be referred to the modern Baikalian genus *Benedictia*, for the poor degree of preservation of shells from RAMMELMEYER's materials renders exact definition absolutely impossible.

All Benedictiinae species are so stenotopic that even when entering the Angara they remain only in the uppermost part of it, not far from the outflow. They never enter the sors and gulfs.

The Baikalian Valvatidae from the subgenus *Megalovalvata* do not show close relationships either with modern or fossil Valvatidae. True, North America has several species of very small costate valvates (*Valvata tricarinata* and others) having a certain outer resemblance to the costate Baikalian *Valvata piligera*. But there are no sufficient anatomical grounds to regard these American species as close relatives of the Baikalians; by the structure of the radula (WALKER, 1918) they are closer to the widespread *Valvata aliena* WEST. and the European *V. piscinalis* MÜLL. than to the Baikalian species. Fossil forms unquestionably close to the subgenus *Megalovalvata* have not been found yet, either.

The numerous group of species of the Baicaliidae family was incorrectly placed by THIELE (1929—1934) and the author in the family Micromelaniidae represented chiefly in the Caspian Sea and lagoons of the Black Sea.

As has already been noted above, anatomic studies (KOZHOV,

1951) have shown that the initial opinion concerning the close kinship of the Baicaliidae with the Caspian Micromelaniidae was not justified. The latter are closer to the Hydrobiidae and specifically to the genus *Hydrobia* from which they evolved in middle and post-Tertiary time in derivatives of the Sarmatian Sea, whereas the Baicaliidae are anatomically quite apart and stand far from both the Hydrobiidae and the Caspian Micromelaniidae, as well as from the subfamily Pyrgulinae from Lake Ohrid.

In a fossilised state, unquestionable Baicaliidae have been found only in Tertiary deposits in the south-eastern coast of Baikal. The attribution of the Mezosoic *"Cerithium" gerassimowi* REISS from Jurassic or lower Cretaceous deposits in the Vitim valley and elsewhere in the trans-Baikal area to the genus *Baicalia*, which Y. S. RAMMELMEYER did (1940), is incorrect, because in spite of a certain outward similarity with the modern turriform *Baicalia*, their shell has a structure alien to the latter. Recently G. G. MARTINSON (1949b) has renamed the fossil Mezosoic *"Baicalia"* into *Probaicalia*, but this does not imply a close relationship to the Baicaliidae.

Not long ago an interesting gastropod fauna was found in Tertiary (Miocene-Pliocene) continental deposits in North-West China. The shells of species from these deposits very much resemble the shells of some species of modern and fossil Baicaliidae. G. G. MARTINSON (1955a, b, 1956, 1959a, b, 1960) refers some of them directly to the genus *Baicalia*.

In his latest works this author stresses a possible genetical connection between the Baicalian and Tertiary Pontocaspian ("Balkan") faunae of gastropods. The central place in the freshwater and brackish gastropod faunae of the Pontocaspian range and the Balkans is occupied by Micromelaniidae, Hydrobiinae and Pyrgulinae, which do not have, as has been said above, any close relation to the Baicaliidae of Baikal.

Recently G. G. MARTINSON & S. M. POPOVA (1959) have described several species of turriform gastropods from Tertiary deposits of the south of West Siberia which they also designate to the family Baicaliidae and even to the genus *Baicalia*. But it is possible also that the features of outward similarity in the structure of the shell of some turriform Tertiary gastropod species from West Siberia and the Baicaliidae is a result of convergent development.

It would be natural to suppose that in Tertiary time Central Asia had its own centre of formation of a distinct and rich mollusk fauna analogous to the Pontocaspian and Balkan centres, but the development of this fauna occurred in an earlier epoch.

Bryozoa

According to G. G. ABRIKOSOV (1924, 1927, 1959), Baikal is inhabited by the following species of bryozoans: *Plumatella emarginata* ALLM., *Plumatella fungosa* PALL., *Cristatella mucedo* CUV., *Plumatella repens* L., *Paludicella articulata* EHRB., *Fredericella sultana* BL., *Hislopia placoides* KOROTN.

The first four species have an extensive European and Asian distribution and are found in the Baikal area very frequently; in Baikal they live only in the depths of gulfs and bays deeply indenting the coast (the Mukhor, Chivyrkui and other gulfs). They do not enter open Baikal and can not therefore be regarded as characteristic representatives of the Baikalian fauna.

We have no information about the habitat of *Paludicella articulata* and *Fredericella sultana*, which are widespread in Europe, and their presence in Baikal requires confirmation. In any case, they do not live in open Baikal. As regards *Hislopia placoides* (fig. 72), this Baikalian endemic described by A. KOROTNEV (1901) is widely distributed throughout Baikal but lives predominantly in its more or less open parts, from 1—2 to 100 m in the coastal belt. It also enters the Angara and spreads down it up to the Yenisei inclusive. It has also been found in the Arctic lake Taimyr in the lakes of Norilsk region (Pyasina drainage fig. 94) and also in the lakes Munduiskoye and Nalimye lying in the lower reaches of the rivers Kureika and Lower Tunguska in the Yenisei drainage, (GREZE, 1953, VERSHININ, 1960).

Living in different conditions in Baikal *H. placoides* includes several forms of which f. *ripariensis* ABR. (fig. 72 1,1a) has the widest distribution. It lives on stones in very small colonies having the form of variously shaped brown patches or reticula noticeable only at very close observation. Viewed through the magnifying glass, these colonies are found to consist of thin small trailing branches composed of zooecia densely disposed one upon another; separate zooecia are of a sack-like form, about 0.5 mm long and 0.3 mm broad, with four well-developed pointed spines projecting from the edges of the mouth. Settling on stems of underwater plants, *H. placoides* covers them in small but sometimes massive growths, with the zooecia acquiring a more elongated form and bigger spines (*m. sabulosa* ABR. (fig. 72,3). At times colonies of this bryozoan settle on sand grains, forming minute trailing branches.

A. A. KOROTNEV (1901) established a special genus for the Baikalian bryozoan, *Echinella*, while N. ANNANDLE (1911—12) attributed it to the genus *Hislopia*. Several species of this genus are known from the fresh waters of India, South China and the Malay Peninsula. Some authors who studied the position of *Hislopia* in the system of Bryozoa believe it has comparatively recently diverged from the marine Ctenostomata (PELSENEER, 1905). ANNANDLE (1911—12) sees an affinity between the genus *Hislopia* and the genus *Victorella,* which is represented chiefly in the brackish waters of the coasts of the Indian and Atlantic oceans and also in the lakes Tanganyika in Africa and Issyk-Kul and in the Aral Sea.

Fig. 72. Bryozoan *Hislopia placoides* KOROTN. 1 - forma *ripariensis* ABRIK.; 2 - forma *intermedia* ABRIK.; 3 - forma *sabulosa* ABRIK., length of zooecia up to 0.5 mm. After ABRIKOSOV 1924.

In the opinion of L. S. BERG (1937, 1949a), *Hislopia placoides* in Baikal is a relict of the fresh-water fauna which was widespread in Europe in the Pliocene. G. Y. VERESHCHAGIN (1940a, b) considers the genus *Hislopia* to be a young immigrant to the fresh waters of the lake from the sea. According to G. G. ABRIKOSOV (1927, 1959), the Baikalian *H. placoides* is a relict of Tertiary fresh-water bodies of water of a more southern nature, but not a marine element.

Pisces

A start on the study of the Baikalian fish population was made by PALLAS and GEORGI, but the foundations of our knowledge were laid by B. DYBOWSKY (1876) and L. S. BERG (1900, 1903, 1907, 1948—49). The Soviet period has seen numerous new studies of the systematics and biology of the Baikalian fish, as well as of fishing in the lake, described in detail in the collection of articles "Fish and Fishing in the Basin of Lake Baikal" (1958), edited by K. I. MISHA-RIN and the author. It also contains the main bibliography on fishing in Baikal.

At the present time Baikal with its sors and gulfs is known to be inhabited by 50 species of fish belonging to 9 families of which the

Fig. 73. Fishes: 1 - *Comephorus dybowskii* KOROTN., body length up to 16 cm; 2 - *Procottus jettelesi* DYB., body length up to 18 cm. After TALIYEV, 1955.

family Comephoridae (Cottoidei) is endemic to the lake. It has one genus, *Comephorus*, with two species:

C. baicalensis PALLAS and *C. dybowskii* KOROTN. (fig. 73,1). The first of these species was described long ago by PALLAS (1776, 1811), the second by A. A. KOROTNEV (1905). Both of them are very characteristic dwellers of the mass of water of open Baikal, where they live at 0—500 m and occur deeper. The body of the fish, which is not more than 18—20 cm long, is absolutely scaleless, glassy-dull and very much translucent in living individuals. The ventral fins are absent, while the pectoral are very long, reaching almost half the length of the body; the head is very large, with well developed cavities of organs of the lateral line.

It was established long ago by B. DYBOWSKY (1873a) that both *Comephorus* species do not spawn but give birth to living fry, numbering 2,000 to 3,000 per female. According to observations by G. Y. VERESHCHAGIN (1926, 1937), VERESHCHAGIN & SIDORYCHEV (1929) and E. A. KORYAKOV (1955, 1956, 1958), the mating of *C. dybowskii* takes place in September—November, and the mass outcrop of fry occurs in February—April. *C. baicalensis* mates in April—June, with the fry appearing in September—October. The opinion regarding the asexual propagation of *Comephorus*, expressed in its time by D. N. TALIYEV (1949, 1951), has not found confirmation.

Comephoridae are spread all over Baikal, keeping the year round in the mass of water of open deep-water regions with maximum density at 100—300 m (at day time) and avoiding gulfs, sors and other shallow and sheltered sections. Adult *Comephorus* feed upon the amphipod *Macrohectopus*, the larvae and young of *Cottocomephorus* and their own larvae. The juveniles feed upon *Epischura*.

The Cottidae family is represented in Baikal by the subfamilies Cottocomephorinae and Abyssocottinae (TALIYEV, 1955). The former comprises 5 genera: *Procottus* (one species, fig. 73,2), *Metacottus* (1 species), *Batrachocottus* (4 species), *Paracottus* (4 species, fig. 74,2), and *Cottocomephorus* (2 species, fig. 74,4).

The subfamily Abyssocottinae includes the following genera: *Cottinella* (2 species), *Asprocottus* (5 species, one of which has been described from Lake Baunt in the Vitim drainage, fig. 74,1), and *Abyssocottus* (5 species, fig. 74,3).

Two species of the genus *Paracottus* are known: *P. kneri* DYB. (fig. 74,2) and *P. kessleri* DYB. Besides Baikal, these two species live in some lakes of the Baikal basin, in the Angara and occur also in the drainage of the Lena.

Both species of the genus *Cottocomephorus*, *C. grewingki* DYB. (fig. 74,4) and *C. inermis* JAKOWL., are pelagic. As distinct from *Comephorus*, the *Cottocomephorus* species prefer the coastal belt of Baikal. They spawn at shallow places near the shores, mostly on

140

Fig. 74. 1 - *Asprocottus kozhowi* TAL., length 10—15 cm; 2 - *Paracottus kneri* DYB., length 12 cm; 3 - *Abyssocottus pallidus* TAL., body length up to 18 cm; 4 - *Cottocomephorus grewingki* DYB., female, body length up to 12 cm. After TALIYEV, 1955.

rocky bottom. The number of eggs in the females ranges from 900 to 2,400.

Despite their small size (11 to 18 cm in length), these species are commercially important and are easily caught during spawning and wintering. For spawning they converge near shores in thick shoals. The spawning migration is observed in March and lasts till May— June. Sometimes adult females with roe are found later. After spawning the males protect the roe for two or three weeks, after which most of them and part of the females perish. In July— August considerable masses of cottids stay near the shores, feeding upon plankton and devouring their own hatchlings. Some cottids of all ages remain at the shores till late autumn. In autumn they gradually depart to deeper layers (50—100—200 m), where they winter often together with the omul near extensive shallows.

Fig. 75. Omul *Coregonus autumnalis migratorius* GEORGI, length of adult females (5—7 years) up to 40 cm, weight of old specimens (15—17 years) up to 3—4 kg. After MISHARIN, 1958.

Young *Cottocomephorus* constitute an essential component of the diet of the omul, while adults are consumed by the seal.

All other Cottidae genera are represented by benthonic species living at various depths but concentrating chiefly in the coastal belt.

The Coregonidae are represented by the omul (*Coregonus autumnalis migratorius* GEORGI, fig. 75) and subspecies of *Coregonus lavaretus* L. Baikal is populated by at least 4 races of the omul: the North-Baikalian, which spawns in the affluents of North Baikal; the Selenga, spawning in the River Selenga; the Chivyrkui, which breeds in the rivers emptying into the Chivyrkui Gulf; the Posolsky, spawning in the rivers of the Posolsky Sor (TALIYEV, 1941; MUKHO-MEDIAROV, 1942; MISHARIN, 1953, a,b, 1958).

The omul is an exeptionally valuable commercial fish both in its quality and abundance. It matures in the 5th to 7th year, reaching an average of 30 cm in length and 300 to 450 g in weight.

In spring and summer the omul concentrates, in the main, in extensive life-rich shallows. It winters at a depth of 200—300 m, mostly near the same regions which serve as spring convergence and summer feeding grounds. In winter the omul feeds inactively, consuming planktonic crustaceans, which survive in winter in deep waters, and partly bottom-dwelling gammarids and cottids. As it moves towards spring convergence grounds it feeds more intensely, with near-bottom gammarids and imaginal stages and pupae of caddis flies and chironomids beginning to play a certain part in its diet.

The period of the mass shoreward run of omul shoals in spring is regulated by the degree of the warming up of the littoral waters, and therefore the intensity and period of the spring convergence also vary with years depending on meteorological conditions.

Towards the middle of July, when the maximum in the development of pelagic crustaceans, the most important component of the diet of the omul, shifts to more open regions, the omul leaves the shores and feeds in the upper layers of open Baikal.

The same period sees the formation of large shoals of adult omuls which proceed for spawning to the affluents of Baikal. The spawning grounds present sections of rivers with a rocky-pebbly or sandy-pebbly bottom and a moderately rapid current. The number of eggs in one female ranges, depending on the age, from 10,000—15,000 to 30,000—40,000. The development of the roe lasts for 180—200 days and ends in April—May. The hatchlings are passively carried by the current into Baikal, where they begin to feed independently.

In October—November the fattened omuls and the spent shoals migrating downstream leave for deep layers for wintering.

Some of the Baikalian forms of the gwyniads *Coregonus lavaretus* L. are lacustrine, i.e., they spawn in the lake itself. They include, for example, *Coregonus lavaretus baicalensis* DYB. and *Coregonus lavaretus baicalensis natio Dybowskii* KROG. (KROGIUS, 1933; MISHARIN, 1947). The other Baikalian gwyniads are fluviatile and spawn in rivers. The fluviatile forms, according to F. V. KROGIUS (1933), are closer to the European form *Coregonus lavaretus pidschian* GMEL. The Baikalian gwyniads reach 5 to 8 kg in weight and 60 to 77 cm in length at an age of 15 to 20 years.

The lacustrine gwyniads spawn in shallows protected against strong turbulence and rich in underwater vegetation. The spawning takes place in November. The males of *Coregonus lavaretus baicalensis* reach maturity in the 5th to 6th year, the females in the 7th to 8th year. Individual fertility varies from 20,000 to 90,000 eggs.

In August the gwyniads gather in shoals and migrate towards the shores, while the adults go to the spawning grounds abounding in bottom vegetation and concentrate there in autumn (October—November).

The gwyniads feed upon benthos, chiefly mollusks and gammarids, and therefore keep near benthos-rich sections of the bottom, chiefly at 20—120 m depth, in winter evidently much deeper. In spring the gwyniads approach the shores together with the omuls, but in comparatively small amounts (predominantly young fish). The shoreward run of fluviatile gwyniads occurs in August.

In the genus *Thymallus*, two varieties of the widespread Siberian species *T. arcticus* PALL. are known in Baikal: *T. arcticus baicalensis* DYB. and *T. arcticus brevipinnis* SWETOW. (DOROGOSTAISKY, 1923a; SVETOVIDOV, 1931; TUGARINA, 1956a, b). Both forms are endemic. The body coloration of *T. arcticus baicalensis* (the black grayling) varies, but on the whole it is darker than that of *T. arcticus brevipinnis* SWETOW. (the white grayling), which also has a more compressed and bigger body and high fattiness. The white grayling lives throughout Baikal, but most often along the eastern shores, preferring sandy sections of the bottom rich in food. The black grayling occurs everywhere, populating chiefly rocky soil at small depths along the coastal belt.

The black grayling spawns in small affluents of Baikal in early spring. The white grayling prefers bigger affluents for spawning. It enters such rivers as the Selenga beginning with August, but spawns in rapid currents in the Selenga and its tributaries in spring.

Both races of the grayling feed chiefly upon gammarids, larvae of aquatic insects and also mollusks and young Cottidae.

Such representatives of the family Salmonidae as *Hucho taimen* PALL., *Brachymystax lenok* PALL. and *Salvelinus alpinus erithrinus* GEORGI occur very rarely in Baikal and live predominantly only near river mouths. *Salvelinus* can sometimes be observed near the mouth of the River Frolikha which empties into the northern part of Baikal.

The family Acipenseridae is represented in Baikal by the Siberian sturgeon *Acipenser baeri stenorhynchus* NIK., which lives in extensive shallows; big gulfs and the mouths of big rivers serving as its spawning grounds. In summer and particularly in August, when the nearshore waters of Baikal become warmed up to 10—15° C, the sturgeons spread in search of food along the shores far from their usual places of habitation and occur individually near the western shore. They begin to enter rivers in early spring before the breaking up of ice (YEGOROV, 1947). Old sturgeons weigh up to 115 kg with a length of 120 to 180 cm.

The family Cyprinidae has in Baikal general Siberian fluviolacustrine species living mostly in sors, gulfs, pre-estuarine regions and shallow littoral areas of the lake. The particularly numerous cyprinids are the roach *Rutilus rutilus lacustris* PALLAS and the dace *Leuciscus leuciscus baicalensis* DYB. The ide *Leuciscus idus* L. also occurs.

In 1943—44 the Posolsky Sor was artificially populated by the carp from the Amur, *Cyprinus carpio haematopterus* TEMMINCK & SCHLEGEL. Since then it has spread widely in the coastal zone of the Selenga shallow and penetrated into the system of lakes of the Selenga drainage.

The family Percidae is represented by the perch *Perca fluviatilis* L., which lives in shallow sections of sors and in sheltered gulfs, but in summer and sometimes in winter emerges into the open regions of Baikal, concentrating principally near river mouths, sors and gulfs. In August it can be observed along the open coast.

The family Esocidae is represented in the sor system of Baikal by *Esox lucius* L.

The family Gadidae has two races of the burbot (*Lota lota* L.) which have not been studied sufficiently yet. One race lives along the open coast, the other is connected with rivers and pre-estuarine regions.

The other fish that occur in Baikal are *Nemachilus barbatulus toni* DYB., *Cobitis taenia sibirica* GLADKOV, *Phoxinus percnurus* PALL., *P. phoxinus* L., possibly *P. lagowskii* DYB., rarely *Gobio gobio cynocephalus* DYB. and even *Carassius carassius* L.

In recent years, with man's help, the Amur sheat-fish *Parasilurus asotus* L. has penetrated into Baikal through the Khilok-Selenga system.

In the last 20 years the recorded annual catch of fish in Baikal varied between 8,000 and 12,000 tons, with the omul accounting for up to 70%. Considering that at least 30% of the catch escapes statistics, the total annual catch in Baikal, adjacent lakes and lower sections of affluents in ten years between 1946—55 evidently approached 12,000—13,000 tons, 75% of which consists of omul. Counted for the entire area of Baikal (roughly 3 million hectares) the per-hectare catch amounts to 4 kg. But if it is taken into account that fishing in Baikal keeps, as a rule, within 250—300 m depths and that the area of shallows up to these depths does not exceed 600,000 hectares, the amount of fish obtained from the commercial fishing grounds equals 18—20 kg per hectare.

Now for the congeneric ties of the Baikalian fish with fish in other countries.

As pointed out by A. I. BEREZOVSKY (1927) and F. B. MUKHO-MEDIAROV (1942), the Baikalian omul *Coregonus autumnalis migratorius* is very closely related to the Arctic *Coregonus autumnalis* PALL. It is generally believed that the omul penetrated into Baikal through the Yenisei-Angara system from the Arctic Ocean in the Quaternary period (CHERSKY, 1877).

Of special interest among the groups of the Baikalian fish is the endemic family Comephoridae and all other Cottoidei of Baikal.

D. N. TALIYEV (1946, 1955) considers that the few parental forms of the Baikalian Cottoidei lived in the seas washing the eastern shores of Asia. They penetrated into Baikal through rivers at the end of the Tertiary period. Their ancestors originated from the genera *Mesocottus* and *Trachidermis* from the Far-Eastern seas. *Paracottus kneri* and *P. kessleri*, in TALIYEV's opinion, also originate from Far-Eastern forms. L. S. BERG thought the Baikalian Cottoidei to be relicts of the Tertiary fresh-water Siberian fauna. There are no grounds to suppose that *Limnocottus* and the Baikalian Cottocomephorinae in general are descendants of *Mesocottus* (BERG, 1949a). As far as the Comephoridae are concerned, in BERG's opinion they do not have close relatives either in fresh or sea water.

Mammalia

The sole representative of Mammalia in Baikal is a species from the order Pinnipedia, the Baikalian seal *Phoca sibirica* GMEL. (fig. 76). It is a big animal reaching 1.65 m from the tip of the nose to the end of the hind flippers and weighing from 50 to 130 kg. Adult seals have a silvery-grey back and yellowish white belly (B. DYBOWSKY, 1873b).

Fig. 76. Baikalian seal *Phoca (Pusa) sibirica* GMELIN, body length up to 165 cm, weight up to 130 kg. Drawing by N. KONDAKOV.

The Baikalian seal lives all over Baikal but prefers the northern and central parts of the lake.

The breeding takes place in May—June, when the seals gather in sections of the coast difficult of access. Pregnancy lasts for 9 months, and in February—March usually the single pup is born. After 1.5 to 2 months of suckling the cow accustoms the pup to fish. Maturity is reached at the age of four, both by males and females (SVATOSH, 1925; IVANOV, 1938). The seal feeds upon fish, mostly *Comephorus* and *Cottocomephorus*. The summer rookeries are boulder-strewn beaches, predominantly along the south-eastern coast and in the Chivyrkui Gulf. In winter contact with the air is maintained through holes in the ice made during the formation of the ice cover and sustained throughout the winter. The young is born on the ice, in a lair made by the cow amidst hummocks and snowdrifts near an ice-hole. On sunny days in March—April the seals often appear on the surface of the ice and bask in the sun, presenting targets for the seal hunters.

By its morphological characters the Baikalian seal, as S. I. OGNEV (1935) points out, is close to the Caspian seal *Phoca caspica* GMEL. But most researchers consider that *Phoca (Pusa) sibirica* has diverged from the Arctic *Phoca (Pusa) hispida* SCHREB. and adhere to I. D. CHERSKY's (1877) supposition that the seal penetrated into Baikal from the Arctic Ocean through the Yenisei-Angara system in the glacial epoch, simultaneously with the omul.

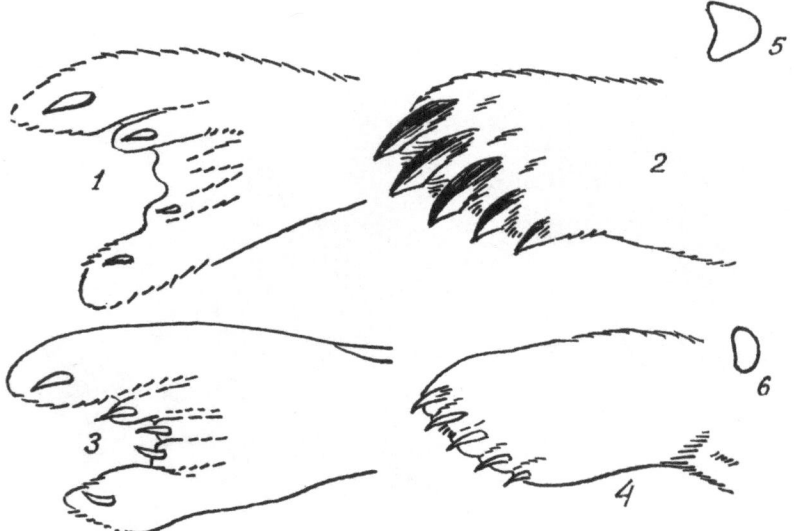

Fig. 76a. Flippers of seals from Baikal, (1—2), from the Caspian *(Ph. caspica* GMELIN (3—4); 5—6 - transverse section through the claw of forefins of the seal. After KONDAKOV, 1960.

This supposition is corroborated by parasitological studies. The louse *Echinophthirius horridus baicalensis* Ass parasitizing the Baikalian seal is very close to the species which is parasitic on the northern seals (Ass, 1935). The nematod *Contracaecum osculatum baicalensis* Mosgov. & Ryzik. parasitizing the Baikalian seal belongs to the species common on the seals of the northern seas but absent on the Caspian seal (Mozgovoi & Ryzhikov, 1950).

According to Taliyev (1940), who based his conclusions on the precipitation reaction, the Baikalian seal shows a closer affinity to the Arctic forms of seals (the Novaya Zemlya, the White Sea) than to the Caspian form.

The hypothesis on the origin of the Baikalian seal from one of the forms of the Arctic species *P. hispida* Schreb. seems to be more creditable.

N. N. Kondakov (1960) notes that the Baikalian seal differs considerably from its relatives of the subgenus *Pusa* in the structure of the flippers. The fore limbs in *P. sibirica* are comparatively long and serve for movement on hard substratum, whereas other seals prefer to move with the help of the muscles of the body. The front fins "are armed with powerful long claws of triangular transverse section, which is observed in none of the representatives of·the genus *Phoca*." The limbs of the Arctic *P. hispida* and the Caspian *P. caspica* are shorter, weaker and armed with thin claws of semi-circular section (fig. 76a). *P. sibirica* differs from other seals also in the structure and armament of the hind limbs. N. N. Kondakov (1960) considers that the peculiar features of *P. sibirica* are indicative of a long period of its independent evolution in Baikal.

General Conclusion on the Composition of the Fauna of Baikal and Its Position in the System of Biogeographical Divisions

As is seen from the review given above, the number of animal species known in Baikal today exceeds 1,200. They are distributed among the following groups:

Table XIV.

Systematic groups	Total number of species and genera		Residents of open Baikal		Number of endemics in open Baikal		
	Species	Genera	Species	Genera	Species	Genera	Fam. and subfam.
1	2	3	4	5	6	7	8
Protozoa	317	80	110	45	90	13	3
Spongia:							
Spongillidae	4	2	—	—	—	—	—
Lubomirskiidae	6	3	6	3	6	3	1
Coelenterata							
(Hydra)	2	1	2	1	1	—	—

Table XIV. (continued)

Systematic groups	Total number of species and genera		Residents of open Baikal		Number of endemics in open Baikal		
	Species	Genera	Species	Genera	Species	Genera	Fam- and subfam.
1	2	3	4	5	6	7	8
Turbellaria	90	15	90	15	90	13	1
Trematodes	17	10	17	10	6	—	—
Cestoda	12	10	12	10	—	—	—
Nematodes:							
Free	10	2—3	10	2—3	5—6	—	—
Parasitic	8	7	8	7	3—4	2	—
Acanthocephala	3	3	3	3	2	—	—
Rotatoria	48	21	12	6	5	—	—
Bryozoa	5	3	1	1	1	—	—
Polychaeta	1	1	1	1	1	—	—
Oligochaeta	62	23	48	20	45	2	—
Hirudinea	17	10	10	4	10	1—2	—
Copepoda-Calanoida	5	4	3	3	1	—	—
Cyclopoida	25	7	19	6	16	—	—
Parasitica	13	7	13	7	2—3	1	—
Harpacticoida	43	9	43	9	38	—	—
Ostracoda	33	3	31	3	31	—	—
Cladocera	10	7	2	2	—	—	—
Bathynellidae	2	1	2	1	2	—	—
Isopoda (Asellus)	5	1	5	1	5	—	—
Gammaridae	240	35	239	34	239	34	—
Acari	6	4	5	3	3	—	—
Tardigrada	1	1	1	1	?	—	—
Trichoptera	36	8—9	16	3	13	2	—
Plecoptera	2	2	1	1	?	—	—
Chironomidae	60	20	22	10	11	(1 subgenus)	—
Anoplura-Parasita	1	1	1	1	—	—	—
Gastropoda	72	12	55	8	53	6	3
Bivalvia	12	3	3	3	3	—	—
Pisces:							
Cottoidei	25	9	25	9	23	8	3
Other fish	25	18	25	18	—	—	—
Mammalia	1	1	1	1	1		
Total	1,219	346	842	253	708	87	11
Species living in open Baikal in % to the total number of species inhabiting the whole of Baikal	100	100	69	73	—	—	—
Endemics in % to the species living only in open waters	—	—	100	100	82	34	—

Not taking the general Siberian species inhabiting the gulfs and bays of Baikal into account, the endemic species living in open waters comprise more than 80% of the total number. In all, 87 genera and 11 families and subfamilies are endemic.

The exceptional wealth of animal life in Baikal becomes especially striking when compared with other lakes and with brackish-water seas. For instance, in the whole of the Caspian Sea 323 autochtonous species (without Protozoa) are known (F. D. MORDUKHAI-BOLTOVSKOI, 1960), whereas in Baikal (also without Protozoa) they number 627. Shown in table XV is the number of species from the best-studied faunal groups in the Caspian, Azov and Baltic seas and the same groups in Baikal. As we see, Baikal is much richer in species than the Baltic Kiel Firth and the Azov Sea. In the number of animal species Baikal is much richer than ordinary fresh-water lakes.

Table XV.

The Number of Species in Some Groups of Invertebrates in Brackish Seas (after MORDUKHAI-BOLTOVSKOI) and in Lake Baikal (after KOZHOV).

	Kiel Firth in Baltic Sea, salinity about 15%	Azov Sea, salinity up to 12% (beyond strongly freshened regions)	Caspian Sea (autochtonous fauna), salinity 12—14%	Lake Baikal	
				All species	Only endemics
Spongia	13	1	6	10	6
Hydrozoa	15	6	4	2	1
Polychaeta	43	38	3	1	1
Prosobranchiata	17	12	32	46	43
Bivalvia	23	14	19	12	3
Amphipoda	18	34	71	240	239
Decapoda	9	8	2	0	0
Total:	138	113	137	309	293

In view of the profound peculiarity of the Baikalian fauna one is at a difficulty in determining its position in the existing zoogeographical zonation. L. S. BERG (1909), basing himself on the fish fauna, designated Baikal as a special Baikal subregion of the Holarctic. G. Y. VERESHCHAGIN (1940a, b) considered that the inclusion of Baikal into the system of zoogeographical divisions existing for ordinary continental waters was essentially incorrect, for the distinctions of the fauna of Baikal from that of the surrounding country are not only very pronounced, but also rest on a different genetic basis. These distinctions are analogous to the differences between the fauna of seas and that of the continental waters around them.

Referring to EKMAN, who singled out the Caspian fauna from the existing system as a special Sarmatian fauna, G. Y. VERESHCHAGIN thinks it more correct to regard the Baikalian fauna as a particular biogeographical unit intermediary between the marine and fresh-water units.

This opinion emphasises a very important point in the history of the fauna of Baikal (just as that of the Caspian Sea and other similar great old lakes), its place of origin. Nevertheless, the removal of Baikal from the existing zoogeographical divisions in the capacity of an independent higher zoogeographical unit, as G. Y. VERE-SHCHAGIN does, will hardly be correct. As will be shown further on, despite the profound endemism of the Baikalian fauna, its history is nevertheless closely connected with the history of the modern fauna of the Palearctic (Holarctic), and there are no grounds to oppose it to the latter as an equal zoogeographical units.

2. The Flora of Baikal.

The flora of Baikal has been studied by K. I. MEYER (1930), S. VISLOUKH (1924), V. N. YASNITSKY (1931, 1936, 1956), V. SKVORTSOV (1937; SKVORTSOV & MEYER, 1928), A. P. SKABICHEVS-KY (1929, 1936, 1952, 1954), N. L. ANTIPOVA, (1955, 1956a, b), and others. Today Baikal and its gulfs, sors, pre-estuarine regions, etc., are known to have 569 species of algae with 162 varieties described as subspecies, forms, variations, etc. About a half of all species are observed in open Baikal with its bays and extensive gulfs poorly protected against winds and the other half live in sors and well-sheltered shallow sections of gulfs and bays. Among the algae, 35% (more than 100 species and 80 varieties) are known to inhabit Baikal only.

Given below are data on the specific composition of various groups of the hydroflora of Baikal.

Table XVI.

The Number of Species and Forms of Algae Found in Baikal and Its Sors, Bays and Gulfs (Compiled by O. M. KOZHOVA)

Floral groups	Number of known species	Varieties of the same species
Cyanophyta:		
Chroococceae	15	2
Chamaesiphoneae	2	—
Hormogoneae	56	5
Total Cyanophyta	73	7

Table XVI. (continued)

Floral groups	Number of known species	Varieties of the same species
Rhodophyta	2	—
Chrysophyta:		
Chrysomonadineae	9	2
Chrysocapsineae	1	—
Chrysosphaerineae	1	—
Chrysotrichineae	1	—
Total Chrysophyta	12	2
Bacillariophyta (Diatomeae):		
Centricae	19	4
Pennatae	315	139
Total Bacillariophyta	334	143
Xanthophyta (Heterocontae)	5	—
Peridineae	7	1
Euglenophyta	3	—
Chlorophyta:		
Volvocineae	5	—
Chlorococcineae (Protococcineae)	37 ⎫	
Ulothrichineae	32 ⎭	6
Siphonocladineae	15	2
Conjugatae	41	1
Total Chlorophyta	130	9
Charophyta	3	1
Grand total	569	162

In addition to algae, about 20 species of flowering water plants have been found in Baikal: species of *Potamogeton, Myriophyllum, Ceratophyllum, Sagittaria, Scirpus, Lemna, Phragmites, Sparganium. Nuphar, Polygonum,* and many other ordinary lacustrine species. But all of them populate only sors, sheltered bays and other such areas densely. In the open waters of Baikal the higher flowering hydroflora is represented only by one or two species of *Potamogeton* and *Myriophyllum,* dispersed in separate small shrubs and accumulations along sandy bottom in coastal sections more or less protected against strong turbulence.

In the open littoral one can also come across the lichen *Collema*

Ramenskii ELENK., which forms small curly shrubs consisting of branching strongly jagged lobes on stones.

Let us proceed to a brief summary of separate groups of algae inhabiting Baikal.

Cyanophyta are represented by more than 70 species, but only 30 of them appear in open regions, and even these live mostly near river mouths and shallow gulfs and bays. Among the planktonic Cyanophyta occurring in open Baikal mention should be made of *Anabaena flos-aquae* (LINGB.) BREB., *A. lemmermannii* P. RICHT., *A. spiroides* KELB., *Aphanizomenon flos-aquae* (L.) RALFS, *Gloeotrichia echinulata* (I. S. SMITH.) P. RICHT.

These algae are often responsible for the summer water "bloom" in coastal areas, pre-estuarine regions, bays and gulfs, and after a long period of warm calm days, also in more open regions near shallows.

Among the benthonic Cyanophyta the most important part is played by *Tolypothrix distorta* (FL. DAN.) KÜTZ. and *Stratonostoc verrucosum* (VAUCH.) ELENK., densely populating the littoral of big gulfs. In some bays a colossal density is attained by *Sphaeronostoc pruniforme* (AG.) ELENK.

Only two Rhodophyta species are known, and even those only from sors and river mouths.

Twelve species of Chrysophyta have been found in Baikal, among them 2 *Mallomonas* species, 4 *Dinobryon* species, and one each of the genera *Uroglena*, *Chrysosphaerella*, *Synura*, *Tetrasporopsis*, *Chrysothallus* and *Epichrysis*. The latter three genera were formerly considered to be endemic to Baikal, but in recent years representatives of these genera have been found in some European bodies of water.

A species common to the open regions of Baikal is *Dinobryon cylindricum* IMH., large amounts of which appear in the spring-summer period in some years. In summer *Dinobryon divergens* IMH. also occurs in open regions, but it is more common only in shallows. The other species of *Dinobryon* (*D. sociale* EHRB., *D. sertularia* EHRB.) are found in gulfs only.

Epichrysis melosirae K. MEYER, an endemic species, is common in the pelagic zone, where it appears in large numbers on the filaments of the widespread planktonic diatom *Melosira baicalensis* (K. MEYER) WISL. *Mallomonas* sp. is often observed in the littoral of open Baikal.

Other species of the planktonic Chrysophyta have been found only in pre-estuarine areas and shallow and sheltered bays, although in summer they are often carried to the adjoining open regions of the lake.

Among the benthonic Chrysophyta, mention should be made of *Tetrasporopsis reticulata* K. MEYER, an alga "having the form of a broad (5 to 15 cm), very thin film of indefinite shape, slippery and slimy to the touch, with torn fimbriate edges. Its colour is dark- or light-brown, sometimes olive-brown. It either floats freely among stones or attaches itself to them with a short, rather thick pedicel" (MEYER, 1930). This alga is found on rocky bottom throughout the littoral.

Chrysothallus baicalensis K. MEYER lives on the stems of the alga *Didymosphenia*.

Bacillariophyta (Diatomea) rank first among other groups in the abundance of species. It has been established that in Baikal this group has 334 species, or 60% of the entire number of species of algae. About 33% of them are endemic.

Diatoms play an exceptionally important part in Baikal's biolo-

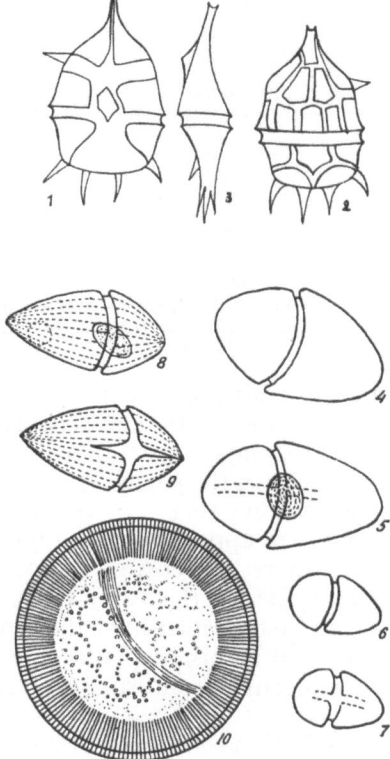

Fig. 77. 1,2,3 - *Peridinium baicalense* KISSEL. & ZWETK., length 0.05—0.07 mm; 1—2 - ventral and dorsal sides; 3 - side view. After KISELYOV & TSVETKOV, 1935; 4—5 - *Gymnodinium baicalense* ANTIP., length of cells 0.05—0.09 mm; 6—7 - *G. baicalense minor* ANTIP.; 8,9 - *G. coeruleum* ANTIP., length of cells 0.04—0.05 mm, after ANTIPOVA; 1955; 10 - planktonic diatom *Cyclotella baicalensis* SKW., cell diameter 0.095—0.113 mm. After SKVORTZOV.

154

gical productivity. In the mass of water of open regions they comprise the main part of phytoplankton. In the littoral they settle on bottom algae, on stones and other kinds of soil, constituting the main part of the diet of many benthonic animals.

Of exceptional importance among the planktonic diatoms in open Baikal are *Cyclotella baicalensis* SKV. (fig. 77,10). *C. minuta* ANT., *Melosira baicalensis* (K. MEYER) WISL. (fig. 78,1,2), *M. islandica* subsp. *helvetica* O. MÜLL. (fig. 78,3,4), *Stephanodiscus Binderanus* (KÜTZ). KOLBE, *Synedra ulna* var. *danica* (KÜTZ.) GRUN. and *S. acus* var. *radians* (KÜTZ.) HUST.

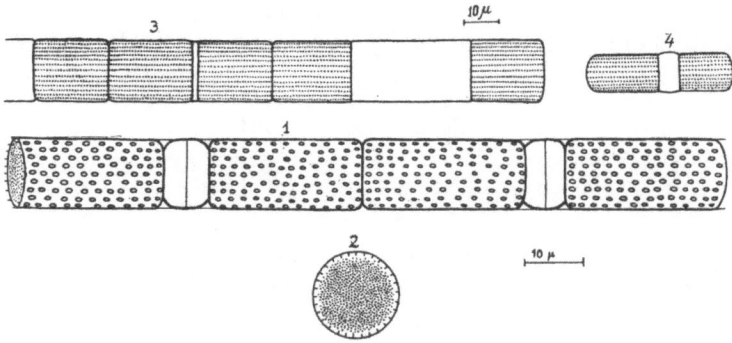

Fig. 78. 1,2 - Planktonic diatom, *Melosira baicalensis* (K. MEYER) WISLOUCH, cell diameter 0.006—0.037 mm; 1 - viewed from the side of the girdle; 2 - from the side of the valve; 3 - *M. islandica* subsp. *helvetica* O. MÜLL.; 4 - the spore of this species. Original.

Besides, in summer, open-water plankton contains *Asterionella formosa* HAAS., but it is more abundant in pre-estuarine areas, gulfs and other shallow regions.

In summer several species of the genus *Fragillaria* are also observed in shallow regions. Among them, *F. crotonensis* KITT. and *F. capucina* DESM. sometimes appear in the adjacent open regions, having been brought there by currents. Species of the genus *Tabellaria*, such as *T. fenestrata* (LYNGB.) KÜTZ. and *T. flocculosa* (ROTH.) KÜTZ., also can be observed in shallow regions in summer.

These are practically all forms of planktonic diatoms living in open Baikal. All others occur only in sors, well-sheltered bays and pre-estuarine areas.

The vast majority of diatom species are epiphytes living on algae, stones, sand grains, etc. Breaking off from the ground, they often appear in vast amounts in the mass of water of the littoral and even in the neighbouring deep-water regions. Those of them which are found in the plankton net especially frequently are *Ceratoneis arcus* (EHR.) KÜTZ., species of the genera *Gomphonema*, *Navicula* and others.

Epiphytes populate the bottom not only in the littoral, but also spread, gradually growing sparser, far beyond its limits to a depth of 60—70 m and perhaps deeper. Of special importance among them are species of the genera *Gomphonema* and *Didymosphenia* which are spread throughout Baikal, covering the stones of the coastal belt in a continuous tomentose yellowish-white layer.

About two-thirds of the epiphytic and bottom diatoms living in Baikal are widely distributed in the fresh waters of Europe and Asia. But there are species with a rather discontinuous distribution. Some species living in Baikal also live in brackish basins.

It should be noted that the species most numerous in open Baikal are, as a rule, endemic *(Melosira baicalensis, Cyclotella baicalensis, C. minuta)*.

Five species of Xanthophyta have been found in Baikal, living in sors and river mouths (2 species of *Ophiocytum*, 1 species of *Tribonema*, etc.).

Pyrrophyta are represented by peridineans. Seven species of peridineans are known so far, a greater part of which are endemic and live in open waters, playing a major role in early-spring phytoplankton. Of greater importance among them are several peculiar species from the genus *Gymnodinium* (*G. baicalense* ANT., fig. 77,4—7; *G. coeruleum* ANT., fig. 77,8, 9), large numbers of which appear in open waters in the sub-ice period (March—April). In spring and autumn in open waters one can also often observe *Peridinium baicalense* KISSEL. & ZWETK. (1935). (fig. 77,1—3). *Ceratium hirundinella* O. F. MÜLL., a widespread species, occurs in open waters only in summer, being more common in shallow gulfs, bays, pre-estuarine areas, etc.

N. S. GAYEVSKAYA (1933) noted the presence of one of the species of the genus *Glenodinium* in the coastal zone of Baikal.

Three Euglenophyta species have been found in Baikal, living only in sors.

Chlorophyta are represented by approximately 130 species. Owing to their abundance they comprise the main part of the phytobenthos of open regions, which are inhabited by more than 40 species from this group; about 30 of them are endemic.

Five species are known from Volvocineae, one of them belonging to the endemic Baikalian genus *Swartschewskiella*: *S. hemisphaerica* (K. MEYER) JASN. This microscopic alga settles on various bottom diatoms of the open littoral. The other four species occur only in well-sheltered gulfs, bays and pre-estuarine areas.

Representatives of Protococcineae are common components of the plankton of the littoral-sor zone, occurring rarely in that of open Baikal. Mention should be made of *Bothryococcus* sp., which usually appears in the plankton of open waters in summer. In some years considerable amounts of it are observed there in spring and winter.

It is possible that the open-Baikal species is not identical with *B. brauni* KÜTZ., which can be seen in summer in shallows, gulfs, bays etc. Fairly often open Baikal also contains *Ankistrodesmus falcatus* RALFS, which reaches the highest numerical density in the summer period. Frequently one can find there *Sphaerocystis (Gloeococcus) Schroeteri* (LEMM.) KORSCH. and also representatives of the genus *Scenedesmus* and some others, which are brought there by currents from shallow regions.

The benthonic Protococcineae include *Sykidion gomphonematis* K. MEYER, an interesting endemic species which lives in vast amounts on the diatom *Gomphonema*. It should be noted that the genus *Sykidion* is represented otherwise only in seas.

An alga common for the open waters of Baikal is *Chlorella vulgaris* BEYER *(Zoochlorella)*, which lives symbiotically in the tissues of sponges and is responsible for their bright-green colour.

The Tetrasporacea of Baikal also deserve attention. In open waters they are represented by species of the genus *Tetraspora*: *T. cylindrica* var. *bullosa* K. MEYER and *T. lubrica* AG. The latter, a bottom alga, plays a more important part. It has the form of a long cylindrical light-green tube which swells strongly with age and turns into an irregularly-shaped sack with wavy swollen corrugated walls. The height of this tube reaches 100 cm, with the diametre equalling a third of the height, but sometimes specimens higher than 1.5 m and more than 50 cm in diametre are found (YASNITSKY, 1952). This alga grows densely on stones in the open littoral at a depth of from 1 to 3 metres.

Of great importance in open Baikal are Ulothrichineae and Siphonocladineae, represented mostly by endemic species. *Ulothrix zonata* KÜTZ., a widespread species, abounds at the edge of water on rocky shores. This species is common in big lakes and rivers outside Baikal.

An exceptionally important part in the phytobenthos of Baikal is played by species of the genus *Draparnaldia* (family Chaetophoraceae). Nine endemic species of this genus are known, comprising a separate group called *Baicalia* (fig. 79). The species most widely distributed is *Draparnaldia baicalensis* K. MEYER (fig. 79,1), which forms small dark-green shrubs 15 to 35 cm high attaching themselves to stones. Its branches reach 3—4 cm in length and are covered with a thick layer of transparent slime. The shrubs of other species are still smaller. Settling on stones, *D. baicalensis* and other species form thick darkgreen growths, whole meadows of which often cover the bottom along open shores. In sors and well-sheltered bays the species of the *Draparnaldia* do not occur.

The same family includes the endemic genus *Ireksokonia*, represented by the species *I. formosa* K. MEYER, which forms thin threads 10 to 12 cm long growing in tufts on stones in the open littoral.

Fig. 79. Baikalian algae: 1 - a branch of *Draparnaldia baicalensis* MEYER, height of branches up to 30 cm; 2 - *Draparnaldia Arnoldi* MEYER; 3 - a branch of *Draparnaldia* sp. Original.

One member of the Ulothrichineae, *Binuclearia tatrana* WITTR., is a planktonic species which is fairly widely distributed in large gulfs and also occurs in open Baikal.

Siphonocladineae are known to be represented in Baikal by 5 genera and 15 species, of which only 5 species can be found in other bodies of water. The rest are endemics, and one species has served for the establishment of the endemic Baikalian genus *Geminiphora*.

In the genus *Cladophora*, the greatest distribution is attained by *C. floccosa* K. MEYER, which grows in small (up to 1 cm) shrubs or globular flakes of numerous thread-like sprigs with a diameter of 1—1.5 cm attaching themselves to stones in the littoral.

The genus *Aegagropila* is represented in Baikal by four endemic species of which *A. compacta* K. MEYER has the widest distribution. It forms minute very compact hemispheric cushions 2 to 5 cm high and 2 to 10 cm in diameter which densely overgrow littoral stones.

In the genus *Chaetomorpha*, an important part in the phytobenthos of the littoral is played by *Ch. pumila* K. MEYER, which grows on littoral stones in dense tufts of thick rigid threads 2 to 4 cm high and about 0.5 cm in diametre, standing more or less erect.

Forty-one species are known from the Conjugatae, but they all occur only in sors, isolated gulfs, bays, and other such littoral-sor areas.

Charophyta are represented in open Baikal by two species of the genus *Nitella*, which often forms thick growths in bays, gulfs and sors.

The studies of bacteria in Baikal (NECHAYEVA & SALIMOVSKAYA-RODINA, 1935; KUZNETSOV, 1951, 1957; RODINA, 1954; O. KO-ZHOVA, 1956b; ROMANOVA, 1958) have shown that the microflora of open waters is far from being quantitatively poor. The mass of water contains up to 1 million cells of bacteria per cubic centimetre, and hundreds of millions of bacteria per square centimetre have been found on stones in the littoral. There are especially many bacteria in macrophyte growths and in tissues of sponges. Various physiological groups have been found among microbes: putrefactive and nitrifying bacteria, yeast fungi, as well as nitrogen-fixing bacteria: *Azotobacter* and *Clostridium pasteurianum*.

DISTRIBUTION OF FAUNA AND FLORA IN BAIKAL

1. Siberian and Baikalian Faunae and the Problem of their Relative Immiscibility.

Baikal is populated by two genetically and ecologically different complexes of fauna and flora. One of them consists of Euro-Siberian species and genera; the other of indigenous Baikalian. The latter are responsible for the very distinct character of the Baikalian fauna and flora.

This fact was noted long ago by W. DYBOWSKY (1912), who distinguished the Baikalian fauna into the complex populating the coastal waters and the indigenous Baikalian complex residing in the deep-water region, or Baikal proper, which is inhabited by an old fauna having nothing in common with the modern fauna of the Palaearctic.

Other students of Baikal have also pointed frequently to this genetic and ecologic heterogeneity of the Baikalian fauna (KOROT-NEV, 1901; KOZHOV, 1931, 1936b, 1947; VERESHCHAGIN, 1935, 1940b), with G. Y. VERESHCHAGIN, suggesting that the general Siberian fauna of Baikal be called the Siberian complex; the indigenous Baikalian fauna the Baikalian complex; and the group of forms belonging to widespread Palaearctic genera that have adapted themselves to life in Baikal and formed new subspecies and species there, the Siberian-Baikalian complex.

The totality of biotopes populated by the Siberian complex of forms has been singled out by us as a separate littoral-sor zone, as distinct from the more or less open waters of the lake inhabited chiefly by Baikalian elements. To this zone we refer the sors, and enclosed sections of large gulfs and bays, the Mukhor Gulf (the Maloye More), the southern part of the Chivyrkui Gulf to the south of the Baklany Island and other similar areas of the lake.

The population of Baikal's sors does not differ, in fact, from that of any Siberian lake having similar conditions and consists of ordinary representatives of the Siberian-European hydrofauna. The waters of shallow gulfs, too, are penetrated by almost all representatives of this fauna, but some forms are absent or rare there; this applies, for instance, to *Gammarus lacustris*, the mollusks *Limnaea stagnalis*, *Stagnicola palustris*, *Physa fontinalis*, and other species. As the conditions grow more Baikalian, more and more general Siberian elements disappear, being gradually replaced by indigenous Baikalians.

A careful study of the distribution of the benthonic fauna in various sections of Baikal has shown that the lacustrine-sor fauna consisting of Siberian-European species is confined to the littoral belt of Baikal as isolated colonies rarely occupying considerable areas and being dozens of kilometres apart. In regions where these colonies are not widely separated, owing to the large indentations of the coastline, a superficial study may lead one to the conclusion that the coastal belt as a whole is populated by a motley mixture of some general Siberian and purely Baikalian elements. But after a closer look at the distribution of the fauna in such regions it becomes clear that the lacustrine-sor species which inhabit some enclosed gulf and also spread outside it in both directions, become increasingly scarce as the distance grows and do not venture beyond the 15—20 m depth, huddling along the shores; nearer to a similar enclosed gulf their number increases again, and so on. In regions with a less dissected coastline the colonies of common Siberian fauna are dozens of kilometres apart. The same holds good for the flora of Baikal: endemic species do not live in the sors and sheltered shallow bays, while ordinary lacustrine flowering plants avoid the open waters of the lake.

In the frontier belt lying between sections of the littoral-sor zone and the more open regions of Baikal, i.e., where the environment already becomes more Baikalian but at the same time preserves some features of ordinary lacustrine conditions, we find a rather motley population comprised partly of some Siberian elements and partly of Baikalian endemics. Among the Baikalian elements, the mollusks *Gyraulus gredleri* and *Radix auricularia* penetrate farther than others into the open regions of Baikal along the coast. Only active swimmers among the Siberian fish species and also planktonic organisms, given favourable conditions, temporarily spread beyond the gulfs and bays.

The indigenous Baikalian fauna consisting of endemic species does not live in the sors and shallow gulfs deeply indenting the coast, with the exception of very few relatively eurytopic species. Those which are found more frequently in the waters of bays and gulfs and in the sors are the oligochaetes *Limnodrilus arenarius*, *Propappus volki* and *Peloscolex inflatus*; the gammarids *Gmelinoides fasciatus*, *Micruropus wahli*, *M. possolskii*, *M. talitroides*, *M. litoralis* and *Eulimnogammarus viridis*, the bryozoan *Hislopia* and more rarely the mollusk *Baicalia korotnewi* and the polychaete *Manayunkia baicalensis*.

A more detailed account of the population of the littoral-sor zone and the conditions of life in it will be given further on.

Much has been written on the problem of the relative immiscibility of Siberian and Baikalian elements in Baikal. It is generally known that living conditions characteristic for every ecologically distinct group are one of the major factors determining the distribution of

Fig. 80. Surf on open Baikal. Drawing by B. LEBEDINSKY.

Fig. 81. The littoral in the Bolshiye Koty area. Photo by N. TYUMENTSEV.

organisms. Such groups thrive only in a quite definite specific complex of conditions to which they have adapted themselves in the course of a very prolonged period of struggle for existence, developing definite physiological and morphological characters that correspond to these conditions. The general Siberian fauna, inhabitants of comparatively shallow bodies of water, and the fauna of such a great and peculiar lake as Baikal, have been developing in basically different ways, each in different ecological conditions. In the sors and ordinary lakes, the considerable warming of the whole mass of water in summer owing to their shallowness intensifies biological processes there, which reduces the oxygen content of their waters; in winter, under the ice cover, the oxygen content of these waters is still lower. This complex of conditions, typical, on the whole, of shallow waters, is unsuitable for Baikalian endemics, but lacustrine forms are well adapted to it. On the other hand, the complex of conditions of a typically Baikalian type is unfavourable for lacustrine forms (KOZHOV, 1936b).

Let us review some of the factors of the water environment which may determine the distribution of fauna in Baikal. In the open regions of the lake, the temperature of the bottom layers in summer rarely exceeds 12—13° C even at a depth of 3—5 m in the littoral and remains at this level only for 10 to 15 days (August). Outside of the littoral it is lower still (fig. 14). In sections with a narrow inshore shoal, during strong winds the water temperature even in summer often drops to 5—4° owing to the emergence of hypolimnetic waters to the surface and the driving of warm superficial layers away from the shores.

An entirely different regime exists in the sors and shallow sheltered gulfs (fig. 15). In summer the bottom temperature at a depth of 3—5 m there reaches 22—23° and even more on some days, and the temperature exceeding 12—13° (i.e., higher than the maximum Baikalian temperature) is sustained for at least 3.5 to 4 months. In winter, thanks to thick silt deposits, the bottom temperature does not drop below 2—3°. The average number of day degrees in such littoral-sor regions in the 0—5 m layer, expressed as the product of temperature multiplied by time, is twice the number of day degrees at the same depths in the open regions of Baikal, as has already been noted (fig. 22).

It is perfectly obvious that the low water temperatures in the open regions of Baikal present a powerful obstacle to their colonisation by ordinary Siberian-European fauna and flora, the reason being, not the individual capacity for survival of any particular specimens of these species penetrating there from rivers, adjacent lakes and sors, but mainly the adverse effect of low temperatures on the processes of propagation and embryonic development of these species. It seems very probable that in Baikal the heat-loving litto-

162

Fig. 82. The littoral in the Ushkany Islands area. Photo by M. Kozhov.

Fig. 83. The littoral on a scoured moraine along the northeastern shore of Baikal.
Photo by M. Kozhov.

ral-sor species do not find enough warmth for the normal ripening and development of the genital products — eggs and spermatozoa — and also for the normal progress of the first stages of the embryo's development, even if fertilisation has taken place. This proposition has found indirect proof in the following observations.

In early spring, in the southern and central parts of the shallow Mukhor Gulf (the Maloye More) we find egg clusters and teeming young of the ordinary lacustrine mollusk *Limnaea auricularia* var., but closer to the range of the gulf and particularly outside of it the number of the young drops sharply.

A similar picture is observed in other gulfs and bays. For instance, in the Boguchan Bay (North-East Baikal) the egg clusters and young of ordinary lacustrine mollusks occur only in shallow and well-sheltered sections, where one can also find adult specimens, inhabiting marginal lakelets and bogs connected with Baikal. But very few egg clusters and juveniles can be found in more open parts of the bay. Notably, adult specimens of such ordinary lacustrine-sor species of mollusks as *Radix auricularia*, *Valvata aliena* and others are much smaller there than in the neighbouring Slyudyanka lakes.

It should be mentioned that ordinary Siberian-European forms of Trichoptera do not develop in the open regions of Baikal and that there are no Plecoptera and Ephemeroptera there, although some species undoubtedly deposit eggs in the water near the sors and adjacent lakes. We have caught some females of lacustrine species of chironomids, such as *Chironomus* gr. *plumosus*, over the water in the open regions of Baikal, but we have never seen larval stages of these species in the waters of these regions. The amphipod *Gammarus lacustris* does live in lakes and sors, but it never occurs in the open regions of Baikal, although vast numbers have been spread in the inshore belt of Baikal as a bait in ice-hole fishing for decades.

The low temperature, which adversely affects the propagation and development of species of the Siberian fauna and flora, also inhibits the development of ordinary lacustrine plankton in the open regions of the lake, which will be dealt with further on in the review of the plankton.

It is necessary to stress that the ecological barrier which prevents the heat-loving dwellers of eutrophic waters from settling in open Baikal is not, contrary to G. Y. VERESHCHAGIN's opinion (1940b), impervious to some cold-loving of eurytopic species, dwellers of oligotrophic bodies of water, brooks and cold deep mountain lakes. For instance, cold-loving species of chironomids from the groups Sergentia and Orthocladiinae survive in the open regions of Baikal, while the mollusks *Gyraulus gredleri*, a highly eurytopic species, and *Radix auricularia*, which forms small varieties sometimes venturing beyond sheltered bays and gulfs, can also be observed in open

Fig. 84. The littoral in the Ajaja Bay, North Baikal. Photo by M.Kozhov.

Fig. 85. The littoral in the Kharsagay cape area on the western shore of Baikal.
Photo by M. Kozhov.

waters, but not far from the sors. Oligochaetes from the genus *Nais*, the fresh-water *Hydra*, the fishes grayling, gwyniad, burbot, etc., have also succeeded in establishing themselves.

While the low temperatures of the waters of Baikal are an obstacle to the lacustrine-sor species of the Siberian fauna in its open regions, the thermal and chemical regime of the sors and ordinary shallow lakes, in its turn, is obviously unfavourable for the indigenous Baikalian fauna. This is explained by the relatively high temperatures as well as the low oxygen content and the pollution of waters with organic substances, to which the Baikalians are not accustomed and which inhibits the normal course of physiological processes in them. The importance of this factor becomes clear after a study of the peculiarities of the distribution of Baikalians in the zones directly adjacent to the sors and other littoral-sor sections. Of special interest in this respect is the behaviour of mollusks. For instance, in the Chivyrkui Gulf almost none of the Baikalian species of mollusks occur to the south of the Baklany Island in the vast shallow stretch lying between this island and the southern shore of the gulf; on the other hand, this section is colonised with lacustrine-sor species. True, one can find there adult living specimens of *Baicalia dybowskiana carinatoides*, but more often one comes across dead shells of this species. Evidently the individual specimens penetrating there perish without leaving offspring.

In the Mukhor Gulf (the Maloye More), only *Baicalia korotnewi* and *Sphaerium baicalense* of the Baikalian mollusks reach the central part of the gulf, but the nearer the middle, the fewer of them can be found, with the absence of the young of *B. korotnewi* testifying that this species evidently does not propagate there, either. In addition, the specimens of Baikalian species in such areas have a poor appearance, and form dwarf varieties, etc.

The water in the upper course of the Angara, as in Baikal, is very cold, and prior to the construction of the dam its rapid flow prevented it from getting heated to more than 10—12° C over 80 to 100 km. Due to the similarity of thermal and chemical conditions it was inhabited by a fairly large number of Baikalian species, primarily from among the littoral fauna. A certain number of Baikalian species, especially gammarids, live in the Angara and reach the Yenisei, the Yenisei Gulf and other waters of the Arctic seaboard.

What makes the Angara more comfortable for the Baikalian species to live in is, firstly, the comparatively moderate summer temperatures characteristic of fluvial waters; the lower the summer temperatures (the upper section of the Angara), the more Baikalian species survive there, and, conversely, the higher the temperature (the lower reaches), the lower their survival value. Secondly, it is the similar chemical regime, especially as regards the abundance of oxygen, and the absence of pollution, etc.

The influence of the relatively high summer temperatures on the Baikalian species consists, evidently, in disturbing the normal process of the ripening of the genital products, or the development of the young, if the egg has been fertilised. Our study of the ripening of the genital products of the Baikalian mollusks Benedictiinae (Kozhov, 1928, 1945) has shown that an impetus to the extensive ripening of spermatozoa in the testicles of the males is provided by the water temperature of about 8—10°, which is the maximum value for the bottom layers observed in the littoral waters of Baikal's open regions towards August. As the genital products ripen the mating takes place, usually in August-September, upon which the females deposit cocoons with eggs throughout the autumn, the later part of the winter, and in the spring. Animals placed in the tank with running Baikal water maintained at a temperature of 8—10° C (as in the bottom layers in the littoral of the lake), also mate, and the females deposit egg cocoons. When the temperature is raised to 14—16° cocoons are not deposited, although the animals will live on for months if fresh plants are added as food for them. If the temperature is raised to 18—20°, the animals will perish within several weeks. Similar experiments have been conducted with the mollusk *Valvata baicalensis*. If the tank has running water, in July and August this mollusk deposits cocoons even at a temperature higher than in Baikal (14—15° as against 8—10° in the lake). The eggs in the cocoon begin to develop normally, but after two or three weeks all the embryos die, and the cocoons become covered with a fungus.

All these data add confirmation to the generally known fact that the temperature and chemical regime are powerful regulators of physiological processes and serve, therefore, as a decisive factor in the distribution of animals.

The combined influence of chemical factors and the thermal regime is evinced very graphically in the behaviour of the roach, perch and other fish inhabiting the sors and coastal lakes. In summer most of these fish live in the inshore belt of Baikal's shallows, where they find sufficient food and where the temperature of the water at the end of July, in August and also in September differs little from the temperature of the water in the sors. In autumn, when storms rage in Baikal and the shallow inshore belt open to the winds becomes too uncomfortable to live in, the fish withdraw to the sors, rivers and marginal lakes and live there till January—February. Towards this period the water in the sors becomes very poor in oxygen, and the fish return to Baikal, which is now covered with ice. There they remain till spring and after the ice break-up they re-enter the sors for spawning and feeding and live there till July—August, when they again leave the sors for Baikal. But whereas such active organisms as fish can change their environment, appearing now in the sors and now in Baikal, depending on changes in the water, poorly mobile and especially sessile animals die when adverse conditions set in. That is precisely why we find, for instance, accumulations of dead shells of mollusks in the transition zone, i.e., in the zone of contact (in the narrow sense) between elements of the Siberian and Baikalian faunae. Both of them die out there if they fail to endure the new conditions or cannot propagate normally.

In discussing the possible causes of the immiscibility of the Siberian and Baikalian faunal complexes in Baikal we could also turn to differences in the sources of food, if only in the case of vegetable-feeders, since algae of endemic species and genera predominate among the macrophytes in Baikal, while ordinary lacustrine plants prevail in the sors. But there are indications that both the Baikalian and Siberian species are more or less indifferent to the systematic composition of the vegetable food which they consume, even during a prolonged period of time.

There is no doubt that constant turbulence and strong surf in the inshore belt of Baikal present a formidable obstacle to the settling of dwellers of sors, marginal lakes or quiet sheltered bays in this section of the bottom.

Thus it is to be admitted that the thermal and gas conditions of the waters and partially perhaps the disturbances caused by the surf present important limiting factors in the distribution of aquatic organisms; the causes of the relative immiscibility of the Siberian and Baikalian faunae should evidently be sought in the fact that they have been developing in different environments. The extremely small number of even relatively eurytopic species in Baikal shows that there have been no stages in the history of Baikal favourable for their formation.

Contrary to this, the ordinary continental fresh-water fauna (that of lakes, bogs, rivers, etc.) has been developing in conditions of extremely variable environmental factors, both in the seasons of the year and in the course of the long history of these bodies of water. The rapid warming through of the waters in spring, the high temperatures in summer, the long duration of the warm period, followed by an abrupt cooling, sometimes even complete freezing (in Siberia, at least), and the sharp changes in the gas conditions have induced this fauna to develop a number of essential adaptive physiological characters, especially as regards the processes of propagation and embryonic development. The propagation period of the vast majority of these animals is timed for early spring, thus allowing their young to grow bigger and stronger during the short and hot summer. The wide diversity of these bodies of water themselves and of life conditions in them leads, in turn, to the development of eurytopic forms. Stenotopic elements can evolve only in cold mountain waters, springs, cave waters, thermal springs and other such places. But in the duration of existence and in the constancy of life conditions they have no comparison with the deep zones of Baikal.

Somewhat different views on the causes of the immiscibility of the Baikalian and Siberian faunae were held by G. Y. VERESHCHAGIN. Back in 1935 he pointed out in his work "Two Types of the Biological Complexes of Baikal" that the specific features of the Baikalian fauna and its ecological distinctiveness could not be explained away by the old age of Baikal and the constant influence of such environmental factors as temperature, gas conditions, etc. Rejecting this explanation, G. Y. VERESHCHAGIN pointed, instead, to the presence of water "with a changed molecular structure" (MENDELEYEV, 1935). This is what he wrote on this score (1940b, p. 211): "It is possible that precisely owing to this peculiarity that, since the formation of great depths, the water of Baikal has been detrimental to many organisms, and that is probably why this admixture of abyssal water in the superficial layers of Baikal in some regions and in some seasons of the year presents one of the obstacles to the penetration of the Siberian complex into the Baikalian one... It seems very likely that, not only for Baikal, but for Lake Tanganyika as well the presence of great depths with a peculiar composition of the water is a factor which prevents the modern population surrounding these lakes from mixing with the more ancient population which has gradually, with the progressive formation of great depths of these lakes, accommodated itself to these specific features of the water." Similar views were held by Academician L. S. BERG. In his last work devoted to Baikal (1949a, p. 291—292) we read: "The population of the Baikal waters hardly mixes with that of the local rivers and lakes supplying their waters to Baikal. Aliens usually perish in this lake. This is probably explained by the chemical composition of the Baikal waters, which have not been sufficiently studied yet as regards micro-elements, i.e., elements which occur in very insignificant amounts but are vital for certain organisms."

In his work "The Origin and History of the Baikalian Fauna" (1940b), G. Y. VERESHCHAGIN categorically objects to laying emphasis on such environmental factors as the temperature and chemistry of the water.

He maintains that if temperature was an important factor of the environment, Baikalian elements could spread to cold Siberian lakes, whereas in reality they are not to be found there. It would be wrong, however, to view the thermal factor in isolation from the other environmental factors, especially the gas conditions, which in such cold Baikal mountain lakes as Frolikha or Kulinda differ considerably from those in Baikal proper. The bottom layers of these lakes are poor in oxygen even in summer, the water shows acid reaction, and autotrophic organisms are very scarce. Therefore their trophic value is very low (KOZHOV 1950b).

G. Y. VERESHCHAGIN considers also that the Baikalian species are more or less indifferent towards thermal and gas conditions. In doing so he refers to A. Y. BAZIKALOVA's experiments with gammarids (BAZIKALOVA, 1941), which have supposedly shown that not only littoral, but also deep-dwelling gammarids of Baikal withstand a drop in the oxygen content to 0.5—1.0 cm^3 per litre. But it follows from BAZIKALOVA's experiments that this drop is only endured by a few, relatively eurythermal species, such as *Gmelinoides fasciatus*, while most of the Baikalian species only survive when the oxygen content is comparatively high. The most important thing, however, is that these experiments were conducted in tanks and their duration did not exceed 100 hours, whereas in natural conditions it is a question of surviving throughout the animal's entire life cycle, from birth till maturity, which lasts for several years, and therefore the same species' demand for oxygen will be entirely different and certainly much higher. Instances of this have already been cited above. Evidently the researcher failed to take full stock of the time factor in connection with the life cycle, a factor vital in natural conditions.

Special investigations devoted to the problem of immiscibility have been conducted in recent years (LEVANIDOVA, 1948; BEKMAN, 1952; KORYAKOV, 1959a).

M. Y. BEKMAN tried to find out experimentally whether the Baikal water had a pernicious or inhibitive effect on alien organisms not adapted to living in the lake. The experiment was conducted with *Gammarus lacustris*, dozens of adult specimens of which were put in pails and deposited in Baikal, from the littoral zone to a depth of 1,000 m. They lived there for many months and evinced no signs of being inhibited. They deposited eggs, but the development of the eggs was greatly retarded. For its embryonic development this crustacean needs 360 day degrees; the effective temperatures of the growth and development of its young do not exceed 25° C (the optimum higher than 10°); and the sum of temperatures up to the moment of reaching maturity should not be less than 1,800°. In the case of Baikal, the thermal regime even of the littoral region rules out the possibility of normal growth and development of juveniles.

D. N. TALIYEV and Y. A. KORYAKOV (1947, 1948) conducted experiments to ascertain the degree of sensitivity of the Baikalian cottids and *Comephorus* to the thermal and gas regime. These experiments revealed an extremely low heat resistance of the latter, which perish at a temperature higher than 8—9°. The highest heat-resistance was observed among species of the genus *Paracottus* (*P. kneri, P. kessleri*), as well as in the pelagic cottid *Cottocome-*

phorus grewingki (19—21°) and in *Batrachocottus baicalensis.*

I. M. LEVANIDOVA (1948) undertook a comparative study of the distribution of the fauna in some Baikalian sors and bays and came to the conclusion that among the abiotic factors influencing the distribution of organisms in the "zone of contact" the predominant part was played by the thermal regime, then the seasonal and diurnal gas regimes, the salt content and the turbulence of the surf.

At the same time I. M. LEVANIDOVA advances, as one of the causes of immiscibility, the biocoenotic factor, although she does not develop this proposition on the basis of any special observations. G. Y. VERESHCHAGIN (1940b) also maintains that the biocoenoses which have formed in Baikal from native Baikalians leave no place for new immigrants.

The biocoenotic factor does play a part, of course, but it should not be over-estimated. Even in Baikal we have instances of semingly alien faunal elements easily colonising biocoenoses that have developed in the course of centuries, if conditions favour this invasion (the omul, the seal, etc.).

Thus, all the facts and observations available today point to the untenability of hypotheses reducing the ecological distinctiveness of Baikal proper merely to the influence of some peculiar properties of the water itself, whether it is an admixture of water with a changed molecular structure or peculiar micro-elements still unknown to us. Such hypotheses lead us away from the search of real causes of immiscibility, from ascertaining the mechanism of the complicated centuries-long influence of a complex of such important factors of the water environment as temperature, chemistry, illumination, soil, biotic relationships, etc.

2. Sections of the Littoral-Sor Zone and Their Population.

As has been said above, we refer to the littoral-sor zone, the sors and enclosed and sheltered shallow bays and gulfs.

In degree of isolation from Baikal's open waters, the sors can be divided into the following groups (KOZHOV, 1947):

1. Lake-sors, i.e., bodies of water connected with Baikal proper only by narrow gullies (outlets) cutting their way through the more or less broad and low sand spits separating them from Baikal.

To this group belongs the large sor-lake Rangatui situated on the south-eastern coast of the Chivyrkui Gulf.

The waters of Baikal exert little or no influence on the hydrological and hydrochemical regime of the waters of the lake-sors. In the course of long fluctuations of the water level of Baikal, sors can become fully detached from it and turn into ordinary lakes; in this instance their water level may turn out to be higher than in Baikal, as is the case, for instance, with Lake Kotokel, which takes up about 16,000 hectares on the eastern coast of Baikal in the drainage of the River Turka. Once this lake was undoubtedly a gulf, then it became a sor, and now it is 7 to 8 km away from Baikal. It lies 10 to 12 metres higher than the level of Baikal and is already connected with it by a whole system of small rivers. Moreover, such bodies of water can not only get detached from Baikal, but even lose all connection with it and turn temporarily or permanently into lakes without outflows, as has happened, for instance, with large lakes on the north-western coast of Baikal.

2. Typical sors, i.e., bodies of water connected with Baikal by a more or less broad gullet. The Baikal waters may have a certain influence on such sors when they penetrate into them during inshore winds. The Posolsky and Upper Angara sors belong to typical sors.

3. Gulf-sors, which are not fully detached from Baikal. They are connected with it by channels which are broad enough to allow a considerable hydrological influence of Baikal's open waters on them, especially during inshore winds. The large Proval Gulf situated to the north of the Selenga delta may serve as an example (fig. 15).

The total area of sections of the littoral-sor zone is inappreciable compared to the territory of the lake as a whole. It does not exceed 40,000 to 50,000 hectares. Nevertheless, they unquestionably play a positive part in the biological productivity of the regions of the lake adjacent to them.

As an illustration, we shall describe the distribution of benthos in the Posolsky Sor and in the gulf-sor Proval.

The Posolsky Sor, as has been mentioned in the first part of this work, lies to the south of the delta of the River Selenga. Covering an area of 3,500 hectares, it is 3 to 3.5 metres deep and receives four small rivers.

The bottom consists of more or less clean sands lying predominantly in the coastal belt, in particular opposite to river mouths and along open shores, silted sands in the southern part of the sor, and viscous grey silt with a large admixture of organic substances, occupying the central deep part of the sor.

At the beginning of June vegetation is still scant in the Posolsky Sor. Its vigorous development begins in mid-June. In summer areas protected from waves become almost fully overgrown with the macrophytes *Potamogeton*, *Sparganium*, *Nymphaea*, *Nuphar*, *Polygonum amphibium*, *Myriophyllum*, and other common lacustrine plants. In the central part of the sor *Potamogeton* can be found to a depth of 2—3 metres on separate patches of sandy-silty bottom sometimes covering an area of several dozen hectares.

The predominant part in the fauna of the Posolsky Sor is clearly played by lacustrine Siberian species, but also conspicuous is the presence of Baikalian species, especially gammarids, among which particularly numerous are *Micruropus possolskii*, *M. cristatus*, and also *Gmelinoides fasciatus*; one can also come across *Micruropus fixeni*, *M. wahli*, *M. talitroides*, *M. glaber*, *Pallasea grubei*, and others. The bryozoan *Hislopia placoides* has also been found. All these Baikalian species are widespread primarily in weed beds and on sandy bottoms.

Table XVII gives data on the quantitative composition of the population of various types of bottom. The biomass of the zoobenthos on all types of soil increases from spring till autumn, attaining the maximum in August and the beginning of September. This increase is due chiefly to chironomids, whose biomass increases greatly towards the autumn, gammarids and partly mollusks, whereas the number of oligochaetes remains, on the whole, at the same level throughout the vegetation period. In the course of the whole summer a particular numerical density of living organisms is observed on silty soils and especially near river mouths. In August and at the beginning of September the biomass of the zoobenthos reaches 160—386 kg/ha (1943) in the sor as a whole and 1,000 kg/ha at the mouth of the River Bolshaya, chiefly thanks to the mollusks *Sphaerium corneum* (up to 114 specimens, about 85 g per m²), and also *Pisidium subtruncatum*, *Valvata aliena*, Trichoptera, Hirudinea *(Herpobdella octoculata)*, Chironomidae, etc. In the centre of the sor the biomass reaches 263 kg/ha. The bulk of the biomass on silts in the central trough is composed of chironomids (5 g/m²) and mollusks (10 g/m²), with *Chironomus gr. semireductus*, the mollusks *Valvata aliena* and also *Pisidium henslowanum*, *P. amnicum* and others predominating. The gammarids play a secondary part. The gammarid species that occur more frequently are *Micruropus possolskii* and *Gmelinoides fasciatus*.

Life is not so abundant on sandy-silty soils, where the biomass of zoobenthos in August varies, on the average, from 118.7 to 97.1 kg/ha. But even on these soils it reaches 470 kg/ha opposite to the mouth of the River Bolshaya. The predominant groups are mollusks, among which a high numerical density is attained by *Sphaerium corneum* and species of the genus *Pisidium*, and gammarids, among which the main part is already played by the Baikalian species *Micruropus possolskii* and *Gmelinoides fasciatus*. Gammarids, mollusks and oligochaetes predominate on sandy-silty soils in the central part of the sor, with the same *Micruropus possolskii* and *Gmelinoides fasciatus* being especially numerous among the gammarids, *Pisidium henslowanum*, *P. subtruncatum*, *Sphaerium corneum* and *Valvata aliena* among the mollusks, and *Enchitraeoides aliger* and *Rhyacodrilus coccineus* among the oligochaetes.

The sands are more scantily populated, with the biomass of the zoobenthos ranging, on the average, from 47 to 55 kg/ha and reaching 80 kg/ha in some sections influenced by rivers. The predominant part there is played by gammarids and mollusks of the same species that are characteristic of sandy-silty soils.

Table XVII.

The Biomass of Zoobenthos in the Posolsky Sor in July-August 1938 (raw weight in g/m², number of specimens per m²).

Soil	Sand		Sand & Silt		Silt	
Number of bottom samples	18		14		10	
	Number of specimens	Weight	Number of specimens	Weight	Number of specimens	Weight
Oligochaeta	150	0.630	328	1.585	229	0.953
Hirudinea						
Helobdella stagnalis	—	—	8 ⎫	—	7 ⎫	—
Herpobdella octoculata	—	—	9 ⎬	0.932	23 ⎬	2.052
Glossiphonia complanata	—	—	1 ⎭	—	6 ⎭	—
Total	—	—	18	0.932	36	2.052
Mollusca						
Pisidium subtruncatum	88	—	166	—	197	—
Pisidium henslowanum	29	—	81	—	20	—
Pisidium amnicum	1	—	9	—	13	—
Pisidium Sp.	22	—	95	—	61	—
Sphaerium corneum	2	—	71	—	154	—
Limnaea auricularia	1	—	2	—	—	—
Planorbis gredleri	1	—	1	—	—	—
Planorbis leucostoma	—	—	4	—	1	—
Acroloxus lacustris	—	—	—	—	1	—
Valvata aliena	2	—	20	—	26	—
Valvata sibirica	1	—	2	—	12	—
Total	147	0.660	451	5.782	485	13.032
Gammaridae						
Micruropus possolskii	254	—	246	—	20	—
Micruropus fixeni	1	—	19	—	3	—
Micruropus wahli	1	—	—	—	—	—
Gmelinoides fasciatus	104	—	124	—	8	—
Pallasea grubei	1	—	1	—	—	—
Gammarus lacustris	—	—	6	—	21	—
Juvenis	76	—	64	—	86	—
Total	437	4.062	460	2.836	138	0.773
Chironomidae						
Glyptotendipes	4	—	—	—	—	—
Orthocladiinae	1	—	—	—	—	—
Cryptochironomus	1	—	4	—	5	—
Procladius	—	—	1	—	—	—
Paratanytarsus	—	—	22	—	99	—
Psilotanytarsus	—	—	—	—	19	—
Semireductus	—	—	5	—	2	—
Cladopelma	—	—	—	—	1	—
Not determined	1	—	—	—	—	—
Total	7	0.026	32	0.358	125	0.973
Trichoptera	2	0.030	14	0.428	2	0.200
Other insecta	—	—	—	—	5	0.137
Bryozoa, *Hislopia*	—	—	2	—	—	—
Grand total	743	5.408	1,305	11.921	1,020	18.120
In kg/ha	—	54.08	—	119.21	—	181.20

The total average biomass of the zoobenthos changes accordingly with the seasonal change of the benthos and in different periods is expressed in the following average values:

Table XVIII.

Average Biomass of Zoobenthos in the Posolsky Sor in Different Months of 1943.

Time of the year	Biomass in kg/ha
End of May—beginning of June	34
Beginning of July	41
Beginning of August	80
End of August—beginning of September	144

Now for the fish inhabiting the sor. *Rutilus rutilus lacustris* is widespread in the sor itself and in the shallows of the neighbouring region of Baikal. *Perca fluviatilis* also occurs throughout the sor. *Leuciscus idus* is not numerous. *Esox lucius* lives in the sor the year round. *Leuciscus leuciscus baicalensis* enters the sor for spawning in the second half of April and the first half of May. In summer it returns to Baikal. In autumn its migrations to the sor are observed again, but in winter it prefers to live in Baikal. *Lota lota* is not to be found frequently. *Thymallus arcticus baicalensis* occurs rarely. *Coregonus autumnalis migratorius* (the omul) enters the sor on the way to river spawning grounds from the end of August till autumn. In spring large numbers of omul juveniles descend through the sor from the spawning rivers. Periodically, sturgeons, chiefly young ones, also can be found. In 1944 the Amur carp *Cyprinus carpio haematopterus* was let into the sor. In recent years both adult and young carps have been found in the Posolsky Sor and in the lakes of the Selenga drainage.

The gulf-sor Proval, having an area of 18,500 hectares, lies to the north of the Selenga delta (fig. 15). Depending on the direction of the winds, the waters of the Baikal can freely enter and leave it through broad channels. The arms (gullets) of the Selenga flowing into the gulf carry large masses of suspended material with them which settles on the bottom. Owing to constant winds, insignificant depths and the stirring of the bottom mud, in summer the water in the gulf is always turbid and its transparency very low. The depth of the gulf rarely exceeds 4.5—5 m.

A narrow strip of clean sandy soil lies along the islands separating the gulf from Baikal; such soils also abound near the eastern and southern shores open to waves from Baikal.

The composition of silt varies. In some places it consists of particles of peat washed out of the old peat deposits composing the shore. In the central trough it is predominantly of a mineral nature, being composed of the material brought in by the Selenga. Closer to Baikal it grows richer in organic substances. The thermal regime, as has been noted, is characterised by rapid warming through in spring. In summer the temperature of water in the depths of the gulf reaches 20—26° C and in the shallows part, 13—16° (see table IV, V p. 33—34). The inshore and emergent vegetation is developed chiefly in the pre-estuarine and southern parts of the gulf.

The Proval Gulf, owing to the diversity of its biotopes, has almost all representatives of the general Siberian fresh-water fauna. Exceptional numerical density is attained by the bivalve mollusk *Anodonta*, which very thickly populates sandy-silty soils rich in organic substances, Ordinary Siberian species of *Pisidium*, Chironomidae, Oligochaeta (*Stylaria lacustris* and others) are also widespread.

Conspicuous in areas adjacent to the channels is the abundance of Baikalian species of Gammaridae, particularly *Micruropus possolskii*, *M. wahli*, *M. talitroides*, *Crypturopus pachytus* and *Cr. inflatus*, which occur almost throughout the gulf. In all, up to a score of Baikalian species of Gammaridae are found in the Proval, which is not the case in any other sor. Large numbers of the Baikalian polychaete *Manayunkia baicalensis* and the presence of the Baikalian bryozoan *Hislopia placoides* and the oligochaetes *Rhyacodrilus coccineus*, *Limnodrilus arenarius* var. *inaequalis* and *Propappus volki* have also been noted.

There is no doubt that the presence of a considerable number of Baikalian elements in the Proval is connected with the year-round high oxygen content of its waters and the lower summer temperature of water in the regions adjoining Baikal as compared with other sors.

In 1939—43 the average summer biomass of zoobenthos on silty soils in the gulf-sor Proval amounted to 370 kg/ha, with *Anodonta* accounting for 230 kg/ha.

The fish found in the gulf include *Rutilus rutilus lacustris*, *Perca fluviatilis*, *Esox lucius*, *Leuciscus leuciscus baicalensis*, *L. idus* and *Acipenser baeri stenorhynchus*, for which the Proval presents the main feeding ground. In the autumn most of the sturgeons leave for Baikal and winter there, returning towards the summer.

In addition to these species of fish, the gulf has also been found to contain *Thymallus arcticus brevipinnis* and in August also *Coregonus autumnalis migratorius*, which in this period enters the Selenga for spawning. It does not linger in the gulf and after spawning returns to Baikal. In September and October *Lota lota*, *Phoxinus* and Cottidae can be found in the Proval.

3. Distribution of Benthos in Open Baikal.

Horizontal Division

While studying the distribution of mollusks in Baikal on the basis of material collected in all regions and parts of the lake, we came across a number of facts pointing to clear distinctions between the fauna of its northern and southern parts, with the boundary lying in the region adjoining the mouth of the River Turka on the eastern coast and the region to the south of the Olkhonskiye Vorota Strait on the western coast, i.e., almost in the central section of Baikal. The differences between these parts of Baikal as regards the fauna of mollusks are quite clear; there are many species and varieties whose habitat is restricted either to the northern or southern part of the lake. For instance, such species as *Baicalia variesculpta* (fig. 67,1), *B. jentteriana, B. angigyra, B. macrostoma* and *Choanomphalus annuliformis* live only in the northern part, whereas *Liobaicalia stiedae* (fig. 66,3), *Baicalia herderiana* (fig. 67,2) and *B. turriformis* live only in the southern part (KOZHOV, 1936, 1947).

Similar differences between the northern and southern parts of Baikal in the case of gammarids were noted by V. Č. DOROGOSTAISKY (1923b). Moreover, he pointed to approximately the same boundaries: the area of the Anga on the western coast and the fringe of the Selenga shallow on the eastern coast.

Taking the distribution of gammarids as the basis, A. Y. BAZIKALOVA (1945) divides Baikal, not into two, but three parts in accordance with the morphological depressions (southern, central and northern) that comprise Baikal today. But she, too, points out the transitional character of the gammarid fauna in the central part of the lake, noting at the same time that the differences between the faunae of the three depressions are observable chiefly in the coastal zone and inconspicuous in the deep zones. According to A. Y. BAZIKALOVA, the southern trough is characterised by such deep-dwelling gammarids as *Poekilogammarus lydiae, Leptostenus leptoceras, Garjajewia dogieli, Eulimnogammarus parvetiformis* and others, and the northern by *Eulimnogammarus rachmanowi, Lobogammarus latus, Abyssogammarus gracilis, Ceratogammarus acerus, Pallasea meissneri* and *Axelboeckia carpenteri profundalis*.

The differences between the nearshore bottom faunae of the northern and southern parts of Baikal can be partly explained by differences in the physio-geographical conditions obtaining in them today. For instance, the ice in the northern half breaks up, as a rule, a week or two later than in the southern part, and the annual cycle of temperatures in the north is slightly different. But these differences are not great enough to be held solely responsible for the differences in the fauna. For instance, in the water regime the Maloye More differs more sharply from the open regions of North Baikal than from the southern part, but its fauna is of a distinctly northern nature.

Consequently, the causes of this differentiation of the generally uniform Baikalian fauna should be sought not only in present-day ecological conditions, but also in the geological history of Baikal, which will be dealt with in greater detail in the chapter devoted to the evolution of the Baikalian fauna.

According to L. S. BERG (1909), Baikal presents a sub-region of the Holarctic having only one province, that of Baikal. For this reason North and South Baikal can be regarded, zoogeographically, as the northern and southern districts of the Baikal province.

A more detailed acquaintance with the distribution of Baikal's bottom-dwelling fauna has made it possible to establish a number of sections differing in the nature of their fauna within the confines of both parts of Baikal. Of special interest among them are the Maloye More, Selenga and Ushkany sections.

In the Maloye More section, which incorporates the whole of the Maloye More, almost all Baikalian species are represented. There are even endemic forms, such as the mollusk *Baicalia werestschagini*, which has so far been found only in the Maloye More, and *B. wrzesniowskii olchonensis*, the gammarids *Axelboeckia carpenteri elegans*, and others. In recent years A. Y. BAZIKALOVA (1959) has described five new gammarid species from the Maloye More and established a new genus, *Metapallasea*, for one of them. Future investigations will show whether all these species belong to the Maloye More alone. The so-called malomorsky gwyniad *(Coregonus lavaretus baicalensis)* dwells almost exclusively in the Maloye More. Similar examples can be cited for the area of the Ushkany Islands, as was pointed out by V. Č. DOROGOSTAISKY (1923b), A. Y. BAZIKA-LOVA (1945) and D. N. TALIYEV (1955). According to TALIYEV, this area is inhabited by the Cottoidei species *Batrachocottus uschkani* and *Paracottus insularis*, which have not yet been found in other regions of Baikal, whereas *Batrachocottus baicalensis*, *Paracottus kneri* and the mollusk *Baicalia ciliata*, which are widespread in other regions, are absent there.

There are some peculiarities in the composition of the fauna of the shallows bounding the Selenga delta (the Selenga section), which has its own endemic forms, such as the Mollusk *Baicalia korotnewi gracilis,* a number of species of gammarids, and other faunal groups.

Vertical Zonation in the Distribution of Benthos

Various factors of the water environment, such as temperature, chemistry, light intensity, movement or calmness of the mass of water and properties of the bottom soil are known to depend very much on the depth of the given body of water, i.e., the depth of the water mass between the bottom and the surface. For this reason the differences in the bottom population connected with the alteration of environmental factors make possible vertical zonation of every more or less large body of water. The zones established in the lakes present, as it were, a generalisation of the influence of the complex of environmental factors and particularly light, which is essential for the life of plants, on the organisms inhabiting them. These zones and their boundaries, however, are to a considerable degree conditional, serving to facilitate the description of biocoenoses. A distinguishing feature of biological zonation in Baikal is that this lake is exceptionally deep and the contrasts in life resulting from the change of environmental factors from the coastal shallows to the extreme depths are uncommonly great. Therefore the patterns of zonation for ordinary lakes are not fully applicable in the case of Baikal.

It is known that the properties of the bottom soil are an important environmental factor bearing on the distribution of benthos. The distribution of soil, in turn, depends on many factors, such as the degree to which the nearshore section is protected against the prevailing winds, as well as currents, turbulence, the steepness of the underwater slope, the composition of the shores subjected to erosion, the nearness of the mouths of big rivers filling pre-estuarine areas with outwash, etc. These conditions create a rather motley picture of the distribution of soils in the coastal region, which cannot but influence the distribution of benthos. In view of this, separate patterns of biological zonation should be worked out for every area with characteristic features (extensive shallows, gulfs, open sections with rocky shores and a steep underwater gradient, and so on). But since there is not yet enough material for this, in describing the distribution of benthos we are compelled to use a general pattern based primarily on the peculiarities of the distribution of the bottom flora and fauna connected with the increase in depth and the changes in the properties of the soil, in light intensity, temperature, and other essential environmental factors (KOZHOV 1931a, 1934a, b, 1947; YASNITSKY 1928).

The bottom vegetation in the open regions of Baikal is distributed as follows:

The belt of *Ulothrix zonata*, a green alga which thickly covers beach stones in the area of the most violent turbulence. It extends from the edge of water to a depth of 1—1.5 metres, at times somewhat deeper. Many various species of bottom and epiphytic diatoms are found in *Ulothrix* growths. In its lower part this belt is often adjoined by the algae *Tetraspora cylindrica* and *Didymosphenia geminata*.

Then follows the belt incorporating the bulk of the bottom vegetation. It is broken only where the composition of the bottom is unfavourable for algae—mostly areas with sandy or silty soils opposite to the mouths of big rivers. Rocky bottom in this belt is populated chiefly by the algae *Cladophora* and *Aegagropila*, which spread to a depth of 10—15 m, then species of the genus *Draparnaldia* (fig. 79), which reach the same depths, *Chaetomorpha* species and other algae, spreading to 15—30 m and deeper.

In Baikal's bays and sections protected from the prevailing winds some of these algae, which are characteristic of open Baikal, disappear and are replaced by species of *Nostoc, Potamogeton, Myriophyllum, Ceratophyllum, Chara*, and others. Baikal also abounds in bottom diatoms, which grow very sparse with increase in depth but still can be traced to a depth of 50—70 m.

Now for the laws which have been found to govern the vertical distribution of soils in open Baikal. The edge of the water is adjoined by a strip of clean-washed sands, while a belt of boulders, pebbles and unrounded stones stretches along rocky shores. Farther on lies the belt of comparatively clean coarse sands covering large areas or alternating in patches with ridges of rounded pebbles. From 8—12 m the sands begin to be silted and at 15—20 m they are replaced by considerably silted fine sands. Strong silting near the shore, not far from the edge of water, is observed only opposite to the mouths of big rivers and in areas isolated from currents.

The sharp drop in the density of vegetation, the marked change in the properties of the soil and in the thermal regime of the bottom layers, and other combined factors are evidently responsible for the qualitative and quantitative increase in the distribution of the fauna which is usually observed along open shores in the zone of depths of about 15—20 m. Thus there is outlined a coastal zone analogous to the littoral of other lakes. But even within the confines of this zone the fauna and flora gradually alter in accordance with changes in the life conditions. In this zone we can distinguish the surf subzone, with fairly distinct life conditions, then the subzone of small depths stretching to a depth of 4—6 m, where the waves already lose their destructive force and the bottom plants can develop on a large scale, and deeper still, the lower subzone of the littoral.

Beyond the 20 m depths begins the region of transition to deeper zones. The upper part of this region extends to a depth of approximately 70 m (the limit of the spread of bottom diatoms). We call it the sublittoral. There the effect begins to be felt of the lower (in summer) bottom temperatures, the obviously insufficient light intensity, the low content of organic matter in the soil, and also the predomination of silted sands over rocky and sandy soils. In this connection, along with littoral species, we can already find there

Table XIX.

Zone	After Kozhov (1931, 1947)	After Sovinsky (1915)	After Dorogostaisky (1923)	After Bazikalova (1945)	After Vereshchagin (1949)
Littoral	0—20 m Upper section or surf subzone, 0—1.5 m Middle section, 1.5—5 m Lower section, 5—20 m	0—10 m	0—5 m	0—5 m	Surf zone, 0—1.5 m Zone of small depths, 1—1.5—10—15 m
Sublittoral	20—70 m	10—15 m	5—50 m	5—70—150 m	10—15—100 m
Supra-abyssal (transitional)	70—250 m	Transitional, 50—200 m	Transitional, 50—300 m	Transitional, 70—150—300 m	Transitional, 100—250 m
Abyssal	Deeper than 250 m. Upper section, 250—500 m Lower section, deeper than 500 m	Subabyssal, 200—500 m Abyssal, deeper than 500 m	Abyssal, deeper than 300 m	Deep, 300—500 m Abyssal, deeper than 500 m	Subabyssal 250—500 m Abyssal, deeper than 500 m

supra-abyssal and sometimes abyssal species of fauna. The zone of depths beyond the sublittoral reaching the 250—300 m isobath is called by us the transitional, or supra-abyssal zone. Its lower limit lies in the layer of water where the temperature remains practically unchanged the year round. The bottom is composed of strongly silted sands or silt, while bare rocks predominate in sections with a steep gradient.

The abyssal, or deep zone embraces the underwater slope and the whole of the main trough of Baikal beginning with 250—300 m depths. The bottom of the abyssal zone is covered, with rare exceptions, with deep-water, predominantly diatom ooze; the water regime is characterised by constancy; the temperature stays at about 3.4—3.6° C the year round; the oxygen content ranges from 75 to 90% of the saturation point; light is practically absent. The fauna is represented almost exclusively by deep-dwelling species.

Two subzones can be distinguished in the abyssal zone: the upper, to a depth of 500—600 metres, and the lower, extending to the extreme depths, the domain of typically deep-water species.

The pattern of zonation in the distribution of benthos in Baikal outlined by us is very similar to the pattern elaborated by S. STANKOVIČ (1960) for Lake Ohrid, which is vertically divided into the littoral (0—18—20 m), sublittoral (20—40—50 m), and deep zone (from depths greater than 50 m to the bottom). Evidently the deep zone of Lake Ohrid has a corresponding division in Baikal's supra-abyssal zone, with depths from 70 to 250 m. There is no true abyssal zone in Ohrid, its maximum depth being 286 m.

Our version of vertical zonation in open Baikal is not a generally accepted pattern. Attempts to outline zones in the distribution of bottom fauna in Baikal were made by V. A. LINDHOLM (1909), V. SOVINSKY (1915), DOROGOSTAISKY (1923), VERESHCHAGIN (1949), BAZIKALOVA (1945), TALIYEV (1955), BEKMAN (1959).

Without going into a detailed analysis of all these patterns, we give a table (XIX) containing those of them which have been more thoroughly elaborated.

Let us proceed to a closer review of the repartition of benthos in the zones of open Baikal outlined above.

The Littoral Zone (0-20m)

The width of the upper subzone of the littoral affected by the surf strongly varies in different sections of Baikal; in some sections with rocky shores it does not exceed 10—15 metres, whereas in other places it reaches hundreds of metres. The bottom is composed chiefly of pebble and large cobble roundstone with frequent huge slightly rounded fragments of coastal rocks. Sands predominate near river mouths and sand beaches and in bays.

Conditions of life in this subzone are very peculiar. Strong surf rolls stones and pebble and rolls and shifts sands; only heavier stones remain more or less motionless (fig. 80, 81, 82). Sand, debris, remains of plants previously washed ashore and bodies of insects and other small animals are washed away from the shore and are either carried off far away, settling on the bottom, or again brought to the shore. During strong storms the water in the surf zone becomes turbid, especially in the area of sand shores. On clear and calm days the water at a depth of 0.25 m is heated to 5—7° C at the beginning of June and 12—16° C and more at the end of June and in August. But during strong winds from the mountains the temperature of water near the leeward shores drops sharply and rapidly.

In summer diurnal fluctuations of temperature are observed at the edge of water, with the difference in temperatures in the course of the same day sometimes exceeding 2—3° C.

The water chemistry also varies considerably throughout the 24 hours; at daytime in summer the water can be found to be over-oxygenated (up to 120—150%), whereas at night a slight oxygen deficit may be observed (YASNITSKY, BLANKOV & GORTIKOV, 1927), with an oversaturation with free CO_2.

The surf belt constantly shifts owing to fluctuations in the level of water in Baikal, which rises and falls within a range of 1 metre and more in the course of the year. In winter the inshore belt freezes through to a depth of 1 m.

Macrophytes are not to be found on the sands of the surf belt in open sections, for the constant rolling of the soil prevents them from getting fixed. The animals are represented by the tiny yellow oligochaete *Mesenchitraeus bungei*, large numbers of which populate the sands of the surf belt, especially at the edge of water and on the surf-wetted shore.

In summer the stones on the surf belt are covered with a thick green mat of the alga *Ulothrix zonata*, which appears in spring at the very edge of water and later on, with the rise of the level of Baikal, covers the bottom to a depth of 1—1.5 m.

The most characteristic animal inhabitants of rocky sections of the surf belt are the oligochaete *Mesenchitraeus bungei*, which forms thick accumulations between and under stones, the large gammarus *Eulimnogammarus verrucosus*, the tiny light-blue *E. cyaneus*, then *E. viridis*, *Gmelinoides fasciatus*, *Brandtia lata* (fig. 59,1) and some others. In *Ulothrix* growths one can usually find masses of the ento-mostracans Harpacticidae. The turbellarians are represented by *Baicalobia guttata* (fig. 39,1), *B. copulathrix*, *B. variegata* (fig. 37,5). The algal beds are very densely populated by larvae of Orthocladii-nae and Tanytarsini chironomids and the oligochaete *Nais* sp. During strong surf the animals usually hide between stones, in algal growths or in the sand.

In early spring vast numbers of mating Trichoptera in the imaginal stage accumulate in the surf belt at the very edge of water. There are no sponges in the surf subzone, and only individual mollusks enter it on quiet calm days.

Towards the autumn the *Ulothrix* algae disappear from the surf belt and the animal population moves to the deeper adjacent zone. Only the oligochaete *M. bungei* stays on, burrowing deep in the soil.

The surf subzone with the complex of population characteristic of it is well shown along the shores of all open Baikal and in its open bays and gulfs.

In summer it is fairly rich quantitatively, especially on rocky bottom, among algae. Life is particularly abundant in spring and in the first half of summer, when masses of caddis-flies appear. In this period the graylings *Thymallus arcticus baicalensis* appear there at night and often keep at the very edge of water, attracted by plentiful food, and Cottoidei are also usually present. The surf belt of sandy soils in strongly open and shallow regions, especially along the eastern shores (the Selenga shallow), is thinly populated.

Life conditions in the littoral beyond the surf subzone are more favourable for Baikalian organisms. The influence of surf waves is greatly weakened there and the range of seasonal fluctuations of temperatures in the bottom layers is less than in the surf subzone.

The predominant soils along the open shores and far from big rivers are pebbles and comparatively clean coarse sands which, beginning with the 5—6 m depths are replaced by fine sands with an admixture of mud. In bottom hollows one can often come across considerable accumulations of coarse detritus brought by rivers, dead weeds washed away from the shores, etc.

In the pre-estuarine regions of big rivers the bottom of the littoral is usually composed of sand, often with rows of pebbles or strongly silted; in sheltered sections of pre-estuarine regions silted sands sometimes directly follow the surf subzone.

Vegetation is concentrated chiefly on rocky patches of the littoral bottom. On sandy soils, especially in areas protected by long promontories, one can now and then come across separate small growths of *Myriophyllum*. In better sheltered areas *Potamogeton, Polygonum amphibium*, as well as Characeae and other water plants are found.

Now for the main biotopes of the population of the littoral.

Rocky bottom. Predominating in the littoral along the open shores of Baikal, this bottom is populated by a fauna which is very rich and varied. As a rule, any stone taken from a depth of 2—10 m is covered on top, on the sides and sometimes even from below, in hollows, cracks, etc. entirely or in patches, by the algae *Didymosphenia, Cladophora, Chaetomorpha, Draparnaldia* (fig. 79), *Tetraspora*, and others, as well as with lichens and colonies of the

Baikalian sponge *Lubomirskia baicalensis*. Often one species prevails over other plants, and sometimes the stone is covered almost entirely by sponges.

The Baikalian sponge *Lubomirskia baicalensis* (fig. 32) populates the rocky bottom in particularly large numbers. At a depth of 4—10 m whole thickets of this bright-green branching sponge can be seen. In the area of the Listvennichny Gulf sponges occupy at least one-fourth of the area of rocky bottom of the littoral (GAVRI-LOV, 1950a, b). Oligochaetes, chiefly *Lamprodrilus nigrescens*, live under the base of the sponge; hundreds of tubes of the Baikalian polychaete *Manayunkia baicalensis* (fig. 41) often stick out around the edge of the base and from oscular openings; larvae of caddis flies, which construct their cases from coarse grains of sand, live in the hollows of stones. On the exterior of the sponge we always find vast amounts of very attractive minute emerald-green gamma-rids *Brandtia (Spinacanthus) parasiticus*, which tenaciously cling to the sponge with their feet and do not part with it even when it is lifted from the water. In addition the exterior of the sponge is richly inhabited by large dark-red gammarids *Eulimnogammarus cruentus*, tiny reddish *E. grandimanus*, and many other species. One can also always find there Harpacticidae, which move along the branches in the form of whitish dots. They account for the presence of large numbers of the gammarids *Brandtia parasitica* and other species.

A very rich fauna is found on overgrown stones at a depth of 2—5 m. The bare parts of stones are inhabited by many turbella-rians, among which *Baicalobia copulathrix, B. guttata* (fig. 39,1) and *B. variegata* (fig. 37,5) are more common; the oligochaetes *Propappus glandulosus, Lamprodrilus nigrescens* and others; the Baikalian leeches *Trachelobdella torquata* (fig. 43); large numbers of small Rhabdocoelidae; the isopods *Asellus angarensis* and *A. baicalensis* (fig. 57); larvae of the caddis fly Baicalinini and chironomids; masses of gammarids hiding in cracks, algal growths, under stones, etc. Especially numerous among them are *Eulimnogammarus grandi-manus, E. cyaneus, E. viridis, E. maacki, Brandtia lata* (fig. 59), *Gmelinoides fasciatus, Baicalogammarus pullus*, and others. We also find there a wealth of mollusks, especially *Choanomphalus amauro-nius, Ch. maacki* (fig. 66,8) *Baicalia herderiana* (fig. 67,2), *B. bithy-niopsis, B. ciliata* (fig. 68,2) and, in the north of Baikal, *B. macrosto-ma* and *B. variesculpta* (fig. 67,1). On stones and between them one can often see Baikalian Cottoidei fish, particularly *Batrachocottus baicalensis, Paracottus kessleri* and *P. kneri* (fig. 74,2) and in spring also *Cottocomephorus grewingki* (fig. 74,4), which spawns there.

These biocoenoses are quantitatively rich, as is seen from the results of studies conducted by us in the area of the Bolskiye Koty (table XX).

Table XX.

Population of Stones Covered by Algae and Sponges per 1 m² at a Depth of 3—4 m in the Area of Bolshiye Koty (Kozhov, 1931)

Group	Number of species	Number of specimens	Raw weight, in g	Predominant forms
Spongia	2	15 (colonies)	—	*Lubomirskia baicalensis*
Turbellaria	6	5	1.3	
Polychaeta	1	200	0.2	*Manayunkia baicalensis*
Oligochaeta	4	100	0.7	*Lamprodrilus nigrescens*
Hirudinea	2	77	0.2	*Trachelobdella*
Asellus	1	55	0.1	*Asellus angarensis*
Gammaridae	up to 20	972	25.0	*Eulimnogammarus grandimanus*
Mollusca	9	2,066	80.0	*Baicalia herderiana, Choanomphalus amauronius, Ch. maacki*
Trichoptera	2—3	785	7.8	Baicalinini
Chironomidae	2—3	143	0.3	—
Total:	up to 50	4,508	115.6 (weight without sponges)	

* Ostracods and other small animals were discounted.

It is believed that the biomass of zoobenthos on alga- and sponge-covered rocky grounds in the littoral at a depth of 2—3 m should average 1,000 kg/ha, not counting the vast amount of sponges, whose weight often exceeds 1 kg/m².

A comparison of the wealth of life in various belts of the bottom with rocky sections shows that the depths directly adjoining the surf zone are more scantily populated than depths of 3—8 m, where life is particularly abundant. Wherever the bottom is composed of whole rock, life is also poorer than on coarse pebbles and large stones. The variety of the fauna of the rocky littoral can be judged by the fact that in the area of the Listvennichny Gulf the following number of species were found on one hectare of the bottom at a depth of 0—4 m: Spongia 3, Turbellaria 14, Bryozoa 1, Polychaeta 1, Oligochaeta 24, *Asellus* 3, Gammaridae 75, Trichoptera 6, Chironomidae 12, Mollusca 18, Pisces 15—172 species in all (Gavrilov, 1950a, b).

Rocky grounds of the littoral of open Baikal are favoured by the Baicalian black grayling *Thymallus arcticus baicalensis* and many species of Cottoidei.

Coarse sands of the littoral at a depth of 2—10 m along the open coast are usually embedded in rocky sections in patches of from 1 m² to several hectares or occupy large areas opposite to sand shores and in pre-estuarine regions. In the area of the Selenga delta and on both sides of it sands spread over a vast territory, reaching out into open Baikal far beyond the 10 m isobath to a distance of 8—10 km from the delta. Extensive areas are taken up by sands also in the vicinity of the mouths of the rivers Upper Angara, Kichera and Turka and in many other places.

These sands are populated by a very distinct fauna. The turbellarians are represented by several species, the most characteristic of them being the pale pinky *Archicotylus plana* and *Sorocelis nigrofasciata* (fig. 37,1). Among the oligochaetes, particularly numerous are *Teleuscolex korotneffi*, *Clitellio multispinus*, *Tubifex inflatus*, then *Haplotaxis ascaridoides*, and *Limnodrilus pygmeus*. One can also often find the sandloving form of the polychaete *Manayunkia baicalensis* and the leech Piscicolidae. But a particular numerical density is attained by gammarids, whose motley or yellow-brown colouring blends well with the colour of the sand. Special mention should made of the genus *Micruropus*, most of the species of which are typical burrowing sand-dwellers. The most numerous of them are *M. kluhki*, *M. wahli*, *M. talitroides*, *M. crassipes* and *M. cristatus*.

The genus *Hyalellopsis* (fig. 61,2) is also very typical of the sands, with the tiny compact yellowish crustaceans *H. variabilis* and *H. czyrnianskii* being the commonest species of this genus on coastal sands. The representatives of other groups of Baikalian gammarids found frequently on the sands include the red-eyed dingy-brown crustacean *Carinogammarus rhodophthalmus*, *C. microphthalmus*, *Echiuropus macronychus* and *Poekilogammarus araeneolus*, to name but a few. Among the mollusks typical of the coastal sands we should mention *Choanomphalus dybowskianus*, *Ch. schrenki*, *Ch. gerstfeldtianus*, *Baicalia oviformis*, *B. dybowskiana*, *B. florii*, *B. bithyniopsis*, *B. elata*, *B. semenkewitschi*, *B. pulla*, *B. contabulata*, *Benedictia baicalensis* (fig. 70,3), *Sphaerium baicalense*, *Pisidium maculatum*, *P. korotnewi*, and *Valvata baicalensis* (fig. 66,6,7). The specific composition of the biocoenoses of the coarse sands of the littoral at depths of up to 5 m is very rich, with the number of species reaching 50, not counting small-size forms penetrating through the one-millimetre sieve. There are especially many mollusks (about 20 species) and gammarids (more than 20 species). The number of specimens (without microfauna) per 1 m² of the bottom sometimes exceeds 2,000, with gammarids and mollusks predominating. The biomass is lower than on rocky bottom, but still fairly high, reaching, on the average, 150 kg/ha. Mollusks account for the greater part of it.

On these sands one can also find considerable numbers of the Cottoidei fish, especially *Paracottus kessleri*.

According to MIKLASHEVSKAYA (1935), clean sands at a depth of 5—10 metres in the area of the Selenga shallow average, per 1 m² of the bottom, 14.52 g (145.2 kg/ha) of biomass and 731 specimens of animals, which are distributed as follows:

Gammaridae	384	specimens weighing	4.81 g
Oligochaeta	203	specimens weighing	9.19 g
Polychaeta	116	specimens weighing	0.12 g
Hirudinea	2	specimens weighing	0.01 g
Mollusca	26	specimens weighing	0.39 g
Total	731	specimens weighing	14.52 g
in kg/ha			145.2

In the area of North Baikal, the biomass of zoobenthos on the sands of the littoral is estimated by us to average 150 kg/ha.

It can be inferred from the data available that the biomass of zoobenthos of the sandy soils of the littoral of open regions approximates, on the average, 12—15 g per m² or 120—150 kg per hectare.

Silty fine sands with detritus at a depth of 8—20 m are more richly populated, as regards both quality and quantity, than clean sands. There one comes across not only typical sand dwellers, but also some new forms, which prefer deposits of detritus. For instance, the oligochaete population includes, besides *Licodrilus schizochaetus*, *L. parvus* and *Limnodrilus arenarius* also *Licodrilus dybowskii*, *Lamprodrilus stigmatias*, *Styloscolex baicalensis* and some other forms. Large numbers of the polychaete *Manayunkia baicalensis* are found there. The oligochaetes, in addition to some typical sand-loving forms, include many *Crypturopus pachytus*, (fig. 58,3), *Cr. inflatus*, *Cr. tuberculatus*, *Carinogammarus rhodophthalmus* and *Axelboeckia carpenteri* (fig. 61,6); *Pallasea brandti* and *Carinogammarus cynnamomeus* are frequent at depths of about 20—25 m. The mollusks are abundantly represented by *Pisidium maculatum*, *Baicalia carinata*, *B. contabulata*, *B. florii*, *B. pulla*, *B. pulchella*, *B. wrzesniowskii*, and others. The number of species found there exceeds 50, being more or less evenly distributed among mollusks, gammarids and oligochaetes. Population density reaches 2,500—5,000 specimens per 1 m² of the bottom (without microfauna), and the biomass 150—250 kg/ha, with mollusks outweighing all other groups of benthos.

Opposite to the mouths of big rivers the population of sandy and sandy-silty soils is similar, on the whole, to that of other sections with the same soils, but the biomass of zoobenthos varies strongly depending on the amount of detritus, which, in its turn, is determined by the distance from the shore or the degree of protection against turbulence.

The growths of Characea, *Potamogeton*, *Myriophyllum* and other water plants in areas protected by promontories are, as a rule, very rich in life, which is even more varied than on the soils described above. One can find there such dwellers of overgrown rocky bottom as the mollusks *Choanomphalus amauronius, Baicalia herderiana* (fig. 67,2,), the gammarids *Eulimnogammarus grandimanus, Baicalogammarus pullus,* etc., as well as *Carinogammarus rhodoph-thalmus, Crypturopus tuberculatus, Micruropus litoralis,* and other typical sandloving and detritus-eating forms. Other gammarids found frequently are *Pallasea cancelloides, P. brandti* and *Hetero-gammarus sophianosi, Pisidium korotnewi, P. maculatum, Benedictia baicalensis, Kobeltocochlea martensiana* and *Baicalia oviformis* are very numerous among the mollusks.

The density of population is seen from the figures contained in table XXI.

Table XXI.

Zoobenthos of Hydrophyte Growths on Dark Silty Sands per 1 m² (the Area of Bolshiye Koty, Summer).

	Species	Specimens
Turbellaria	1— 2	12
Oligochaeta	6— 8	290
Gammaridae	8—10	120
Mollusca	15	870
Trichoptera and other insects	5	200
Total	up to 40	1,492

Silted soils or fine strongly silted sands occur at a depth of up to 20 m chiefly opposite to river mouths, in sheltered sections. Sometimes they lie near the shores, bounding surf sands. They are rich in organic admixtures and have a dark colour.

These soils have a rich population of oligochaetes and gammarids. According to L. G. MIKLASHEVSKAYA (1935), the biomass and population density on silted facies of depths of 10—25 m in the area from the Selenga shoal to the southern extremity of the lake (opposite to river mouths), is expressed in the following average figures per 1 m² of the bottom:

Gammaridae	1,900 specimens weighing 22.8 g
Oligochaeta	150 specimens weighing 3.6 g
Hirudinea	20 specimens weighing 0.18 g
Mollusca	80 specimens weighing 0.12 g
Total	2,150 specimens weighing 26.7 g (267 kg/ha)

Conditions of life in bays differ markedly from those along open shores, chiefly owing to their being relatively well protected against

Fig. 86. Peschanaya Bay. Photo by M. Kozhov.

Fig. 87. Babushka Bay on the western shore of Baikal.

strong turbulence. For this reason the bottom layers of water are in a calmer state, the soil is less subjected to overturning, and the water is more or less isolated from nearshore currents. Therefore the predominant part in bays is played, as a rule, by sandy, sandy-silt and silted soils deposited in troughs, whereas rocky grounds are of secondary importance, although they form accumulations opposite to rocky promontories. The surf belt is not so clearly defined there as in open Baikal, and bottom plants have more favourable conditions for development on sandy soils due to their relative stability and insignificant turbulence. The wash of the rivers and streams flowing into the bays is not carried far into Baikal, but is deposited in their troughs, providing good food for bottom animals.

There are many such bays in Baikal such—the Peschanaya Bay (fig. 86), Babushka Bay (fig. 87), Anga Bay (fig. 88), Zavorotnaya Bay (fig. 89) and others. As an example, let us review the benthos of the Boguchan Bay situated in the north-west of Baikal 45 km to the south of the northern extremity of the lake (KOZHOV, 1934a).

The surf belt in the bay is clearly defined only along the promontories, and the animal population of sands and rocky soils covered with the alga *Ulothrix* is the same as in the open regions of Baikal. Rocky ground composed of pebbles, large rounded stones and unrounded boulders occupy comparatively large areas near the promontories, reaching depths of 5—10 m and more. The spaces between stones are filled with sand. On stones we find the common Baikalian macrophytes *Draparnaldia*, *Aegagropila* and others (to depths of 5—6 m), and also *Cladophora* and *Chaetomorpha*, which occur up to depths of 20—25 m. They are distributed in belts in the same order as in the open regions.

The animal population of these stones abounds with a typical littoral fauna: sponges, vivid multi-coloured gammarids as well as oligochaetes and those species of turbellarians which are to be found on stones in the open regions. *Baicalia variesculpta* (fig. 67,1), *B. macrostoma*, *Choanomphalus amauronius* and *Ch. maacki* (fig. 66,8) are numerous among the mollusks.

The density of the population of stones in the Boguchan Bay is not lower than in the open regions.

Yellow and grey sands occupy more than three-fourths of the bottom area of the bay, reaching a depth of 10—15 m. The flora is represented everywhere by growths of *Myriophyllum*, *Potamogeton* and sometimes *Cladophora floccosa*, which is found on sand grains in the form of small shaggy balls, then *Chaetomorpha*, *Tetraspora*, etc.

The specific composition of benthos in these sections is rich and population density high. The predominant forms are the gammarids *Micruropus cristatus*, *M. talitroides*, *M. klukhi*, *Crypturopus tuberculatus*, *C. inflatus*, *Baicalogammarus pullus*, *Pallasea cancellus*,

Fig. 88. Anga Bay. Photo by M. Kozhov.

Fig. 89. Zavorotnaya Bay, Baikal mountain ridge in the background. Photo by
M. Kozhov.

Poekilogammarus araneolus, Pallasea brandti, Gmelinoides fasciatus, Carinogammarus rhodophthalmus microphthalmus. The more numerous mollusks are *Baicalia florii* and *B. oviformis*, which populate these soils in vast numbers; *Pisidium amnicum* var. *baicalense* and *P. korotnewi* are also very numerous; one often finds *Valvata piligera, Baicalia carinata* and *Benedictia baicalensis; Limnaea auricularia* var. also occurs frequently. The predominant oligochaetes are *Tubifex inflatus, Propappus volki, Teleuscolex korotneffi, Lamprodrilus satyriscus, Clitellio korotneffi, Limnodrilus arenarius, Lamprodrilus pygmeus.* The tubellarians that can be found in the bay are *Archicotylus* sp. and *Sorocelis nigrofasciata.* The polychaete *Manayunkia baicalensis* is also frequent.

The belt of sands richest in life lies at a depth of 3—7 m in the bay (table XXII). At greater depths farther from the shore and at a depth of less than 3 m the population of the sands is noticeably poorer.

The littoral of extensive gulfs will be described in a section specially devoted to these regions.

The studies of the microbenthos of the littoral of Baikal conducted in recent years have shown that it contributes greatly to enriching the soil with organic substances. Overgrown stones and sands in the littoral are inhabited by many bottom diatoms as well as by Harpacticidae and Cyclopoida; rotifers, tardigrades and nematodes also occur. Among the copepods, a significant part is played in the microbenthos by *Cyclops serrulatus* and species of the genus *Acanthocyclops.* Regrettably, the quantitative aspect of microbenthos in the littoral of Baikal's open waters has not been studied.

Table XXII.

Biomass of the Zoobenthos of the Sands in the Boguchan Bay at a Depth of 3—6 m (the Mean of 11 Samples Taken by the Bottom Grab in the Summer of 1931)

	Species	Specimens	Raw weight in g/m^2	% of weight
Vermes	18	521	11.147	22
Mollusca	19	741	42.100	70
Gammaridae	16	386	2.352	4.3
Insecta	5	111	2.214	3.7
Total	58	1,759	57.813	100

An essential part in the diet of the benthic animals of the littoral is evidently played by bacteria. According to A. G. RODINA (1954), bacteria, which develop in vast numbers on stones in the littoral, participate in the destruction of these rocks and in the cycle of certain elements, silicon included. They cover stones in a continuous film, being particularly numerous in areas with rich vegetation.

Large numbers of bacteria have been found in the tissues of sponges (up to 563 million per 1 cm²), which leads A. G. RODINA to suppose that bacteria not only serve as food for sponges, but also bear a relation to their excretory processes.

In summer (July—August) the biomass of microbes in the rocky littoral reaches, according to A. G. RODINA, an average of 8 g/m². Considerable numbers of microbes have also been found on the sandy soils of the littoral, especially in detritus-rich sections. The greatest numerical density of microbes established on sandy sections reaches 62 million per gram of the soil.

It is probable that the raw biomass of zoobenthos (without microbenthos, sponges and fish) per square metre of the bottom of the littoral in the open regions of the lake and in its open bays averages not less than 25—30 g/m², i.e., 250—300 kg per hectare. The area of the littoral in Baikal (depths of 0—20 m) minus the Maloye More, the Chivyrkui and Barguzin gulfs and the sors is approximately 120,000—150,000 hectares.

In the abundance and variety of life the littoral of Baikal can be compared to that of Lake Ohrid (S. STANKOVIČ, 1960). The littoral of that lake is also most rich in life, but in Baikal the vast majority of species of the littoral are endemics, whereas in Ohrid the proportion of endemics in the littoral population is very low. As distinct from Baikal, the soil in the lower zone of the littoral in Ohrid is enriched with shells of dead mollusks, accumulated in a belt along the fringe of the lower zone. There is no such belt in Baikal, where dead mollusks shells are rapidly dissolved owing to the thinness of their walls and the presence of carbonic acid in the water.

The Sublittoral Zone (20—70m)

The sublittoral of the open regions of Baikal spreads over the upper sections of the bottom slope, the gradient of which varies greatly. Along the open western coast of Baikal the gradient often reaches 30—45° and sometimes even more. On a more gentle slope the rocky bottom of the sublittoral is covered with a thin layer of sand, in most cases slightly silted. Along the eastern coast and especially opposite to the mouths of big rivers the gradient is still less, and the bottom is covered with silted sands, often with an admixture of detritus giving them a dark colour. Large areas are covered with sand in the Selenga shallow. Over more than 90 kilometres along the delta and on both sides of it the belt of silted sands is in some places 5 to 10 kilometres wide. Silted sands take up extensive areas also in North Baikal, near the mouths of the rivers Kichera, Upper Angara and Tiya, where they lie at a distance of 1—2 km from the shore.

But in the sublittoral as well we can often find very clean sands, primarily in areas with near-bottom currents. Silted sands in sections strongly influenced by fluvial outwash contain large admixtures of organic substances.

The water of the bottom layer in the sublittoral is subjected to much weaker seasonal fluctuations of temperature than in the littoral. At a depth of 50 metres the annual amplitude of fluctuations of temperature does not exceed 5—6° C. The influence of turbulence is practically imperceptible. Light intensity is insufficient for the development of macrophytes, but in immediate proximity to the littoral zone, on more or less clean or slightly silted sands or on stones one still can come across colonies of some shade-loving green algae, as well as bottom diatoms. A. P. Skabichevsky (1936) points out that some species of *Cladophora* and *Chaetomorpha* can live at a depth of up to 36—60 m, but evidently they do not penetrate beyond these depths, and only a scant evidence of living diatoms is found to a depth of 70 m.

The sublittoral still experiences a strong influence of the nearshore fauna, and quite a few littoral forms can be found there, but there are also many species preferring to live only in the sublittoral. The density of population and the biomass vary strongly, from insignificant amounts on the rocky slope and sands poor in organic matter to very high values on silted detritus-rich sands.

The distribution of animals among the biotopes of the sublittoral has been studied insufficiently, and we can only give a brief and approximate characteristic of the populations of some of them, the most widespread ones.

It is very difficult to study the population of rocky sections of the bottom of the sublittoral, for none of the instruments used by hydrobiologists work satisfactorily there.

The sponge most frequent on the rocky slope of the sublittoral is *Baicalospongia bacillifera* (fig. 33). The pale-green bark-like form of the sponge *Lubomirskia baicalensis* (fig. 32) can also be found sometimes. The gammarids caught there include such large-size species as *Poekilogammarus pictus*, *Carinogammarus cynnamomeus* and *Odontogammarus pulcherrimus*, which, however, are common on sandy soils as well. The isopods are represented by the highly peculiar flat *Asellus dybowskii* (fig. 57c) and the mollusks by *Baicalia godlewskii* (fig. 68,3) and *B. costata* (fig. 66,2).

The population density is evidently very low there, but we have no figures relating to this density.

Slightly silted sands are populated by a fairly varied fauna, although it is somewhat poorer as compared with silted sands enriched with detritus. The animals most frequently found on them are the mollusks *Benedictia baicalensis* (fig. 70,3), *B. limnaeoides*, *Baicalia carinata*, (fig. 68,1), *B. wrzesniowskii*, *B. carinato-costata*, *B. contabulata*, *Pisidium korotnewi*; the gammarids *Pallasea viridis, P. brandti, Hyalellopsis carinata, Axelboeckia carpenteri* (fig. 61,6), *Crypturopus inflatus, Echiuropus macronychus, Poekilogammarus pictus*; the oligochaetes *Tubifex inflatus, Limnodrilus baicalensis, Styloscolex baicalensis, Limnodrilus schizochaetus*; and the polychaete *Manayunkia baicalensis.*

The biomass of zoobenthos ranges there from 50 to 100 kg/ha, but in areas close to the life-rich littoral it reaches 120—150 kg/ha.

As has been noted above, the pre-estuarine regions of rivers and streams are always silted to a certain extent. Usually the sands there contain a considerable admixture of organic substances and are dark-coloured. Silted sands with detritus are found also near the littoral in bottom hollows, where vegetable matter washed out from the littoral is deposited.

The population of silted sands with detritus is rich and varied. The species most often found there are the turbellarians *Sorocelis nigrofasciata* (fig. 37,1) and *Archicotylus plana*; the oligochaetes *Limnodrilus schizochaetus, Limnodrilus arenarius, Tubifex inflatus, Teleuscolex korotneffi*; the gammarids *Poekilogammarus pictus, Odontogammarus pulcherrimus, Crypturopus inflatus, Micruropus wahli, M. litoralis, Echiuropus macronychus, Carinogammarus cynnamomeus, C. rhodophthalmus, Axelboeckia carpenteri* (fig. 61,6), *Pallasea brandti, Acanthogammarus albus*; the isopod *Asellus baicalensis* (fig. 57a); the mollusks *Benedictia limnaeoides, Kobeltocochlea martensiana, Liobaicalia stiedae* (fig. 66,3) *Baicalia pulla, B. contabulata, B. duthiersi, B. turriformis, B. godlewskii* (fig. 68,3), *B. wrzesniowskii, B. carinata*, (fig. 68,1), *B. carinato-costata, B. dybowskiana, Pisidium korotnewi, P. maculatum*, and others.

Besides, one can often find there the polychaete *Manayunkia baicalensis*, larvae of one or two species of caddis-flies and bright-red larva of the chironomids Orthocladiinae (fig. 64, 65).

What is impressive is the number of genera and species of gammarids and mollusks sharing a comparatively small area of the bottom scooped by the bottom-grab. For instance, in 10 samples taken by the Petersen bottom-grab we found 21 species of gammarids and 21 species of mollusks. There were especially many specimens of the gammarids *Odontogammarus pulcherrimus* and *Carinogammarus cynnamomeus* and the mollusks *Pisidium korotnewi* and *Baicalia godlewskii* (fig. 68,3).

The biomass of zoobenthos on silted sands ranges in various sections of the sublittoral from 5 to 30 g/m². The more reliable drag samples yield more often the values of 10—20 g/m², and the average for such sections should probably be 15 g/m², or 150 kg per hectare of the bottom.

The sublittoral is inhabited by a considerable number of Cottoidei fish, among which *Batrachocottus baicalensis, Procottus jettelesi* (fig. 73,2), *Metacottus gurwici* and others are more common. Sandy-silted soils in fishing areas also have gwyniads, burbots and the graylings *Thymallus arcticus brevipinnis*.

The population density on silts and sands with detritus is usually fairly high. For instance, according to MIKLASHEVSKAYA, the biomass in 4 bottom-grab stations taken from silted sands at depths of from 28 to 46 metres ranged from 0.594 to 4.5 g per sample (from 5.95 to 45 g per square metre of the bottom); in 4 stations taken there from silts in various points opposite to the delta at a depth of 30—70 m, the weight of the biomass varied from 1.665 to 4.5 g per station; the average for 4 stations was 3.3 g, i.e., 33 g per square metre of the bottom (330 kg/ha). Several stations sampled in 1931 opposite to the mouths of the Upper Angara and the Kichera (KOZHOV, 1934a) yielded much lower values, although, judging by some stations, even there the biomass of zoobenthos in some sections reaches 30 g/m² (300 kg/ha). Thus, we see that the pre-estuarine regions of big rivers in the sublittoral have a fairly dense and varied population on silted soils rich in organic substances, with an average biomass hardly less than 25—30 g per square metre of the bottom, which means 250—300 kg per hectare.

In its ecological conditions and animal population the sublittoral of deep bays differs little from the sublittoral of open Baikal (table XXIII).

Table XXIII.

Biomass of Zoobenthos in g/m² on Silted Soils with an Admixture of Sand in the Boguchan Bay (the Average of Bottom-Grab Samples Taken from 60—65 m in the Summer of 1931).

Group	Raw weight in g
Vermes	7.473
Mollusca	0.865
Gammaridae	5.520
Total:	13.858
In kg/ha	138.58

Oligochaetes and gammarids constitute the predominant groups in the zoobenthos of sandy soils in the sublittoral of the bays, with mollusks playing a secondary part.

Summing up these data, we can make the following generalisations. Sands of various degrees of silting are the most characteristic and widespread bottom soil of the sublittoral. The fauna and flora of the sublittoral differs markedly in its composition from that of the littoral and has a number of typical forms, but the influence of the littoral zone is still considerably great there. The predominant groups of the bottom fauna are oligochaetes, gammarids and mollusks, with mollusks disappearing and being replaced by oligochaetes with increase in silting. Turbellarians, the polychaete *Manayunkia* and larvae of chironomids are also observed frequently.

The richest populations have been found there in pre-estuarine regions on silted sands with detritus and on silts with sand and detritus (the average: up to 30 g/m² or 300 kg/ha), then on similar soils in the troughs of the bays, where the biomass of zoobenthos approaches 20 g/m² (200 kg/ha), and underwater valleys and hollows opposite to river and stream mouths (15—20 g/m² or 150—200 kg/ha). Population is poorer on clean sands with a low organic matter content, lying far from the littoral or on a steep slope, and on accumulations of pebbles and stones (5—10 g/m² or 50—100 kg/ha). Bare rock with a steep gradient is still more poorly populated.

The average weight of the zoobenthos in the whole of the sub-littoral of Baikal can be put at 20—25 grams per square metre of the bottom, or 200—250 kg/ha, this figure being, of course, a very rough guide. The area of the sublittoral in Baikal (without the Chivyrkui and Barguzin gulfs and the Maloye More) is approximately 150,000 hectares.

Supra-abyssal (Transitional) Zone (70—250—300 m)

The steepness of the bottom slope in the supra-abyssal zone varies. Along the open western shores the slope continues to be very steep. It is composed there usually of rock, bare or covered with a thin layer of silty sand. The width of the supra-abyssal zone in the area is, as a rule, insignificant—rarely more than 200—500 metres.

Opposite to the delta of big rivers and especially along the eastern shores of the southern part of Baikal, the slope is gentler, and the supra-abyssal zone takes up extensive areas. The predominant soils in such regions are fine silted sands often supplanted by silt. Wherever the coastal flat is well developed and the slope is very gentle, the sands are usually strongly silted at a depth of 30—40 m, with wedges of fine viscous silt among them which grow broader as the distance from the shore increases. Silting is particularly strong in pre-estuarine regions, where silts occupy almost the whole of the sublittoral and supra-abyssal zones.

Seasonal fluctuations of water temperature in the bottom layers are perceptible only in the upper parts. At about 100 m in open Baikal the water is heated only to 5—5.5° C; this maximum is usually reached in the middle of October and lasts for not more than 15—20 days. The temperature of the 200 m layers remains practically unchanged throughout the year, at 3.6—4°. The amount of light reaching the bottom of the supra-abyssal zone is so insignificant that bottom-dwelling plants cannot survive there and are completely absent.

The bottom population of the supra-abyssal region of open Baikal has not been studied sufficiently, and we have to content ourselves with brief characteristics of some of the biotopes from which a few bottom-grab samples have been taken.

The fauna of the rocky slope of the bottom is very poor everywhere, both in quality and quantity. From time to time one can find there the sponge *Baicalospongia bacillifera* (fig. 33) and the crustacean *Asellus dybowskii* (fig. 57c). Evidently these two forms are most characteristic of the "rock" of the supra-abyssal zone. Several species of gammarids also can be found there.

In areas subjected to the influence of fluvial outwash and in bays silted sands and silts contain admixtures of detritus and are dark coloured. Far from the rivers and shores these soils are poor in organic debris. We find the following animals on silted sands at about 100 m: the oligochaete *Limnodrilus arenarius*; the gammarids *Eulimnogammarus acheneus*, *Poekilogammarus pictus*, *P. ssukatzewi*, *Ommatogammarus* sp., *Plesiogammarus gerstaeckeri*, *Carinurus ssolski*, *Carinogammarus seidlitzi*; the turbellarians *Archicotylus* sp. and sometimes *Polycotylus* sp. (fig. 38); the mollusks *Benedictia limnaeoides*, *Baicalia godlewskii*, *Valvata piligera* (fig. 66,4); and bright-red larvae of chironomids of the group Orthocladiinae. According to A. Y. BAZIKALOVA (1945), the depths of 70—150 m are characterised by the gammarids *Homocerisca caudata*, *Axelboeckia carpenteri profundalis*, *Carinurus reissneri*, *Garjajewia cabanisi*, *Plesiogammarus zienkowiczi* and *Ceratogammarus dybowskii*. In the same belt of depths Bathynellidae can be found.

Heavily silted sands at 180—200 m have a somewhat different specific faunal composition. The animals found most often on these sands are the oligochaete *Lamprodrilus wagneri* and *Rhynchelmis brachycephala*, the gammarid *Macropereiopus* sp., and the mollusk *Valvata bathybia*.

The density of population is very low there; for instance, 6 samples taken by the Petersen bottom-grab from 180—250 m depths in the area of Bolshiye Koty were found to contain, per 1 m²:

Turbellarians *Archicotylus*	3 specimens
Oligochaetes *Lamprodrilus wagneri*	2 specimens
Isopods *Asellus dybowskii*	2 specimens
Gammarids *Carinogammarus sablozkii* and 9 other species more poorly represented	60 specimens
Mollusks *Valvata bathybia*	2 specimens
Chironomids	7 specimens
Total: 20 species	76 specimens

The biomass does not exceed 3—5 g/m², i.e., 30—50 kg/ha.

The density of population and the biomass in the supra-abyssal zone vary greatly depending on the type of soils. Population is scantiest on silted sands and silts lying on the slope far from rivers, life-rich bays and gulfs or the littoral, from where organic debris could be brought in by currents and waves. The bottom-grab takes there usually not more than 0.1—0.4 g, which corresponds to a biomass of 1—4 grams per square metre of the bottom or 10—40 kg per hectare.

About the same figures have been obtained from the silts extending in some places into the supra-abyssal zone. For instance, the average for 10 Petersen bottom-grab samples taken by us in North Baikal from 100—200 metres was 0.347 g or 3.47 g/m² (34.7 kg/ha), with gammarids accounting for 1.32 g, oligochaetes for 0.5 g and chironomids for 0.1 g.

A more or less rich population can be found only in areas lying opposite to river mouths and in the belt of currents from pre-estuarine regions. There the biomass of zoobenthos on silted sands and silts ranges from 20 to 50 g/m² or 200 to 500 kg/ha, with the predominant part in the biomass and numerical density played by oligochaetes, which usually account for more than 50% of the raw weight. A high biomass is sometimes found in sections of the supra-abyssal zone situated opposite to open bays.

Grey mud has been traced in open Baikal from 80—120 m down to the extreme depths. The gammarid *Macropereiopus dagarskii*, which is dominant there both as regards the number of specimens and the biomass, is the most typical bottom-dwelling animal living on this soil. Other animals common there include the gammarids *Carinogammarus seidlitzi, C. rhodophthalmus, C. r. microphthalmus, Garjajewia cabanisi*, (fig. 58,4) *Crypturopus inflatus* subsp., *Abyssogammarus* sp.; the oligochaetes *Rhynchelmis brachycephala, Tubifex inflatus, Lamprodrilus wagneri* and *Styloscolex* sp. Grab samples are from time to time found to contain the mollusks *Benedictia limnaeoides* and *Valvata bathybia*.

The density of population on deep-water mud in bays at 80—140 m is fairly high, as is seen from the following data (table XXIV).

The silts of the supra-abyssal zone along the open shores of the lake are populated more poorly.

The fish found on silted and sandy-silted soils include a considerable number of Cottoidei species, among them *Batrachocottus multiradiatus, Procottus jettelesi, Asprocottus herzensteini, A. pulver, A. megalops*, and others. Besides, this zone is evidently visited by the deep-water form of the burbot *(Lota lota)* and gwyniads.

As can be seen from these data, sections of the supra-abyssal zone opposite to the mouths of big rivers and also in some bays are populated by a fairly rich fauna. The average values of the biomass

Table XXIV.

Biomass of Zoobenthos on Silted Soil in the Boguchan and Senogda Bays in the Summer of 1931 (after Kozhov, 1934a)

Place	Boguchan	Senogda	
Soil	Silt	Silt	
Depth in m	10—140	82	
	Raw weight in g/m²	Raw weight in g/m²	Number of specimens
Gammaridae	10.760	2.000	340
Oligochaeta	2.220	1.500	320
Mollusca	0.140	20.000	190
Chironomidae	—	0.100	10
Total	13.120	23.600	860
In kg/ha	131.20	236.00	—

in these sections on sandy-silty soil would be approximately 15—18 g/m² (150—180 kg/ha) at 80—200 m in pre-estuarine regions and 15—20 g/m² (150—200 kg/ha) at 80—150 m in the troughs of bays and gulfs.

Since these sections are relatively rich in zoobenthos and occupy extensive areas in Baikal, it can be supposed that the average biomass of zoobenthos in the whole of the supra-abyssal zone will be about 10—15 grams per square metre of the bottom or 100—150 kg/ha, as against 200—250 kg/ha in the sublittoral. The total area of the supra-abyssal zone (without the Maloye More and the Barguzin and Chivyrkui gulfs) is approximately 150,000 hectares.

Abyssal Zone
Upper Subzone (250—500 m)

The upper subzone of the abyssal zone extends over the slope into the main trough of Baikal. In many places the slope is still very steep, and therefore the subzone is not broad (0.5—1 km), especially along the western shore. But in some regions depths of up to 500 metres occupy considerable areas. For instance, opposite to the Selenga delta they take up a vast territory between both coasts of Baikal. They also spread over large areas in the northern extremity of the lake and in the central part along the Olkhon Island-Ushkany Islands line (the underwater Akademichesky Range).

The temperature of the bottom layers of the abyssal zone remains at about 3.4—3.6° C throughout the year. The oxygen content of the bottom layers is somewhat reduced (up to 75—90% of saturation). Light is practically absent.

Most of the bottom soil is composed of viscous silts of greyish or pale-blue colour, often with ferro-manganese nodules. But in the vicinity of the shores the silts still often contain a considerable admixture of clay. Bare rock with patches of silted sand or brown mud accumulating in hollows has been detected on the steep slope between underwater valleys along the western shores. Opposite to river mouths we again come across many extensive areas of strongly silted sands alternating with patches of mud, and in some places also rows of pebbles and stones.

The benthos of depths of more than 250 metres has not been studied adequately.

The rocky slope of the bottom with patches of silted sands and mud accumulating in hollows seems to be very scantily populated. Sometimes bottom-grabs and trawls tear away from stones colonies of the sponge *Baicalospongia bacillifera* and bring up *Asellus dybowskii*, some mollusks *Benedictia fragilis* (fig. 70), *B. maxima* and *Valvata bathybia* and deep-water forms of the gammarids *Abyssogammarus sarmatus* (fig. 58,2), *Acanthogammarus albus*, *Garjajewia cabanisi* (fig. 58,4) and *Ommatogammarus flavus*, but in very small numbers. Population is undoubtedly richer on silted sands sometimes occurring along the slope of the bottom. On such sands in the southern part of Baikal we have taken the gammarids *Macropereiopus albulus*, *Brachiuropus Grewingki*, *B. reicherti*, *Ommatogammarus albinus* (fig. 58,1), *Parapallasea puzylli* (fig. 60,2), *Abyssogammarus sarmatus* (fig. 58,2), *A. swartschewskii* and *Garjajewia cabanisi*.

According to A. Y. BAZIKALOVA (1945), the gammarids living there also include *Garjajewia dogieli*, *G. sarsi*, *Poekilogammarus rostratus*, *Polyacanthisca calceolata*, *Abyssogammarus swartschewskii*, and others. Bathynellidae also evidently live there.

In addition to typical deep-dwelling gammarids, the upper subzone is visited by species that belong to the coastal complex, in the broad sense of the word: *Poekilogammarus pictus*, *Eulimnogammarus ussolscewi*, *Parapallasea borowskii*, *Eulimnogammarus acheneus*, *Carinogammarus rhodophthalmus* and *Acanthogammarus godlewskii*.

Conspicuous in bottom-grab and trawl catches at 250—500 m is a comparatively large number of near-bottom forms of gammarids, such as some species of *Ommatogammarus* and *Abyssogammarus*, which are often found also in plankton nets lowered to the bottom layers. The mollusks found there include *Benedictia fragilis*, *B. limnaeoides*, *Valvata bathybia* and *Baicalia godlewskii* and the oligochaetes *Lamprodrilus wagneri*, *Rhynchelmis brachycephala*, *Limnodrilus arenarius* and *Tubifex* sp. One can also find turbellarians, among them *Polycotylus* sp. The fish in the upper abyssal zone are represented only by Cottoidei: *Abyssocottus korotneffi*, *A. godlewskii griseus*, *A. pallidus* (fig. 74,3), *Batrachocottus nikolskii*, *Asprocottus gibbosus*, as well as *Asprocottus megalops*, *A. herzensteini*, *Batracho-*

cottus multiradiatus, Procottus jettelesi and others, which enter from the supra-abyssal zone.

The population of deep-water silts is evidently similar to that of silted sands, but it is still poorer.

The gammarids caught by us on silts with an admixture of sand at 250—400 m depths in the Barguzin Gulf included *Macropereiopus dagarskii, M. florii, Carinogammarus seidlitzi, C. rhodophthalmus, C. microphthalmus, Carinurus belkini, C. platicarinus, Pallasea dryshenkoi, Garjajewia cabanisi, Crypturopus inflatus, Homocerisca perla, Micruropus cristatus* and *Cheirogammarus* sp. *Macropereiopus dagarskii* was more numerously represented than all other species.

Among the mollusks, only individual specimens of *Baicalia korotnewi* (fig. 66,1) were found; *Rhynchelmis brachycephala* occurred more frequently among the oligochaetes and *Archicotylus multiclada* among the turbellarians.

Qualitative data on the density of the faunal population of depths greater than 250 m are very scanty. It can be inferred from the material available that the density of population and the biomass in these depths vary greatly. Opposite to the mouths of big rivers, in hollows where organic fluvial outwash is deposited, and also in the belt of currents from river mouths to open Baikal the biomass can reach considerable values (20—50 g/m^2 or 200—500 kg/ha); evidently the bottom is not poor also in areas favourable for the deposition of organic substances washed out from the life-rich littoral. But far from the shores and rivers, where the rain of dead plankton and pelagic fishes is the only source of food for the bottom-dwellers, the density of population is very low.

These conclusions can be illustrated by the following data.

One bottom-grab station taken at a depth of 319 m in a hollow opposite of the northern mouth of the Selenga revealed this population (counted per m^2):

Gammaridae	550 specimens weighing	6.60 g
Oligochaeta	1,780 specimens weighing	42.72 g
Chironomidae	20 specimens weighing	00.01 g
Total	2,350 specimens weighing	49.33 g
In kg/ha		493,3

Data on the biomass in the Barguzin Gulf at a depth of 250—400 m on silty soil with an admixture of sand are given in table XXV and on that in the Maloye More at 200—300 m, in table XXVI.

In the Barguzin Gulf the influence of fluvial outwash is also evidently felt, although the population of 250—400 m depths in the gulf is much poorer than in the area opposite to the Selenga.

The biomass of 200—300 m depths in the Maloye More approaches 80—90 kg/ha, but it falls to 50 kg/ha with increase in depth and the distance from the shore.

Table XXV.

Biomass of Zoobenthos in g/m² on Grey Mud with a Small Admixture of Sand at a Depth of 250—400 m in the Barguzin Gulf in the Summer of 1931 (4 bottom-grab Samples)

Group	Biomass	%
Gammaridae	2.925	26.53
Oligochaeta	7.900	71.65
Turbellaria	0.200	1.82
Total	11.025	100
In kg/ha	110.25	

Table XXVI.

Biomass of Zoobenthos in g/m² on Deep-water Silts at a Depth of 200—300 m in the Range of the Maloye More in the Summer of 1940.

Group	Depth of 200—300 m, average of 8 bottom-grab samples	Depth of more than 300—400 m, average of 6 bottom grab samples
Turbellaria	0.150	0.080
Oligochaeta	3.600	2.670
Gammaridae	4.780	2.280
Chironomidae	0.020	0.010
Mollusca	0.030	0.020
Total	8.580	5.060
In kg/ha	85.8	50.6

At a still greater distance from the shore the bottom-grab in most cases does not bring up a single organism even when it is full of silt, which indicates that the density of population is exceedingly low there. For instance, 8 bottom-grab stations taken by us in 1931 from silts at depths of from 250 to 400 m in North Baikal resulted in 785 mg of oligochaetes and 324 mg of gammarids. This adds up to 1.109 mg, which means 1.386 g per square metre of the bottom or 13.86 kg/ha.

Lower Subzone (Deeper than 500 m)

The lower abyssal subzone embraces partially the slope of the bottom and the entire central part of Baikal. In some places, where the slope is steep, bare rock can still be sounded there, but the whole of the main trough is covered with viscous blue mud with a large admixture of valves of dead planktonic diatoms.

This area of the bottom is characterised by the unchanging temperature of the bottom layers of water (3.3—3.6° C the year round), absence of light and a somewhat reduced O_2 content (about 10 mg/l).

The fauna consists almost entirely of distinct abyssal forms most of which have been mentioned above.

Data on the biomass and population density of depths of more than 500 metres are very incomplete, although these depths occupy at least five-sixths of the entire bottom area of Baikal.

Trawls sometimes bring up a fairly rich and varied population, which often consists exclusively of gammarids and oligochaetes. But it should be borne in mind that these implements are dragged along the bottom for many hundreds of metres and therefore an impression is created of a relatively abundant population in the abyssal zone. The few bottom-grab samples taken there show that the fauna is very poor quantitatively. The supply of organic substances is extremely limited, and life can be sustained only on dead planktonic organisms and pelagic fish falling on the bottom. These remains attract abyssal gammarids, most of which are excellent swimmers. There is no doubt that the biomass of zoobenthos in the lower abyssal zone equals but a fraction of that of the coastal zones and even the upper abyssal subzone.

The only fish that have been found in this zone belong to the Cottoidei: *Cottinella boulengeri, C. werestschagini, Asprocottus gibbosus, A. herzensteini abyssalis* and *Batrachocottus nikolskii.* It is also inhabited by some Cottoidei species typical of the upper part of the abyssal zone. The abyssal Cottoidei species usually have a flabby body covered with a very tender skin easily gathering in folds. The organs of the lateral line are developed very well; the eyes in the majority of forms are reduced. The body is almost always colourless or light pale-yellow and, as a rule, spotless (TALIYEV, 1955). The eyes of the abyssal gammarids usually lack any pigmentation or are pale pink, but the antennae, as a rule, are very long. The body coloration is usually whitish or pinky-white.

Distribution of Benthos in Big Gulfs and the Maloye More

The water of the gulfs and the Maloye More is warmed up more rapidly than in open Baikal and reaches higher temperatures. As has been noted, the warming begins at the coastal shallows and gradually spreads to the open parts of the gulfs (fig. 16, 17).

In the Chivyrkui, which is Baikal's second biggest gulf, the bottom soil is predominantly composed of sands, and it is only at the outlets of the gulf, at depths of more than 20—30 metres, that they are replaced first by silted sands and then by grey silt. Black or brown silt covers the bottom of the bays. Rocky bottom is developed poorly in the gulf.

The bottom vegetation, consisting of both flowering plants and algae, is very rich and covers the bottom of the littoral almost every-where, reaching depths of 20 m and even more. All this, plus the presence of considerable amounts of organic substances in the soil, greatly favours the thriving of zoobenthos (BUROV, 1935).

As regards the conditions of life and the population, the southern part of the Chivyrkui Gulf with depths of not more than 4—5 m and the middle of the bays should be referred to the littoral-sor zone inhabited by the general Siberian fauna and flora already characterised above.

The Baikalian species abiding there include the gammarids *Gmelinoides fasciatus*, *Pallasea cancellus*, *P. cancelloides*, *Crypturo-pus pachytus tuberculatus*, *Micruropus talitroides*, *Eulimnogammarus fuscus*, *E. verrucosus* var., and *Pallasea grubei*. Large numbers of the mollusk *Baicalia dybowskiana carinatoides* are also found there.

On fine sands in the southern part of the Chivyrkui Gulf the biomass of zoobenthos reaches 33 g/m² or 330 kg/ha. It is lower on black or brown silt in the bays of the gulf. Mollusks, gammarids and oligochaetes usually comprise the predominant groups of zoo-benthos on sands, while species of gammarids account for the greater part of the biomass on silts in the bays.

Flora and fauna are very rich quantitatively on the boundary between the southern and central parts of the gulf, where one ob-serves thick growths of macrophytes and a numerous motley animal population consisting of both Baikalian and ordinary lacustrine species. The mollusks are represented particularly abundantly by the baikalian forms *Baicalia oviformis*, *B. dybowskiana carinatoides*, *B. carinato-costata*, *B. korotnewi*, *Sphaerium baicalense*, *Benedictia baicalensis*, and species of the genus *Pisidium*. There is a wealth of *Radix auricularia*. A dozen gammarid species, ordinary inhabitants of sheltered shallows of the lake, also live there.

The fish permanently residing in this area are gwyniads, the grayling *Thymallus arcticus baicalensis*, the roach, sturgeon, ide, perch, pike and many Cottidae species. Large herds of seals appear there in autumn.

In the central part of the gulf sands at depths of from 1 to 10—15 metres are populated very densely also. The specific composition of benthos is, on the whole, the same as on sandy soils in the littoral of the open regions of Baikal, but the biomass and population density are much higher. The average biomass of zoobenthos approaches 30 g/m² or 300 kg/ha, with the predominant part played by mollusks, gammarids and oligochaetes.

A rich population has been found on silts in the exterior parts of the bays at 2—10 m depths, where the biomass reaches 34—50 g/m² (340—500 kg/ha). Only brown silts in the middle of the bays, which already belong to the littoral-sor zone, are populated more scarcely (up to 120 kg/ha).

Table XXVII.

Biomass of Zoobenthos in g/m² (Raw Weight) in the Chivyrkui Gulf in 1932

Soil	Littoral					Sublittoral and supra-abyssal		
				Bays				
	Sand	Sand	Sand with black silt	Black silt	Brown silt	Sand	Silt	Silt
Depth in m	1—10	10—20	6—17	2—10	1.5—4	20—30	40—50	70—250
Number of bottom-grab samples	27	12	8	5	6	4	4	5
Gammaridae	4.244	3.780	15.600	12.200	5.800	0.850	5.650	2.010
Oligochaeta ⎫ Polychaeta ⎭	5.894	4.260	22.400	7.900	0.600	0.862	20.125	3.240
Mollusca	16.807	21.154	10.400	9.300	2.500	10.500	0.100	0.030
Turbellaria	2.126	0.087	0.200	0.300	0.400	—	0.875	—
Hirudinea	0.041	—	1.500	3.200	2.200	0.080	0.037	—
Insecta	0.159	0.380	—	0.800	0.500	—	—	—
Total	29.271	29.661	50.100	33.700	12.000	12.292	26.787	5.280
In kg/ha	292.71	296.61	501.0	337.0	120.0	122.92	267.87	52.8

Data on the biomass of zoobenthos in the Chivyrkui Gulf are contained in table XXVII.

Sandy soils in the central part of the gulf, especially those with a large admixture of detritus, abound in the mollusks *Pisidium amnicum, P. korotnewi, Baicalia dybowskiana, B. oviformis, B. korotnewi, Valvata aliena ssorensis, Sphaerium baicalense, Valvata piligera,* and others.

Other groups of animals found there in large numbers are the gammarids *Acanthogammarus victorii, Parapallasea puzylli, Pallasea cancellus* and many littoral species of *Eulimnogammarus* and *Micruropus,* as well as multitudes of oligochaetes, the polychaete *Manayunkia,* turbellarians, leeches, caddis-flies and chironomids.

The richest life has been observed in areas adjoining the bays of the central and exterior parts of the gulf; evidently they have the optimum for the existence of Baikalian endemics: the water regime is very close to that of Baikal proper, and the soil is rich in nutrients is the form of decaying vegetable remains constantly delivered there by turbulence and currents from the neighbouring bays. Inside the bays, strong silting prevents the development of a rich fauna of mollusks and gammarids. An analogous picture is to be observed in other regions: life is abundant at the outlets of all sheltered bays while lacustrine-sor forms are absent there, with the exception of *Gyraulus gredleri borealis.*

The population of sands at 20—30 m depths is much less dense than in the littoral, the raw crop of zoobenthos averaging only 12—12.5 g/m² (120—125 kg/ha). The widespread species among the mollusks are *Pisidium korotnewi, Choanomphalus dybowskianus, Baicalia semenkewitschi, B. contabulata* and *B. carinata.* The gam-

Fig. 90. The Barguzin gulf. Drawing by B. LEBEDINSKY.

marids are abundantly represented by *Acanthogammarus victorii* and many other species.

Silts in the sublittoral of the gulf at a depth of 40—50 m are still populated fairly densely. The biomass of zoobenthos there averages 27 g/m² (270 kg/ha).

At depths of more than 70—80 m the bottom soil is composed chiefly of silt. The density of population at these depths falls to 5.3 g/m² (53 kg/ha), although sections with a higher biomass can be found there.

All the above-cited data show that the Chivyrkui Gulf is one of the richest regions of Baikal. The biomass of zoobenthos for the whole of the gulf at depths of up to 50—70 m should evidently be not lower than 30 g/m² (300 kg/ha).

In the Barguzin Gulf (fig. 90,16) the distribution and properties of the soil are strongly influenced by the River Barguzin. Thanks to its influence the bottom of almost the whole of the shallow part of the gulf adjoining the mouth is blanketed with sands and silts enriched with detritus. The bottom soil in the deep part of the gulf is composed of typical deep-water mud.

Rocky ground in the surf zone of the littoral of the gulf is populated by the same species (primarily turbellarians, gammarids, sponges, mollusks, insects, oligochaetes, etc.) that live in the open lake.

The biomass of zoobenthos on sandy and silty soils in the Barguzin Gulf is shown in table XXVIII. The animals most frequently occurring on the sands of the littoral are the gammarids *Micruropus cristatus*, *M. talitroides*, *M. wahli*, *Crypturopus pachytus* (fig. 58,3), *Cr. tuberculatus*, *Echiuropus macronychus*, *Acanthogammarus godlewskii*, *A. flavus* (fig. 61,5), *Hyalellopsis czyrnianskii*, *Pallasea viridis*, *Gmelinoides fasciatus*, *Poekilogammarus araneolus* and numerous species of the genus *Eulimnogammarus*; the mollusks *Sphaerium baicalense*, *Pisidium baicalense*, *P. korotnewi*, *Baicalia oviformis*, *B. elata*, *B. carinata*, *B. semenkewitschi*, *B. costata*, *B. godlewskii*, *B. pulla*, *B. contabulata*, *B. variesculpta* (fig. 67,1) and *Valvata lauta*; the oligochaetes *Lamprodrilus pygmeus*, *L. korotneffi*, *Teleuscolex korotneffi* and *Clitellio multispinus*. Often one can come across the polychaete *Manayunkia* and the turbellarians *Archicotylus plana* and *A. lacteus*. Caddis-flies and chironomids are also frequent.

Highly characteristic of dark silted sands at a depth of about 50 m are the gammarids *Macropereiopus dagarskii*, *Micruropus cristatus*, *Crypturopus inflatus*, *Carinurus platicarinus*, *Carinogammarus microphthalmus*, *Echiuropus macronychus* and *Acanthogammarus godlewskii*. *Macropereiopus dagarskii* and *Micruropus cristatus* show a vast preponderance both in the biomass and the number of specimens. The mollusks most often found on these sands are *Pisidium korotnewi*, *Baicalia semenkewitschi*, *B. korotnewi*, *B. pulla*, *B. carinato-costata* and *Benedictia baicalensis*; the more frequent oligochaetes are *Lamprodrilus pygmeus*, *Teleuscolex korotneffi* and *Limnodrilus dybowskii*, while *Archicotylus plana*

Table XXVIII.

Biomass of Zoobenthos in the Barguzin Gulf (after M. M. Kozhov, 1934 b)

Zone	Place	Soil	Depth in m	Number of bottom grab samples	Raw weight in g/m²							In kg/ha
					Turbellaria	Oligochaeta	Polychaeta	Gammaridae	Insecta	Mollusca	Total	
Littoral	Along the open eastern coast	Sand	2—3	4	—	0.325	—	2.025	—	0.200	2.550	25.5
	The central part of the gulf	Sand with silt	up to 10	6	0.066	1.764	1.783	2.317	0.083	1.633	7.646	76.46
	To the right of the mouth of the Barguzin	Sand	8—15	9	—	8.000	—	3.775	0.174	0.533	12.482	124.82
	Along the south-eastern coast	Sand	10—22	10	0.050	11.750	—	2.650	0.390	1.670	16.510	165.1
	In the bays of the south-eastern coast	Sand	4—10	3	—	26.033	—	9.568	14.468	21.383	71.452	714.52
Sublittoral	The central part of the gulf	Silted sands	up to 50	21	0.324	19.976	3.295	11.076	0.581	1.290	36.542	365.42
	The north-eastern part of the gulf	Dark strongly silted sands	10—60	18	4.100	11.588	1.972	17.402	0.188	0.328	35.578	355.78
Supra-abyssal	The central part of the gulf	Silted sands	50—120	5	—	19.980	—	4.160	—	0.220	24.360	243.6
	Along the south-eastern coast	Silted sands	50—160	7	—	9.171	0.428	6.329	—	0.286	16.214	162.14
		Grey mud with an admixture of sand	100—240	6	0.133	14.450	—	9.767	—	0.166	24.516	245.16

and *Sorocelis* sp. are more common among the turbellarians.

The gammarids more numerously represented on dark heavily silted sands occupying the entire north-eastern part of the gulf at depths of up to 50—100 m are *Macropereiopus dagarskii, Micruropus cristatus, Crypturopus pachytus, Carinurus platicarinus* and *Echiuropus macronychus.* Up to 200 specimens of such species as *Macropereiopus dagarskii* and *Micruropus cristatus* are found in one bottom-grab sample from 0.1 m². Among the mollusks, which are very scantily represented there, we found *Pisidium baicalense, Baicalia korotnewi, Sphaerium baicalense, Benedictia limnaeoides, Baicalia carinata, B. semenkewitschi* and *B. pulchella* with the first two species obviously predominating.

Among the oligochaetes, greater numerical density is attained by *Limnodrilus schizochaetus, L. baicalensis, Lamprodrilus korotneffi, L. pygmeus, L. glandulosus,* and others. The turbellarians have been found to be represented by *Sorocelis hepatizon.* The polychaete *Manayunkia baicalensis* is also fairly common there.

On deep-water silts we have found the same organisms that occur in the other sections of the abyssal zone of Baikal. There are especially many gammarids such as *Macropereiopus dagarskii.* The mollusks are represented only by *Baicalia korotnewi,* the oligochaetes by *Rhynchelmis brachycephala,* and the turbellarians by *Archicotylus multiclada.*

Below we give a summary table of the biomass of zoobenthos in the Barguzin Gulf (table XXVIII).

The fish population of this gulf is fairly rich. In addition to the omul, it includes the gwyniad, sturgeon, roach, perch, pike, ide and other fluvio-lacustrine and Baikalian species.

The Maloye More takes up an area of approximately 80,000 hectares (fig. 91, 92, 93).

Its bottom is gently inclined from south to north. The depth of the southern extremity (the Mukhor Gulf) does not exceed 5 m, whereas in the northern part the depth reaches 200—300 m. The bottom is composed chiefly of silted sands with various degrees of silting, and silts proper (KOZHOV, 1936a, 1947a; PATRIKEYEVA, 1959; BEKMAN, 1959). The Mukhor Gulf is open to north-easterly winds. The winds drive colder waters into it from the open and deeper regions of the Maloye More, which mingle with the water of the gulf and cool them, especially in the outer half. Therefore the summer temperature of the water of the gulf rarely rises above 20° C (fig. 17).

The gulf freezes up at the end of October or in November. It takes a long time for the ice-cover to form because of strong north-westerly storms, which are particularly violent in autumn. The thickness of the ice at times exceeds 1 m.

The bottom of the gulf is almost fully overgrown, with ordinary

Fig. 91. The littoral of Besymyannaya bay (the gulf Chivyrkuisky). Photo by M. Kozhov.

Fig. 92. The southern part of the Maloye More, with the "Lomonosov", the expedition launch of the Baikal Biological Station, in the foreground and Olkhon Island in the background. Photo by M. Kozhov.

flowering plants as well as Characeae; *Nostoc*, etc. predominating. Practically the whole central part is covered with *Potamogeton*, which forms extensive and rich underwater beds there.

The animal population of the southern and central parts of the Mukhor Gulf differs sharply not only from the open regions of Baikal, but even from the fairly shallow southern half of the Maloye More. There, as in the sors, ordinary lacustrine species predominate. The Baikalian fauna is represented by a few species, including 5 or 6 species of mollusks, among which *Baicalia korotnewi* and *Sphaerium baicalense* are more numerous. But these species do not penetrate into the innermost shallow part of the gulf and its sheltered bays. The Baikalian gammarids entering the gulf include a considerable number of species also occurring in the sors, namely, *Crypturopus pachytus, C. inflatus, Micruropus litoralis, M. talitroides, M. possolskii, M. wahli, Gmelinoides fasciatus* and some species of the genus *Pallasea*, such as *P. kessleri* and others. In the range of the gulf one can find the Baikalian polychaete *Manayunkia* which, however, does not penetrate deep into it. Nearer to the neck of the gulf the Siberian species gradually disappear and the number of Baikalian species increases. The biomass of zoobenthos on silted soils averages 88.2 g/m² (882 kg/ha), with the predominant part obviously played by mollusks, including *Anodonta* (465 kg/ha).

On silted soils the fauna is also rich quantitatively. A poorer population is observed on coastal sands, where the biomass of zoo-

Fig. 93. The Shaman Bay in the Maloye More. Photo by M. Kozhov.

benthos is put at 16.3 g/m² (163 kg/ha), on the average, with mollusks ranking first and oligochaetes second.

According to the data available, the average biomass of zoobenthos in the Mukhor Gulf should be 42.1 g/m² (421 kg/ha).

The density of the population of the bottom of the gulf is very high. Gammarids of the genus *Micruropus* are especially numerous there. Counted per square metre, the numeric density of *M. wahli* reaches 16,180 specimens, *M. possolskii* 16,180, *M. litoralis* 15,400, *Gmelinoides fasciatus* 19,200, *Pallasea kessleri* 4,420, *Eulimnogammarus viridis* 4,000 specimens (BEKMAN, 1959).

Owing to the abundance of benthos, the Mukhor Gulf present a rich feeding ground for benthophagous fish. One can always find there perch, dace, roach, ide, pike and species of Cottidae. The gulf is also visited by the grayling, and the malomorsky gwyniad spawns there.

The fauna of the other bays of the Maloye More is also quantitatively very rich. The Zagli and Kharin Irgi bays have been found to contain two species of Baikalian sponges, 15 species of oligochaetes and up to 30 species of mollusks (among which only *Limnaea auricularia* is a fluvio-lacustrine species), about 40 species of gammarids, and so on. The open littoral of the central and northern parts of the Maloye More is populated much more poorly than that of the bays and of the southern part. The biomass of zoobenthos there does not exceed 14—16.5 g/m² (140—165 kg/ha). Conspicuous everywhere on the soft soils of the littoral is the preponderance of oligochaetes, mollusks and gammarids.

The bottom of the sublittoral of the Maloye More is composed chiefly of sands with various degrees of silting. The greatest numerical density of the population of the sublittoral sands has been observed in the southern part, where the biomass of zoobenthos reaches 30—35 g/m² (300—350 kg/ha), almost half and sometimes even three-quarters of which falls to the share of oligochaetes.

Qualitatively, the fauna of the Maloye More sublittoral is highly varied. The animals most frequently found there include the oligochaetes *Limnodrilus schizochaetus* (on strongly silted sands and on silts), *Lamprodrilus satyriscus*, *L. pallidus* and *L. pygmeus*; the mollusks *Benedictia limnaeoides*, *B. maxima*, *Kobeltocochlea martensiana*, *Baicalia nana*, *B. florii*, *B. semenkewitschi*, *B. jentteriana B. pulla tenucosta*, *B. korotnewi*, *B. wrzesniowskii*, *B. carinata*, *B. carinato-costata*, *Valvata piligera* and *Pisidium korotnewi*; the gammarids *Hyalellopsis carinata*, *Crypturopus inflatus*, *Micruropus talitroides*, *Poekilogammarus pictus*, *Carinogammarus cynnamomeus*, *C. rhodophthalmus*, *C. capreolus*, *C. fuscus*, *Baicalogammarus pullus*, *Axelboeckia carpenteri*, *Acanthogammarus godlewskii*, *Pallasea brandti P. kessleri*, and others.

The mollusk fauna in the Maloye More sublittoral grows markedly poorer in connection with the silting of the sands.

Strongly silted sands and silts predominate in the supra-abyssal zone at 70—250 m depths lying in the central and northern parts of the Maloye More.

The richest population has been found on the silts of the central part, where the biomass of zoobenthos reaches 30 g/m² (300 kg/ha). Nearer to the deeper part of the Maloye More life on the bottom grows scarcer, with the biomass of zoobenthos falling to 14 g/m² (140 kg/ha) and in the deepest part itself, at depths of more than 200 metres, to 8.5—5 g/m².

The Maloye More is inhabited by the omul, gwyniad, black and white grayling, perch, pike and, more rarely, roach, dace and some other fluvio-lacustrine species, as well as numerous Cottidae species.

General Conclusions on the Distribution of Zoobenthos in Baikal

The distribution of the biomass of zoobenthos in Baikal is given in table XXIX, which should serve only as a guide. We see from it that the biomass is the highest in sections characterised by transition to Baikalian conditions. They include, for instance, the northern half of the Mukhor Gulf (40 g/m²), the bays of the Maloye More (33 g/m²), the southern part of the Chivyrkui Gulf (33 g/m²), etc. The biomass of zoobenthos in the sors and in the bays with conditions close to those of eutrophic Siberian lakes is considerably lower, ranging from 2 to 10—15 g/m², which accords well with summer biomass in the ordinary lakes of the Baikal area. The coastal belt of the open regions of Baikal is much richer in the biomass of zoobenthos than the sors, especially in the littoral (25—30 g/m²) and the sublittoral (20—25 g/m²), while the extensive shallows of such areas as the Maloye More, the Barguzin gulf, the Chivyrkui Gulf, etc., are richer than the open regions of Baikal at corresponding depths.

The average biomass of zoobenthos (without sponges) in the coastal belt of Baikal to a depth of 250 m is tentatively put at 22 g/m².

As regards the biomass of zoobenthos in the abyssal zone of Baikal (deeper than 250—300 m), it seems to be very low. Only in those regions which fall under the influence of fluvial deposits does the biomass of the abyssal zone reach values comparable to the biomass of the coastal belt, but far from the shores, in the main trough of Baikal occupying up to 85% of all the area of the Baikal, the average biomass evidently does not exceed 1.5—3 g/m². This is understandable, since food for the bottom-dwellers can be provided only by the bodies of big and small pelagic organisms falling from the mass of water of the pelagic zone on the bottom of the lake.

The average biomass of zoobenthos for the entire area of Baikal can be put at approximately 5—6 g/m².

Table XXIX.

Biomass of Zoobenthos in Lake Baikal in g/m².

Littoral-sor zone								
Sors				Sheltered gulfs and bays of Baikal				
Lake-sor Rangatui, August	North-Baikalian Sor, June-August	Posolsky Sor, August	Gulf-sor sor Proval, August-Sept.	Mukhor Gulf (Maloye More)	Malo-ye More bays	Southern part of Chivyr-kui Gulf, south of Baklany Island	Bays of Chi-vyrkui Gulf
Up to 3.0	Up to 3.0—7.0	In August 10.3—14.4	38.0—56.0 without *Anodonta* 12.0—15.0	42.0 without *Anodonta*	33.0	33.0	7.0—17.0

Open Baikal								
Coastal belt, 0—250 m								
Open regions			Extensive shallows			Average for depths of 0—250 m in open Baikal	Abyssal	
Litto-ral, 0—20 m	Sublit-toral, 20—70 m	Supra-abyssal 70—250 m	Chivyr-kui Gulf, 0—50 m	Bargu-zin Gulf, 0—100 m	Maloye More, 0—250 m		Upper sub-zone, 250—500 m	Lower sub-zone, deeper than 500 m
25.0—30.0	20.0—25.0	10.0—15.0	30.0	20.0	22.0	20.0—25.0	3.0—5.0	1.0—2.0

* The area occupied by depths of 0—250 m in Baikal approaches 600,000 hectares; that of the abyssal zone (deeper than 250 m), 2,550,000 hectares. Counted for the entire area of the bottom, the biomass of zoobenthos evidently will not exceed 50—60 kg/ha, or 5—6 g/m².

It should be noted that the biomass of zoobenthos in Baikal remains, on the whole, at one level throughout the year. Its seasonal variations are insignificant. Any marked seasonal variations resulting from consumption by fish have not been detected, either, for such main consumers of zoobenthos as the Cottoidei fish, graylings, gwyniads, sturgeons and others evidently feed in Baikal the year round. The investigations conducted in the Maloye More by M. Y. BEKMAN (1959) have shown that from spring till autumn the total biomass of all the benthonic groups changes very little. This peculiarity of the Baikalian benthos distinguishes it sharply from the zoobenthos of the sors and the lakes of the Baikal area, whose biomass varies with seasons, equalling in spring a half or a third of the autumn value and falling again in winter as a result of both natural dying out and consumption by fish.

There is a gap in the knowledge of the Baikalian benthos owing to the lack of quantitative data on the biomass of microbenthos in different biotopes. As far as we know, this microbenthos is fairly rich, varied and distinct in its composition. The micro-population of the bottom and of the near-bottom layers is known to contain quite a few species of Cyclopoida, Harpacticoida, Cladocera, Rotatoria and Nematodes. In the microbenthos of the Maloye More bays I. K. VILISOVA (1959b) found 13 species of Copepoda, 6 species of Cladocera, Hydra and Ostracoda and 7 species of Rotatoria.

Even the cursory study of the microbenthos of Baikal conducted in recent years has yielded many interesting results, such as elucidation of the important role of the microbenthos in the diet of fish, particularly their young, as well as of mollusks, gammarids and oligochaetes. An investigation of the animal and vegetable microbenthos presents one of the primary tasks of the students of Baikal.

If the biomass of the zoobenthos of Baikal is compared with that of other large basins, both fresh-water and brackish and marine ones, the results will not be in favour of Baikal. For example, in 1935—49 the biomass of zoobenthos in the northern part of the Caspian Sea was found to range from 47.8 kg/ha (1938) to 303.7 kg/ha (1949) and 403.2 kg/ha (1935), with mollusks (mostly Bivalvia) accounting for 4 to 36% (BIRSTEIN & SPASSKY, 1952). These values considerably exceed the averages for Baikal. But the biomass of the deep zone of the southern and central parts of the Caspian Sea also seems to be low. In the biomass of zoobenthos, Baikal bears no comparison with the Barents Sea, the Sea of Okhotsk and the Sea of Japan (ZENKEVICH, 1947—51). But the benthos of such big European lakes as Ladoga and Onezhskoye is evidently much poorer than that of Baikal. Baikal is many times richer in benthos than Lake Balkhash, whose biomass, despite its shallowness, is valued at an insignificant 5—15 kg/ha (BURMAKIN, 1956), but on the other hand, Baikal is much poorer than Lake Ilmen, where the biomass of zoobenthos is counted in hundreds of kg/ha.

It follows from the data stated above that the biomass of zoobenthos in the coastal belt of open Baikal is higher than in the sors and in the eutropic lakes of the Baikal area. But this should not lead one to the conclusion that the annual output of zoobenthos in Baikal is also higher than in these bodies of water.

Owing to the lower temperatures the growth of biomass in Baikal is retarded, which can be noted from the comparison of biological processes in the same groups of animals in ordinary lakes and in

Baikal. For instance, the cycle of the development of caddis-flies in Baikal is sometimes as long as 3 years, whereas in the well warmed shallow lakes of the Baikal area it usually takes not more than one year. Most of the lacustrine forms of mollusks reach maturity a year or two earlier than the Baikalian forms. In view of this, the P/B (production—biomass) ratio for the zoobenthos of Baikal is evidently much lower than for that of lakes and sors.

The study of the actual and potential productivity of species of the Baikalian zoobenthos has been started only in recent years (BEKMAN & BAZIKALOVA, 1951; BEKMAN, 1954, 1959). Judging by the rate of growth and cycle of development of some widespread littoral species of gammarids, their potential productive capacity is high. For example, the ratio of production to average annual biomass (the P/B factor) for the gammarid *Micruropus possolskii* approaches 3.4. The same productivity factor has been obtained for the ordinary lacustrine amphipod *Gammarus lacustris* in the lakes of the flood plain of the Angara.

There is no other data on the productive capacity of the components of the Baikalian zoobenthos.

IV

SPREAD OF MODERN BAIKALIAN FAUNA OUTSIDE OF BAIKAL

Data on the distribution of some Baikalian species outside of the lake have already been cited in the chapter devoted to the fauna. Now we shall try to summarise them and evaluate their biogeographical importance (fig. 94).

As has been noted, several forms have developed among the indigenous Baikalians capable of living not only in the littoral-sor region of Baikal, but also in continental waters outside of it. For instance, some Baikalian species of gammarids enter the lower reaches of Baikal's affluents and settle in adjacent running-water lakes. Among them are *Gmelinoides fasciatus*, *Micruropus possolskii* and *M. wahli*. But none of the Baikalian species spread to any length upstream deep into the continent. Nor can they be found in the moutain lakes around Baikal (ZHADIN, 1937; KISELYOV, 1937; KOZHOV, 1950b).

At the present time immigrants from Baikal are represented especially abundantly in the great waterway linking Baikal with the Arctic Ocean, the system of the Angara and Yenisei rivers. It has been established that prior to damming, the upper section of the Angara was populated by up to 50 species of Baikalian gammarids, the Baikalian species of the genus *Asellus*, *A. angarensis* fig. 57,b, about 10 species of mollusks from the genera *Choanomphalus*, *Baicalia* and *Pseudancylastrum*, a considerable number of species of turbellarians and oligochaetes, the polychaete *Manayunkia baicalensis* (fig. 41), the Baikalian sponge *Lubomirskia baicalensis*, the bryozoan *Hislopia placoides* (fig. 72), Baikalian forms of the Chironomidae and Baikalian species of the Cottoidei, such as *Paracottus kneri*, *Paracottus kessleri* and *Batrachocottus baicalensis* (KOZHOV, 1931b). In Baikal, all these forms live predominantly in the littoral zone and in the Angara, in open section with a relatively rapid flow.

The successful colonisation of the upper course of the Angara by some Baikalians is due to the fact that the thermal and chemical regime of its water very much resembles that of the littoral of the open regions of Baikal. Prior to the building of the Irkutsk hydropower station's dam the water temperature there did not exceed 9—11° C even in August—September, reaching 12—13° at the banks only on hot days; the water was oxygenated to the maximum; the bottom, composed of boulders and gravel, was almost fully overgrown; in some places there were patches of sand, and silted

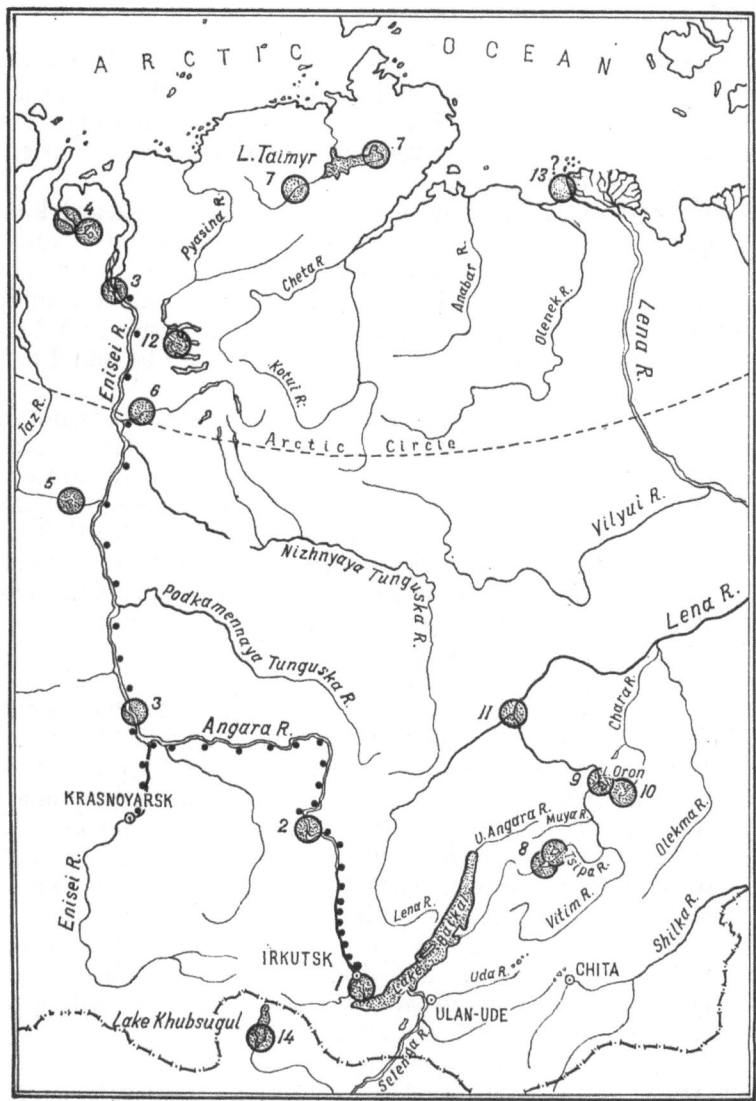

Fig. 94. Outline map of the distribution of the Baikalian fauna outside Baikal.
1 - the upper section of the Angara, inhabited by up to 50 littoral species of Baikal;
2 - the middle section of the Angara, where up to 20 Baikalian species live; 3 - the
River Yenisei, populated by 10 to 15 Baikalian species, including mollusks of the genus
Choanomphalus, gammarids, the bryozoan *Hislopia*, the polychaete *Manayunkia
baicalensis*, Oligochaetes, etc.; 4 - relict lakes of the Gyda Gulf drainage, where the
polychaete, *M. baicalensis*, and Baikalian species of *Eulimnogammarus* have been
found; 5—6 - lakes of the Yenisei drainage, which have been found to be populated
by *M. baicalensis* and the bryozoan *Hislopia;* 7 - relict Lake Taimyr, where two
Eulimnogammarus species, *Manayunkia* and *Hislopia* have been found; 8—10 - big
running-water lakes of the drainage of the Vitim and Olekma, tributaries of the
Lena, where *Manayunkia baicalensis* and the cottids *Paracottus* and *Asprocottus*
have been found; 11 - the middle reaches of the Lena containing the Baikalian
species *Paracottus kessleri* and its parasite *Salmincola cottidarum*, 12 - the Norilsk
lakes, inhabited by *Manayunkia baicalensis*, the bryozoan *Hislopia* and *Eulimnogammarus;* 13 - the Olenek Gulf, where *Manayunkia baicalensis* (?) has been found.
Black circles show the distribution of the Baikalian fauna in the Angara and
Yenisei. Original.

soil occurred only in bays and spots with a quiet flow. In the open sections of the upper reaches of the Angara, general-Siberian lacustrine species were rare, with Baikalians predominating. Thus, ecologically, this part of the river was, as it were, a continuation of the littoral zone of Baikal.

Today, after the building of the hydro-power station near Irkutsk, the upper course of the Angara has virtually become a gulf of Baikal. The fauna of its upper section, as before, is primarily composed of Baikalians, while the gulfs and bays and the middle and lower sections are populated by Siberian lacustrine and fluviolacustrine fauna and flora. No Baikalian species can be found today in the lower section adjacent to the dam.

Downstream of Irkutsk the Angara receives the big tributaries Irkut, Byelaya, Kitoi and others and gradually acquires features of an ordinary Siberian river. But even there, over a stretch of 600 to 800 kilometres from the outlet, rapid-flow sections and particularly the area of the Bratsk Rapids are populated by Baikalian species: the mollusk *Baicalia angarensis*, up to 10 species of gammarids, *Asellus angarensis*, the polychaete *Manayunkia baicalensis*, the Baikalian bryozoan *Hislopia placoides*, Baikalian littoral species of oligochaetes and turbellarians. Some Baikalian species have been able to settle in the lower reaches of the Angara and spread farther on along the Yenisei right up to the Yenisei Bay. P. L. PIROZHNIKOV (1937a, b; 1941) and V. N. GREZE (1954, 1956) offer a fairly extensive list of Baikalian forms found in the Yenisei. The 2,500-km stretch between Krasnoyarsk and the Yenisei Bay contains the gammarids *Eulimnogammarus viridis*, *Micruropus* sp. and *Gmelinoides fasciatus*; in the lower reaches of the Yenisei and in its delta, in addition to the above-mentioned species, such Baikalians have been found as *Pallasea kessleri*, *Micruropus glaber*, *M. vortex*, *Hyalellopsis* sp., *Eulimnogammarus* sp., *E. lividus*, *Brandtia lata*, the mollusks *Choanomphalus amauronius* and *Pseudancylastrum sibiricum*, the Baikalian bryozoan *Hislopia placoides*, up to 10 Baikalian species of oligochaetes, and the fish *Paracottus kneri*.

In the Yenisei's lower course the Baikalian forms of gammarids mingle with marine relict forms and comprise a sizeable part of the biocoenoses of benthos there. P. L. PIROZHNIKOV (1937a) writes: "Occurring frequently and in large numbers, marine and Baikalian forms present a characteristic element of the fauna of the lower Yenisei and lend it a highly distinct colour."

In 1937 G. S. SLASTNIKOV (1940, 1941) found Baikalian species in the drainage area of the Gyda, which pours into the Gyda Gulf of the Kara Sea situated between the mouths of the Yenisei and the Ob. The polychaeta *Manayunkia baicalensis* and the gammarid *Micruropus wahli* have been found abundantly in the fresh running-water lakes of the Gyda drainage, 100 to 150 kilometres from the

Gyda Gulf. SLASTNIKOV, referring to geological data, maintains that the Baikalian forms discovered in these lakes penetrated there from the Yenisei through its ancient link with the Gyda Gulf and the watershed of the Gyda. The same lakes also contain such marine relicts as *Mesidothea entomon* L., the amphipod *Pontoporeja affinis* LINDSTR., then *Mysis relicta* Low., *Senicella calanoides* JUDAY and *Limnocalanus grimaldi macrurus* G.O.S. In view of this SLASTNIKOV regards them as relict lakes.

In Nalimye Lake, which lies on the divide of the Taz and the Yenisei drainage areas, P. L. PIROZHNIKOV (1937a) found *Pallasea quadrispinosa*, which he, after L. S. BERG, considers to be an immigrant from Baikal. *Mysis relicta* has also been discovered there.

According to V. N. GREZE, Baikalian elements also occur in lakes of the drainage of the Lower Tunguska, a tributary of the Yenisei: Lake Munduiskoye, 60 km from the outflow of the Kureika, and Lake Nalimye, 70 km upstream from the outflow of the Lower Tunguska, in which the Baikalian polychaete *Manayunkia baicalensis* and the Baikalian bryozoan *Hislopia placoides* have been found. Lake Munduiskoye also has the amphipod *Pallasea quadrispinosa* and the copepod *Cyclops kolensis*, which lives in Baikal. As V. N. GREZE points out, these lakes are situated in the area of the ancient bed of the Yenisei; they are fairly large (Munduiskoye has an area of 10,450 hectares and Nalimye of 1,190 hectares) and comparatively deep running-water lakes. V. N. GREZE writes that the Baikalian species in them are to be regarded as relicts of the fauna of the ancient Yenisei.

An interesting fauna has been found recently in the system of the large and deep running-water Norilsk lakes in the drainage of the River Pyasina (the Arctic Ocean basin to the east of the Yenisei), between 68° and 70° North Latitude. In addition to widespread Siberian species, they are inhabited by the glacial relicts *Pallasea quadrispinosa, Gammaracanthus loricatus lacustris, Mysis oculata relicta, Pontoporeja affinis*, the goby *Myoxocephalus quadricornis*, as well as Baikalian immigrants, the bryozoan *Hislopia placoides*, the polychaete *Manayunkia baicalensis*, the gammarid *Eulimnogammarus viridis* (VERSHININ, 1960). After L. S. BERG, N. V. VERSHININ considers the Baikalian elements in the Norilsk lakes to be a remnant of the once widespread Tertiary fauna of Siberia. It would be more natural to suppose, however, that Baikalian species penetrated there, as into other Arctic lakes, in Quaternary-time, through the Angara and Yenisei systems and also owing to the Yenisei's communication with the drainages of other northern rivers.

Geological data indicate that in the period of the Arctic Ocean's boreal transgression marine species reached as far as the Norilsk plateau and skirted it from north and west, extending into the

Norilsk valley in a deep gulf (VERSHININ, 1960; p. 753). In that period marine Glacial relics, too, could have penetrated into the Norilsk lakes.

Of great interest is N. V. VERSHININ's statement on the discovery of *Manayunkia baicalensis* (?) in the Olenek Gulf of the Arctic Ocean lying to the west of the Lena mouth. If the presence of *M. baicalensis* in this gulf is confirmed, two suppositions regarding the ways of its migration from Baikal are possible: from the Yenisei and Pyasina system in the west along the Arctic coast, where *Manayunkia* could have spread using the freshened estuaries of Arctic Ocean tributaries, which changed their position throughout the Glacial Period depending on the position of glaciers, or through the Lena system, using the ancient outflow from Baikal into the Lena, which will be dealt with further on.

Remarkable finds have been made by V. N. GREZE in Taimyr, a large Arctic relict lake on the Taimyr Peninsula (V. N. GREZE, 1947, 1953, 1957), among them the polychaete *Manayunkia baicalensis*, the bryozoan *Hislopia placoides*, and gammarids of the Baikalian genus *Eulimnogammarus* (supposedly *E. viridis* and *E. cyaneus*).

Along with Baikalian species, Lake Taimyr is inhabited by the relict forms *Pontoporeja affinis, Mysis relicta, Limnocalanus grimaldi macrurus* and the infusorian Tintinnoidea.

Explaining the presence of Baikalian species in Lake Taimyr, V. N. GREZE suggests an ancient communication between the Yenisei and the drainage of the River Taimyra flowing through Lake Taimyr. It is a noteworthy fact that species of the genus *Eulimnogammarus* still occur 150 to 200 km upstream of the Upper Taimyra. V. N. GREZE dates the communication between the Yenisei and the drainage area of Lake Taimyr to post-glacial time. In that period the eastern part of Lake Taimyr was connected with the Khatanga Gulf, which explains the presence of marine relict forms in the lake. But it is known that still earlier, at the end of the Glacial period, the whole of the Taimyr lowland lay under sea waters, and it is possible therefore that Baikalian and marine (relict) species in general were widespread in the freshened gulfs of the Quaternary sea into which the ancient Yenisei discharged its waters.

Thus, the great system of the Angara-Yenisei serves, as it did in the past, as a route for the travel of many Baikalian species downstream for thousands of kilometres, right up to the Arctic regions. It is a remarkable fact that they can be found in large relict lakes and also in the brackish-water Yenisei and Gyda gulfs, where they form biocoenoses together with brackish-water dwellers of estuaries and pre-estuarine regions of rivers flowing into the ocean. The facts cited here can serve as a vivid example of the spread of highly specific faunae by means of hydrographic connections over vast distances of thousands of kilometres.

It should be noted that in the Angara, the Yenisei and the Arctic lakes of the Yenisei, Gyda and Pyasina drainages, the immigrants from Baikal differ little from the original Baikalian forms. Most of them are evidently identical with the latter. Up till now, only the amphipod *Gmelinoides fasciatoides* GURJAN. has been described as a separate species (GURYANOVA, 1929). But even this amphipod presents merely a variety of the Baikalian *Gmelinoides fasciatus* STEBB. (BAZIKALOVA, 1957).

The comparatively small degree of the divergence of the Angara-Yenisei complex of Baikalian immigrants gives reason to suppose that Baikal's communication with the Arctic Ocean through the Angara and the Yenisei originated in a relatively recent past.

On the question of the time and method of penetration of Baikalian gammarids into the delta of the Yenisei and the polar lakes not connected with it directly, A. Y. BAZIKALOVA shares the opinion of P. L. PIROZHNIKOV (1937b), which consists of the following:

The Quaternary period is known to have been very eventful in the history of North Eurasia. The north of Siberia was repeatedly inundated by sea waters, with the transgression of the Arctic Ocean around the middle of the Quaternary period extending in some places southward to 65—63° N. Lat. Traces of it are detected along the Yenisei up to the mouth of the River Yelogui. That transgression must have coincided with the maximum mid-Quaternary (Samarovo) glaciation, when the southern fringe of glaciers in some places went beyond the Arctic Circle. In P. L. PIROZHNIKOV's opinion, during the maximum glaciation the ice barrier was so thick that it could block up the passage of the great Siberian rivers to the ocean. After merging, the East-Siberian and West-Siberian glaciers formed a continuous ice barrier and cut off a section of the boreal sea which extended to the south in the form of a great gulf. This explains how the inland Quaternary Central Siberian basin could form. By discharging their waters directly into that basin the Ob and the Yenisei raised its level and freshened it. Its fauna must have consisted of several euryhaline and coldwater marine forms and also some Baikalian and fresh-water ones. P. L. PIROZHNIKOV supposes that in the period of the highest rise of its level that sea could have communication with the Aralo-Caspian basin, which made invasion of the Caspian Sea by northern elements possible, including marine relics of the Glacial period close to those which occur today in the coastal waters of the Arctic Ocean and the relict lakes of its seaboard.

In this connection a brief mention should be made of the history of the so-called relics of the Glacial period. This group includes the seal *Phoca hispida* (and the forms related to it which live in Lake

Ladoga and the Caspian Sea), the four-spined cottid *Myoxocephalus (Cottus) quadricornis* L., the amphipods *Gammaracanthus lacustris* G.O.S., *Pontoporeja affinis, Pallasea quadrispinosa* and *Pallasea levis* EKMAN, the isopod *Mesidothea entomon*, the mysid *Mysis relicta*, and the copepod *Limnocalanus grimaldi* GUERNE with forms close to it. They live usually in waters which were or are closely connected with glacial dammed lakes of North Eurasia, freshened sections of sea gulfs, or relict fresh or brackish water bodies, etc. Relict species are widespread in lakes along the Arctic seaboard as well as in the Baltic basin and even in some lakes of Ireland and Scotland. In the Caspian Sea, relict forms are represented by the seal *Phoca hispida caspica* and crustaceans of the genera *Mysis, Pseudalibrotus, Pontoporeja, Gammaracanthus, Limnocalanus*, and also *Mesidothea entomon*, arriving from Arctic glacial dammed lakes in the Glacial Period, they have changed slightly there and produced new forms (MORDUKHAI-BOLTOVSKOI, 1960).

In Siberia, relict forms have been found in the above-mentioned Arctic lakes of the drainages of the Yenisei, the Gyda and the Pyasina and in the Yenisei and Gyda gulfs. *Mesidothea entomon* occurs in the Lena drainage to the south of its confluence with the Aldan and in the upper reaches of the Indigirka; *Pallasea quadrispinosa* has been found in the mouth of the Lena.

Among these relicts, the amphipod *Pallasea quadrispinosa* SARS is of special interest. Its close affinity to the Baikalian species *Pallasea kessleri* DYB. was noted long ago by B. DYBOWSKY, A. Y. BAZIKALOVA (1945) writes: "These two forms have an absolutely identical structure and distribution of cutaneous protuberances, which are slightly less pronounced in the case of *P. quadrispinosa*, and a common structure of the limbs, differing in insignificant detail only." Another Arctic species of the genus *Pallasea, P. laevis* EKMAN, is known to live on Novaya Zemlya only and is merely a variety of *P. quadrispinosa* (SEGERSTRÅLE, 1957).

The genus *Pallasea* can be regarded as having evolved in Baikal long before the Quaternary period. It is represented by more than a dozen species there. According to A. Y. BAZIKALOVA, it became the progenitor of other Baikalian genera: *Parapallasea, Metapallasea, Poekilogammarus, Macrohectopus*. Thus, an immigrant from Baikal related to *Pallasea kessleri*, which still lives in the Angara and the Yenisei, could really have been the ancestor of *Pallasea quadrispinosa*.

Relying on P. L. PIROZHNIKOV's hypothesis, S. SEGERSTRÅLE (1956, 1957, 1958), who has devoted much effort to ascertaining the routes of the migrations of relicts of the Glacial period in Europe, advances the proposition that almost the entire group of European glacial relicts have their birthplace in the "Central Siberian glacial dammed sea-lake" which lay in the area of the middle reaches of the

Ob and the Yenisei. Marine species, ancestors of species of the modern relict group, penetrated there from the Arctic Ocean. Finding themselves in a new environment, they changed, accommodating themselves to fresh-water conditions, and spread from there both to the east (the Lena, the Indigirka) and far to the west of Europe. *Pallasea quadrispinosa* also formed there, though not from marine species but from the Baikalian species *Pallasea kessleri*, which had invaded the above-mentioned central Siberian basin in the Glacial period.

It should be noted, however, that modern investigations in the West-Siberian Lowland reveal no trace of the extensive dammed basin as portrayed by P. L. PIROZHNIKOV and S. SEGERSTRÅLE. Neither are there any indications of an ancient run-off from it towards the Caspian Sea. But there is no doubt that dammed glacial lakes, including very big ones, could exist in North Siberia in Glacial times. We must also take into consideration the possible influence of extensive sea transgressions which stretched into the south of Siberia for more than 1,000 km from the modern coast of the Arctic Ocean. These transgressions undoubtedly left behind relict basins in which marine species could become adapted to brackish and fresh water. The numerous glacial lakes, the readjustment of hydrographic communications, the transition of river systems from one drainage to another, the important changes in the position of the mouths of Siberian rivers—all this must have greatly promoted the spread of relict species from their birthplace far to the east and west along the Arctic coast.

Large fresh water bodies of the Glacial period were colonised simultaneously by immigrants from the northern seas and from Baikal. Thus, at the points of contact of Baikalian and relict forms, distinct biocoenoses appeared which have been preserved in some freshened gulfs of the northern seas and in relict lakes.

It can be supposed that in the Quaternary period not only species of the genus *Pallasea*, but also some other immigrants from Baikal could have spread far to the east and west along the Arctic coast. Among them is, for instance, the polychaete *Manayunkia baicalensis*, which is perhaps to be regarded as the progenitor of the *Manayunkia* forms occurring in some Arctic coastal waters. In their turn, some northern active elements could have worked their way into Baikal and other lakes of the Baikal mountainous regions during the maximum transgression of the sea and the advance of glaciers. They include the omul, the seal, possibly the gwyniad and also *Salvenius alpinus erithrinus*, which still inhabits cold mountain lakes in the regions of Siberia where glacial phenomena were particularly pronounced.

If the modern powerful waterway of the Angara-Yenisei provides, as it did in the Quaternary period, a channel for the spread of some elements of the Baikalian fauna northward right up to the Arctic coast, then in a more remote past Baikal's communication with the Ocean was evidently provided by the drainage basin of another great Siberian river, the Lena, as is shown by interesting finds made in the catchment areas of the Vitim and the Olekma, large tributaries of the Lena, in the last two decades.

Remnants of a Baikalian-type fauna have been found in large running-water lakes of the Vitim drainage (fig. 111), such as Baunt, Busani and Oron, lying in the Tsipa tectonic depression, and Lake Oron-Vitimsky (the Vitim valley), as well as the Lakes Davatchanda, Leprindo and Leprindakan in the drainage of the Chara, a tributary of the Olekma, and in the area of the Muya-Chara tectonic depression (KOZHOV, 1942, 1949, 1956; KOZHOV & TOMILOV, 1949; TALIYEV, 1946; TOMILOV, 1954). In all these lakes the Baikalian polychaete *Manayunkia baicalensis* var. has been found, and in the lakes of the Vitim drainage, also Cottoidei species of the Baikalian genera *Paracottus* (*P. kessleri*, *P. kneri*) and *Asprocottus* (*A. kozhowi* fig. 74,1). A species of the small gwyniad from the group *lavaretus* has been observed in the lakes Baunt, Busani and Oron-Tsipinski. V. N. ANPILOVA (1956a, b) admits that this gwyniad could have diverged from the Baikalian *Coregonus lavaretus baicalensis* DYB.

It is also admissible that *Asellus epimeralis* BIRST. found in the Vitim lakes, originally also lived in Baikal and penetrated into these lakes later. Besides Baikal, the lower reaches of the Lena and the above-mentioned lakes, representatives of the genus *Asellus* have not been found anywhere else in East Siberia.

A Baikalian species of the Cottoidei, *Paracottus kessleri*, is widespread in the Lena drainage area. Recently the parasitic copepod *Salmincola cottidarum* MES., an ordinary parasite of the Baikalian Cottocomephoridae, has been found on this cottid from the middle reaches of the Lena (KORYAKOV, 1959b). There are also indications that this area has other immigrants from Baikal.

We cannot agree with the opinion of L. S. BERG (1949a) that the Baikalian elements in the catchment basins of the Vitim and the Olekma are relicts of the "heat-loving" Euro-Siberian fauna which was widespread in the Pliocene. There is no doubt that they penetrated into these basins from Baikal and stayed on there.

The only explanation for the presence of Baikalian elements in some lakes of the Olekma and Vitim basins is that there was a period in Baikal's history when it was hydrophysically connected with the basin of the Lena (KOZHOV, 1949, 1950b). In that period Baikal's waters flowed into the Lena through the rivers linking Baikal with other large lakes that filled tectonic depressions of the Baikal system: that along which the Barguzin flows into Baikal now, as well as the Tsipa, Muya-Chara, and other depressions, and that is why these lakes were populated by a Baikalian-type fauna, a part of which has survived there. The rise of the north-eastern part of the trans-Baikal area formed a divide which separated Baikal from the other depressions of its system, with its waters breaking through the modern Angara valley and streaming towards the Yenisei.

Geological data also point to the young age of the modern run-off from Baikal via the Angara and to Baikal's possible ancient connection with the Lena basin.

V. V. LAMAKIN (1952, 1957) holds that Baikalian species could have reached the area of the Tsipa and Muya-Chara troughs in the inter-glacial period, when the level of Baikal was very high owing to the thawing of glaciers after maximum glaciation. N. U. DU-MITRASHKO (1952b) considers that such a connection was possible through the valley of the Tiya, a north-western affluent of Baikal. There are suppositions regarding other connections of Baikal with the Lena. But the discovery of living remnants of the Baikalian fauna in the Tsipa, Muya and Chara catchment area shows that this connection was most likely realised through a system of gullies linking Baikal with the above-mentioned tectonic depressions of the Baikal system. It is hardly possible that the roe of such a Baikalian species from the Cottocomephoridae family as *Limnocottus kožowi*, which spawns in Baikal at a considerable depth, could be carried by birds across the Baikal-Tsipa divide. The carriage of the polychaete *Manayunkia* by birds is equally unlikely, for the endemic Baikalian species are highly sensitive to their environment and are still less capable of enduring airborne travel over hundreds of kilometres.

It should be also noted that some representatives of a fauna of the Baikalian type have been found also in the large Mongolian lake Khubsugul, namely, the mollusks *Choanomphalus mongolicus* Kozov and *Kobeltocochlea michnoi* LDH. (LINDHOLM, 1929; KOZHOV, 1946; ANUDARIN, 1953). Their migration from Baikal to Khubsugul up the Selenga and via the complicated system of its tributaries is hardly probable. Evidently they are remnants of a distinct ancient fauna which penetrated into Baikal from Central Asia in the Tertiary and developed vigorously there.

As had been established by P. L. PIROZHNIKOV, V. N. GREZE and others, Baikalian immigrants comprise a considerable part of the highly nutritive biomass of zoobenthos in the lower reaches of the Yenisei. For this reason there arises the question of the possibility and expediency of stocking water bodies in North Siberia with Baikalian species (GREZE, 1951b; PIROZHNIKOV, 1955). This question has assumed particular importance in connection with the construction of hydro-power projects on the Angara and the appearance of large reservoirs where some Baikalian species may constitute the bulk of the biocoenoses of benthos and plankton. Among the pelagic fish, the Baikalian gwyniads and especially the omul may prove particularly valuable in this respect.

V

LIFE IN THE MASS OF WATER

1. Distribution of Plankton, Its Seasonal Changes and Migrations of Pelagic Fish.

The first studies of plankton in Baikal were made by V. N. YASNITSKY (1923, 1924, 1930; YASNITSKY & SKABICHEVSKY, 1957), K. I. MEYER (1930), as well as by S. M. VISLOUKH (1924), A. G. GENKEL (1925), V. A. YASHNOV (1922), B. V. SKVORTSOV (1937; SKVORTSOV & MEYER, 1928), A. P. SKABICHEVSKY (1929, 1935, 1954), A. A. ZAKHVATKIN (1932). Of special importance for the knowledge of the annual and seasonal changes of plankton in Baikal were the stationary investigations conducted by V. N. YAS-NITSKY (1930) in the area of Bolshiye Koty in 1926—28.

Subsequently extensive studies of the Baikalian plankton were undertaken by the author and his colleagues and pupils N. L. ANTIPOVA, I. K. VILISOVA, O. M. KOZHOVA, F. G. MAZEPOVA, L. N. MOGILEV and others.

At the present time Baikal is known to have about 100 species of planktonic algae (O. KOZHOVA, 1956b, 1957, 1959a, b), several dozen species of infusorians, up to 40 species of rotifers and about 20 species of crustaceans.

All planktonic organisms can be divided into two main genetic groups: Baikalian and European-Siberian. The latter is richer in species but inhabits, in the main, shallow enclosed gulfs, bays and other sections of the littoral-sor region. Many of the European-Siberian species, however, sporadically occur in the open waters, while some of them, being permanent residents there, comprise special Baikalian forms. Such are, for instance, the rotifers *Notholca longispina, N. striata acuminata, Keratella quadrata, K. cochlearis* and *Filinia longiseta,* from the Cyclopoida-*Cyclops kolensis* var. *baicalensis,* etc.

Especially numerous in the plankton of the open waters of Baikal, in deep regions in particular, are species of the Baikalian group characterised by profound endemism, such as the diatoms *Cyclotella baicalensis,* (fig. 77,10), *C. minuta, Melosira baicalensis* (fig. 78,1), the peridineans *Peridinium baicalense* (fig. 77,1—3), *Gymnodinium baicalense* (fig. 77,4, 7), *G. coeruleum,* (fig. 77,8, 9), the infusorians *Marituja pelagica,* (fig. 28,2), *Liliimorpha viridis* (fig. 28,1), the Baikalian forms of the *Tintinnidium* (fig. 29), the rotifer *Synchaeta pachypoda* (fig. 40), the gammarid *Macrohectopus branickii* (fig. 60,1) and the copepod *Epischura baicalensis* (fig. 46). The commonest among the European-Siberian species in the open waters of Baikal are the diatoms *Melosira islandica* subsp. *helvetica*, *Synedra ulna*, *S. acus* and *Dinobryon cylindricum*, as well as *Cyclops kolensis* var. *baicalensis* (fig. 47). Among the pelagic plankton-eating fish, the most important part in the open waters is played by the Baikalian omul *Coregonus autumnalis migratorius* (fig. 75), two species of *Cottocomephorus*: *C. grewingki* (fig. 74) and *C. inermis*, and two species of *Comephoridae*: *Comephorus baicalensis*, *C. dybowskii* (fig. 73,1).

As regards the horizontal distribution of plankton (as well as that of temperature), we should distinguish in Baikal 1. sections of the littoral-sor zone already characterised in the chapter on the distribution of benthos, 2. shallow regions adjoining the estuaries of big rivers, 3. extensive gulfs, straits, bays and other relatively shallow sections along the open fringe, 4. deep regions. The first three comprise, in the broad sense, the littoral region which to a certain degree is analogous to the neritic region of the sea. It can be contrasted to the deep-water region removed to a considerable distance from the shore (KOZHOV, 1957; O. KOZHOVA, 1957).

As regards the vertical distribution of life in the mass of water, we distinguish the following zones (KOZHOV, 1954, 1958):

1. The upper, or trophogenous zone, i.e., the zone of intense photosynthetic activity, where most of the primary product is created and the bulk of plankton is concentrated throughout the greater part of the year. The lower limit of this zone changes in the course of spring and summer but evidently does not descend lower than 25—50 metres, although in the period of homothermy the vegetative reproduction of algae takes place also in deeper layers, probably to a depth of at least 100 metres.

2. The middle, or transitional zone, where, as a rule, no more or less thick swarms of plankton are observed throughout the year, with the exception of near-bottom layers in shallow regions. The lower limit of this zone can be conditionally put at about 250 metres.

3. The lower zone, which includes the entire water mass deeper than 250—300 metres. It is very poor in living plankton.

4. The near-bottom stratum, where accumulations of living plankton, especially those of *Macrohectopus*, are observed in winter and early spring.

Table XXX illustrates seasonal changes in the development of plankton in Baikal's open waters. Shown in fig. 95—102 are seasonal and annual changes in the numerical density and biomass of the most important species and groups of plankton (KOZHOV, 1955a, b).

Biological spring in the open regions of Baikal begins very early. Not infrequently individual vegetative reproducing cells and colonies of the vernal algae *Melosira baicalensis*, *M. islandica* subsp. *helvetica*, *Cyclotella baicalensis*, *Synedra ulna*, etc., can be found in

Table XXX.

Biological Seasons in Baikal's Open Waters

Factors	Spring		Summer		Autumn	Winter
	Early spring (sub-ice) period February, March	Late spring (transitional) period May, June	Early, July, first 10 days of August	Late, August-September	October November	December-January
Seasonal average water t° (Centigrade)						
0 m	0.7	2.5	10.0	12.5	6.3	2.0
20 m	0.8	2.8	7.0	9.0	5.8	2.2
Maximum t°						
0 m	1.5	4.0	15.0	15.0	9.0	3.6
20 m	1.0	3.6	10.0	10.0	8.0	3.6
Water circulation	Attenuated circulation	Intense vertical convection and wind-induced circulation	Wind-induced circulation. Horizontal currents	Wind-induced circulation. Strong horizontal currents	Intense vertical thermal circulation. Horizontal currents	Attenuation of circulation at end of period
Compounds of biogenous elements	Decrease; annual minimum towards end of period	Increase; spring-summer maximum towards end of period	Decrease	Decrease; summer minimum	Increase	Increase; winter maximum
Phytoplankton	Mass vegetative reproduction of diatoms and peridineans; annual maximum of biomass in abyssal regions	Beginning of dying-off of vernal forms and their subsidence. High biomass; decrease towards end of period	Mass dying-off of vernal forms. Appearance of aestival forms	Mass development of aestival forms in open shallows	In some years autumnal outburst of diatoms but total biomass diminishes sharply	Scant; annual minimum of biomass

Table XXX. (continued)

Biological Seasons in Baikal's Open Waters

Factors	Spring		Summer		Autumn	Winter
	Early spring (sub-ice) period February, March	Late spring (transitional) period May, June	Early, July, first 10 days of August	Late, August-September	October November	December-January
Zooplankton	Mass outcrop of young *Epischura* of winter-spring generation. Reproduction of *Macrohectopus*	Period of growth of winter-spring generation of *Epischura*. Start of a new explosive reproduction by end of period. Increase in biomass	Mass outcrop of young *Epischura* of summer generation. Emergence of summer forms: *Cyclops, Cladocera, Vorticella*, etc. Annual maximum of biomass	Period of growth of summer generation of *Epischura*. Annual maximum of biomass at outset of period. Decrease in biomass by end of period	Decrease in biomass. Subsidence of adult *Epischura*	Scant; annual minimum of biomass
Vertical distribution of zooplankton	Maximum density in upper layers (0—50 m)	Dispersal in water mass down to 220—300 m, with greater density in upper layers	Maximum density in upper layers, (0—50 m especially 0—25 m.)	Maximum density in upper layers (0—50 m, especially 0—25 m.)	Dispersal in water mass, descent to deep layers	Greater part in deep layers. Ascent by end of period
Intensity of diurnal vertical migrations of zooplankton	small	very small	great	great	small	very small
Plankton-eating fish	Omul hatchlings begin to migrate down at end of period. Movement of omul and *Cottocomephorus* from wintering grounds to shores	Omul shoals converge in shallow fishing grounds. Approach of *Cottocomephorus* to shores for spawning. Omul fry migrate down to Baikal	Movement of omul to open regions and extensive feeding migrations in upper layers. Mass outcrop of *Cottocomephorus* hatchlings and fry in shallows	Extensive feeding and spawning migrations of omul in upper layers. Mass accumulations of *Cottocomephorus* fingerlings along shores	Start of movement to wintering grounds in deep layers	Wintering of omul and *Cottocomephorus* at 200—300 m and deeper, mostly in nearbottom layers close to shallows

January under the just-formed ice cover still free from snow (YASNITSKY, 1930).

In March the numerical density of algae increases considerably; peridineans, especially species of the genus *Gymnodinium*, appear and begin to multiply vigorously. In April the bloom of vegetable life under the ice is observed; water transparency in the photosynthesis zone drops to 8—5 metres, and the zone of mass habitation of algae extends to depths of 30—40 metres (ANTIPOVA & KOZHOV, 1953; O. KOZHOVA, 1959).

Simultaneously with the growth of the biomass of phytoplankton, an increase in the quantity of *Epischura baicalensis*, its chief consumer, and also rotifers and infusorians is observed in the upper zone.

Among the rotifers, especially conspicuous are *Notholca longispina* and *N. striata* f. *acuminata*, the big carnivorous rotifer *Synchaeta pachypoda*, as well as *Filinia longiseta* and species of the genus *Keratella*.

In winter and early spring the hatching of the young of *Epischura* takes place chiefly in deep layers. Beginning with January the young crustaceans rise to the upper layers, where they accumulate and comprise 50 to 80% of the numerical density. But throughout the sub-ice period most of the adult crustaceans remain in deeper

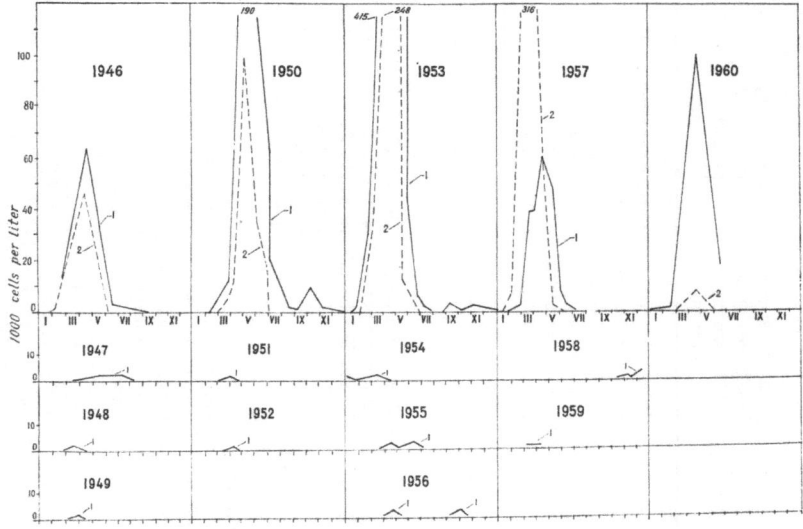

Fig. 95. Seasonal and annual fluctuations in the number of cells of diatom algae in the open water of South Baikal (the Bolshiye Koty). 1 - *Melosira baicalensis;* 2 - *Melosira islandica helvetica* in thousands of cells per litre in the 0—50 m layer (average). After ANTIPOVA.

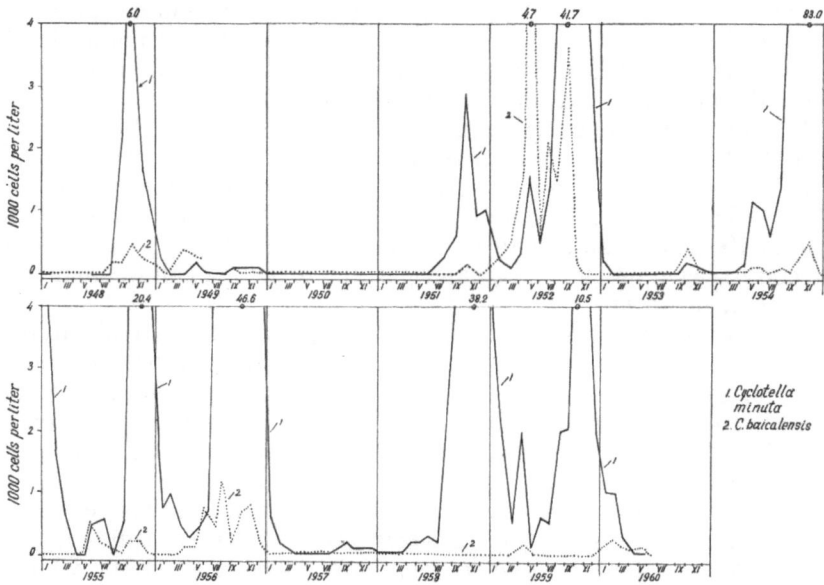

Fig. 96. Seasonal fluctuations in the number of cells of *Cyclotella baicalensis* (2) and *C. minuta* (1) in South Baikal (the Bolshye Koty) in thousands of cells per litre in the 0—50 m layer (average). After ANTIPOVA.

layers. In autumn and winter the amphipod *Macrohectopus* also prefers deep and even near-bottom layers.

In January and February the total raw mass of zooplankton is insignificant. In different years it varies in the neighbourhood of 5, rarely 10, grams under 1 m² in the 0—250 metre layer and between 3.5 and 5 grams under 1 m² (0.07 to 0.1 g/m³) in the 0—50 metre layer.

In April, i.e. at the end of the sub-ice period, the density and bio-mass of zooplankton increase chiefly due to the growth of young *Epischura* of the winter generation.

The melting away of the ice cover is followed in May and June by the mass dying-off of vernal algae, numerical growth of the microbic population (particularly in the 20—50 metres layer), an increase in the oxidation of water, and re-accumulation, in the upper layers, of biogenous compounds, both those which are released as a result of the regeneration of the dying plankton and which rise from the deeper zones. Vernal forms begin to be replaced by aestival ones.

In the transitional period (May—June) adult specimens of the copepod *Epischura* hatch in the deep layers a new (aestival) gene-ration of nauplii, whose number reaches the annual maximum in June. They concentrate in the upper layers.

234

The establishment of direct stratification of temperatures is followed by the restoration of sharply differentiated distribution of plankton. The bulk of the aestival algae replacing the cold-loving vernal complex, and also zooplankton (with the exception of *Macrohectopus*) concentrate in the upper layer. The biomass of zooplankton reaches the annual maximum in August—September: in the 0—50 metres layer it often stands at 40 to 50 grams under 1 m² (0.8 to 1.0 g/m³), and in some sections in peak-crop years 70 to 100 grams and more under 1 m² (1.4 to 2 g/m³). *Epischura* accounts, as a rule, for the greater part (80 to 90%) of all the biomass in the open waters.

In the region of extreme depths, the biomass of aestival phytoplankton is usually very small, but in the littoral regions and in the belt of deep currents, especially after a prolonged calm, aestival algae often cause an abundant water bloom and their biomass may considerably exceed the vernal biomass of phytoplankton. (This will be dealt with later).

In regions with favourable conditions (gulfs, extensive bays, pre-estuarine areas), *Cyclops kolensis* var. *baicalensis* attains maximum development in August—September, keeping, on the whole, in layers above those occupied by *Epischura*.

The end of July, August and September sees an increase in the quantity of rotifers and infusorians in the open waters. Especially abundant among the rotifers are the Baikalian forms of the species

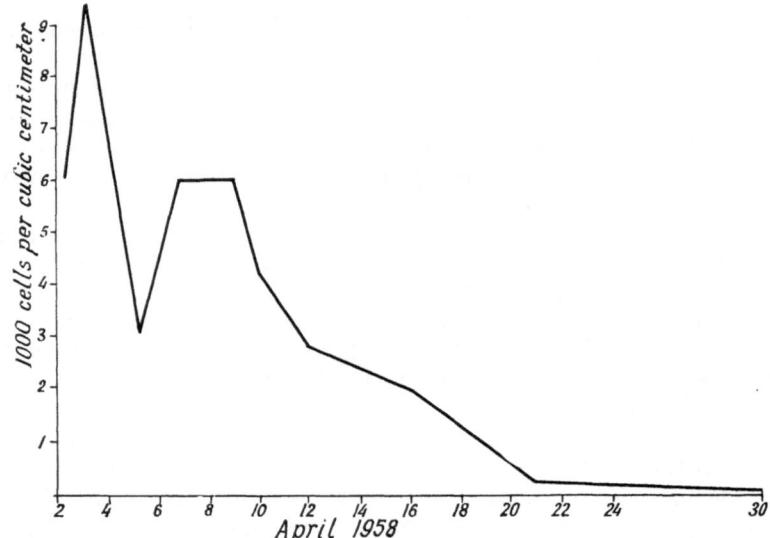

Fig. 97. Fluctuations in the number of cells of *Gymnodinium baicalense* + *G. minutum* in April 1958, in thousands of cells per cm³ in the 0—1 m layer, 10 m from the shore. After O. Kozhova, 1959.

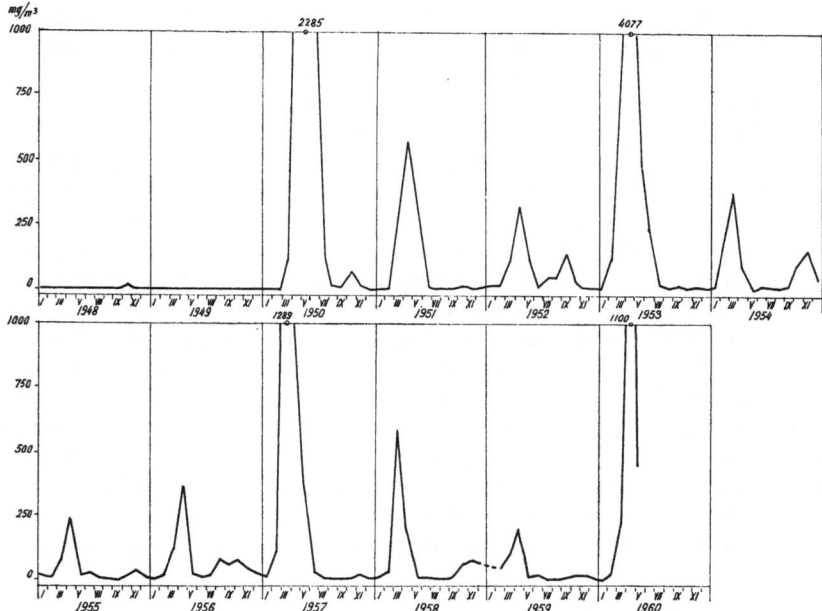

Fig. 98. Seasonal and annual changes in the raw biomass of phytoplankton in the open waters of Baikal in milligrams per m³ in the 0—50 m layer (average). After ANTIPOVA, in litt.

Keratella quadrata, K. cochlearis, Filinia longiseta, Notholca longispina. Vorticellas are sometimes plentifully represented. In summer the main mass of the amphipod *Macrohectopus* keeps between 150 and 250 m of depth, migrating towards the surface during the dark hours.

Seasonal changes in the bacterial plankton of Baikal have been studied by BLANKOV (V. N. YASNITSKY, 1927), S. I. KUZNETSOV (1951, 1957), A. G. RODINA (1954), O. M. KOZHOVA (1953; KOZHOVA & KAZANTSEVA, 1961), A. P. ROMANOVA (1958). These studies have shown that there are two maxima in the development of Baikal's bacterial population: vernal (April) and aestival (August—September), which accords with the maxima of the biomass of vernal and autumnal algae (fig. 99). In April 1954 in the southern part of Baikal the number of bacteria in the 0—25 m layer, which is especially densely populated, ranged between 139,000 and 335,000 per millilitre. In August of the same year the number of bacteria in the upper layers reached 800,000 to 1,000,000 per ml. The lowest density of bacteria was observed in June. The 0—25 m layer is particularly rich in them. Deeper than 100 metres their number is insignificant.

Despite the seeming uniformity of ecological conditions, the pelagic zone of deep-water regions is not entirely uniform as regards the water regime and the composition and numerical density of plankton. Strong horizontal currents and wind-induced movements of water with the emergence of cold water on to the surface often complicate the picture of the distribution of both temperature and plankton. Areas of relatively warm water driven by the wind from the shores can be found in deep-water regions. After strong winds, plankton near the leeward shores is very scarce and more or less evenly dispersed in the 0—50 to 0—100 m layer. In such periods plankton in warmer water in the middle of Baikal is often more abundant than near the shores.

In summer a correlation is observed between temperature and plankton biomass in the upper 10 m layer in deep-water regions. The higher the temperature, the richer the plankton, and vice versa (fig. 101). In August and at the beginning of September the sections with the greatest density of typical Baikalian zooplankton are characterised by a temperature of 14—12° C in the surface waters and 8° at a depth of 10 metres (fig. 101; KOZHOV 1957).

As already mentioned, the plankton of the littoral-sor sections

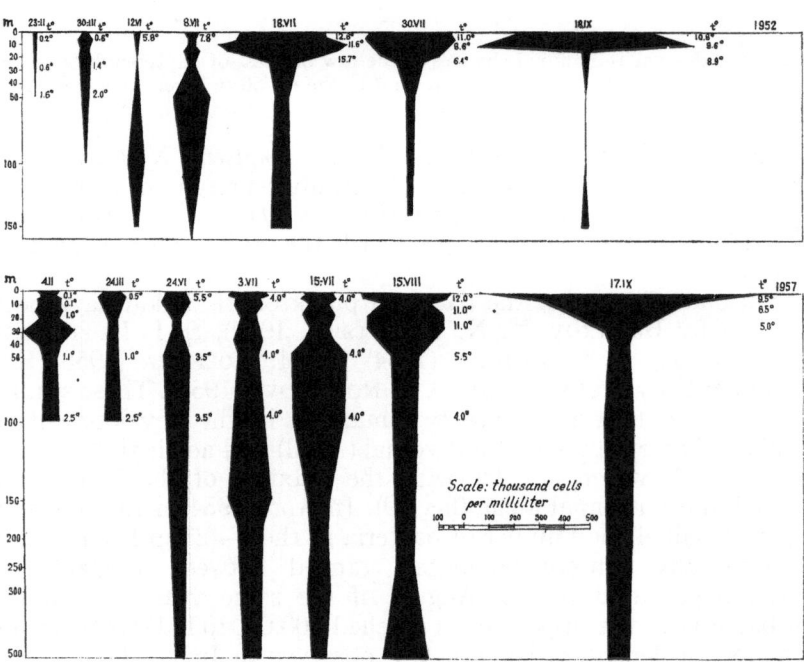

Fig. 99. Seasonal changes in bacterioplankton and temperature in the open waters of South Baikal: A - in 1952 (a Melosira - poor year); B - in 1957 (a Melosira - rich year). After KOZHOVA & KAZANTSEVA, 1961.

(VILISOVA, 1954, 1959a; KOZHOV, 1957) consists of ordinary forms widespread in European-Siberian waters, to which Baikalian species are added only in the cold period of the year. The closer these sections are to the open waters of Baikal, the greater the part played by these Baikalians. In spring, before the ice breaks up, the algae *Melosira baicalensis, Cyclotella baicalensis, Melosira islandica* subsp. *helvetica*, and also *Stephanodiscus Binderanus* and other vernal forms can be observed there. In the zooplankton, small quantities of *Epischura, Cyclops kolensis* and Rotatoria are found. But it is lacustrine species that predominate: *Eudiaptomus graciloides, Cyclops vicinus, Mesocyclops leuckarti, Chydorus sphaericus, Bosmina longirostris, Daphnia longispina,* the rotifers *Polyarthra trigla, Keratella cochlearis, K. quadrata.*

After the freeing of littoral shallows from the ice, vernal Baikalian forms disappear. Gradually, with the warming up of the water, typically lacustrine aestival plankton develops. Among the algae the predominant part is played by *Anabaena flos-aquae, Dinobryon*

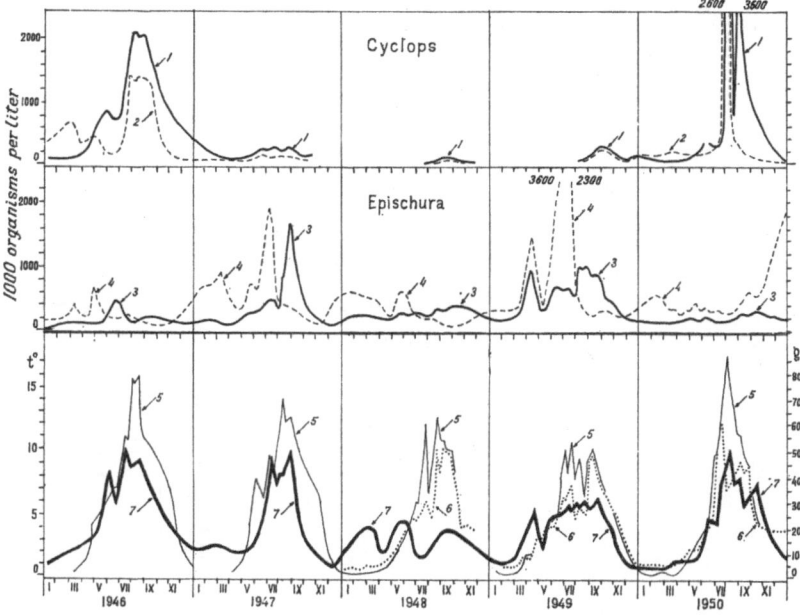

Fig. 100. Seasonal and annual changes in mass species of zooplankton in the 0—250 m layer in the open water of South Baikal. 1—4 - number of crustaceans in the 0—250 m layer in thousands under m²: 1 - copepodit stages of *Cyclops kolensis* var. *baicalensis*; 2 - nauplii of the same species; 3 - copepodit stages of *Epischura baicalensis;* 4 - nauplii of the same species; 5 - temperature of the upper layer of water; 6 - temperature at a depth of 25 m; 7 - raw weight of crustacean plankton in grams under m² in the 0—250 m layer. Original.

238

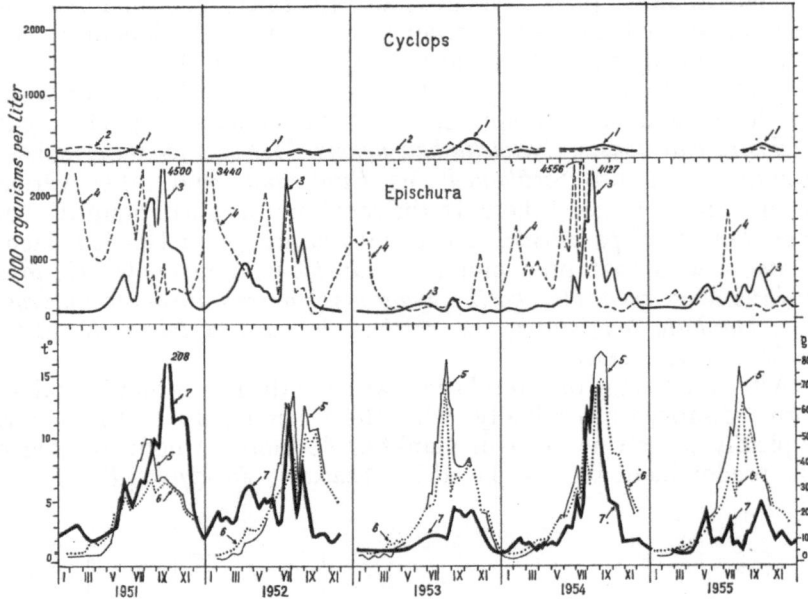

Fig. 100 Continuation 1.

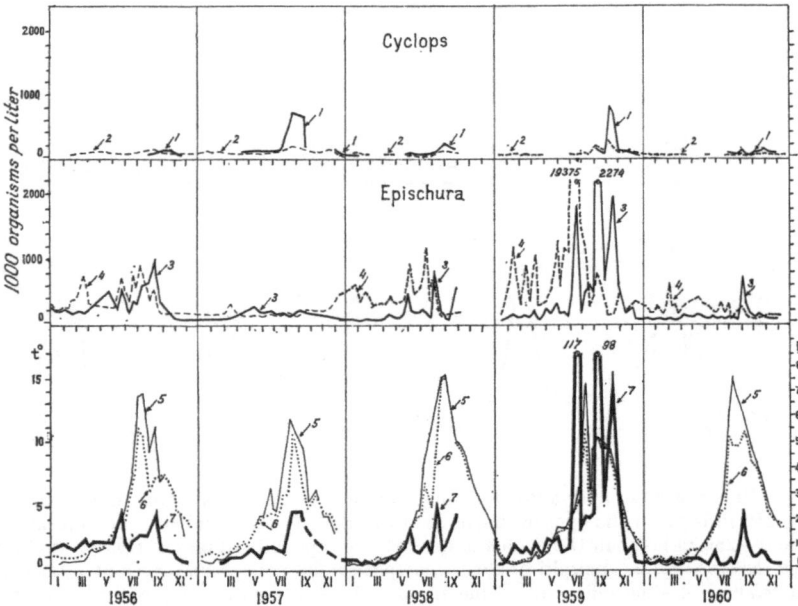

Fig. 100 Continuation 2.

cylindricum, Ceratium hirundinella, Asterionella gracillima, A. formosa, and others. Abundantly represented in the zooplankton are Cladocera, Cyclopoida, Rotatoria, Infusoria.

In spring and summer the concentration of zooplankton in the littoral-sor waters can be high. The biomass of copepods and rotifers often reaches 3—5 g/m³, providing plenty of food

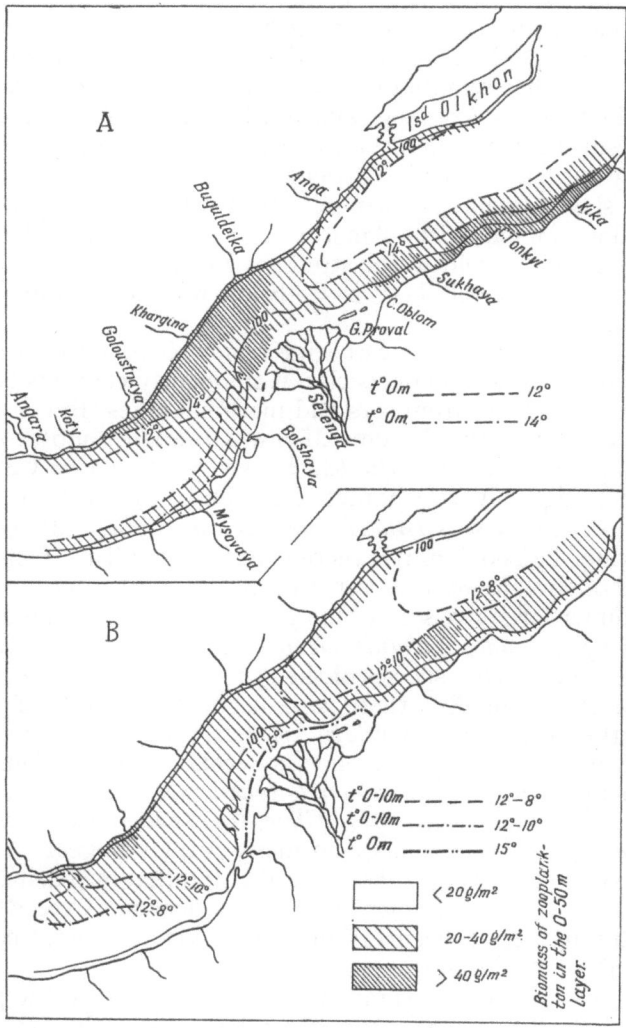

Fig. 101. Distribution of temperature and crustacean plankton biomass in the middle part of Baikal: A - in the second half of July and at the beginning of August 1950; B - at the end of August 1952. After Kozнov, 1958.

for the young of plankton-eating cyprinoids and related fish.

Extensive areas are taken up by the pre-estuarine regions of the rivers Selenga, Upper Angara, Barguzin and Turka. In their water regime these regions can be divided into two zones: internal, with depths not exceeding 10—15 metres, which is constantly under the influence of fluvial waters, and external, influenced by both the internal part and the neighbouring deep-water regions.

As noted above, the water in the sections directly bordering on river estuaries and deltas is warmed rapidly and its temperature in spring is always 2—4° C higher than in the adjacent open zone, causing an earlier seasonal change of plankton. At the same time, its development is adversely affected by the constant turbidity of water resulting from disturbances which raise minute particles of bottom mud. The water is particularly turbid near the river deltas, from the shores to the 10—15 m isobath. The transparency limit there usually does not exceed 1—3 metres. In the area of the Selenga delta the zone of this turbid water stretches along the shores for almost 80 kilometres in a belt 1—2 to 6—8 km wide, depending on wind direction.

The plankton of the pre-estuarine regions is of a mixed composition. Along with Baikalian species, it frequently contains lacustrine species carried away from sors and marginal lakes. But in spring it consists chiefly of the diatoms *Melosira islandica* subsp. *helvetica*, *Melosira baicalensis*, *Cyclotella baicalensis*, *Synedra ulna* var. *danica*, and other algae. Very characteristic for these sections are the diatoms *Stephanodiscus Binderanus*, *Melosira italica*, *M. granulata*.

In spring the zooplankton there contains, as a rule, considerable amounts of *Cyclops kolensis* var. *baicalensis* and rotifers.

In summer large masses of blue-green algae appear, among them *Anabaena flos-aquae*, *Aphanizomenon*, *Gloeotrichia echinulata*, then *Ceratium hirundinella*, species of the genera *Pediastrum*, *Asterionella* and *Dinobryon*; finally, there also occur *Volvox*, *Eudorina*, etc. Among the crustaceans, *Cyclops kolensis* usually predominates in summer, whereas *Epischura* plays a secondary part. Considerable amounts are found of *Eudiaptomus graciloides*, *Daphnia longispina hyalina*, *Bosmina longirostris*, *Chyrodus sphaericus*, *Mesocyclops leuckarti* and *Cyclops vicinus*, and many rotifers appear.

The biomass of zooplankton in the internal pre-delta zone is insignificant even in summer and usually does not exceed 0.3—0.1 g/m³, although its concentration in the surface waters can reach 1—2 g/m³ and more during the night.

In the first half of the summer the external zone of the pre-estuarine regions (beyond the 10—15 m isobath) is characterised by greater water transparency and lower temperatures.

In spring, plankton there is typically Baikalian, while in summer it becomes mixed with a considerable number of forms brought in by currents from pre-estuarine shallows. For instance, abundant development of *Cyclops kolensis* can always be observed there, and *Eudiaptomus graciloides, Daphnia longispina hyalina, Bosmina longirostris, Chydorus sphaericus,* lacustrine species of rotifers, etc., are often found.

Winds and currents carry this mixed plankton far into the adjoining deep-water regions. For instance, from the pre-delta region of the Selenga it penetrates to the south-west, at least to the outflow of the Angara. In North Baikal, it is carried from the pre-estuarine regions of the Kichera and Upper Angara for dozens of kilometres along the western shore to the Boguchan Bay and farther on to the south.

In warm and quiet weather this aestival plankton can also develop in the neighbouring deep-water regions (chiefly in the belt of warm currents, causing water bloom and reducing the transparency limit down to 4—6 metres. But this bloom is usually very short-lived and ends after the first storm.

Fluvial waters enrich Baikal with biogenous compounds, which is evidently one of the causes of summer water bloom in the pre-estuarine sections. But of no less positive significance is the higher temperature both in spring and in summer and the vicinity of the bottom, where the resting stages of algae can be preserved, rising rapidly to the photosynthesis zone in early spring.

The plankton of big gulfs and other extensive regions relatively isolated from the open waters also has certain peculiarities. In early spring the phytoplankton of the gulfs is, more or less, of the same composition everywhere, with the diatoms *Melosira baicalensis, M. islandica* subsp. *helvetica, Stephanodiscus Binderanus, Cyclotella baicalensis, C. minuta* and others comprising the bulk of it, and *Epischura, Cyclops kolensis, Keratella cochlearis, K. quadrata, Notholca longispina* and *Synchaeta pachypoda* predominating in the zooplankton. Owing to the more rapid warming through of the internal sections of gulfs, copepodit stages of *Epischura* and *Cyclops kolensis* develop there already in June, and the biomass of zooplankton reaches, on the average, 1.5—2 g/m³. In the central and external section of gulfs plankton is still poorly developed in June. It is only towards the middle of July that maximum development is attained in the central sections. By this time the predominant part in the internal sections begins to be played by *Cyclops kolensis* var. *baicalensis* as well as Cladocera and lacustrine-sor forms of Rotatoria. The maximum of the biomass of *Epischura* follows the movement of the 10—12—14° isotherm on the surface and the 8—10° isotherm at a depth of 10 metres. When the water temperature exceeds this limit *Epischura* and other cold-loving Baikalian forms gradually disappear.

It should be stressed that an important part in the plankton of big gulfs (Chivyrkui and Barguzin) is almost always played by *Cyclops kolensis* var. *baicalensis* and the alga *Stephanodiscus Binderanus*, for which such regions present a permanent habitat.

The water regime and pattern of plankton development similar to those of big gulfs exist in more or less extensive shallows poorly protected by promontories, and broad and even bays.

Special attention should be paid to the development of plankton in the Maloye More, the physico-geographical characteristic of which was given in the chapter on the distribution of benthos (fig. 17).

The plankton of the Maloye More (VILISOVA, 1959a; KOZHOVA, 1959b) is distinguished by qualitative and quantitative abundance. But in the open regions algae are richly represented by comparatively few forms: *Stephanodiscus Binderanus*, *Melosira baicalensis*, *M. islandica* subsp. *helvetica*, *Cyclotella baicalensis*, *C. minuta*, *Asterionella formosa*, and such aestival forms as *Anabaena flosaquae*, *Gloeotrichia echinulata*, while *Epischura*, *Cyclops kolensis* var. *baicalensis*, *Keratella quadrata* and *K. cochlearis* are usually common in the zooplankton.

In winter the snow cover on the ice in the Maloye More is usually very thin and winds often blow it off. Therefore in January the daytime light intensity in the sub-ice layers is fairly great. This probably explains the unusually early vernal development of algae there. For instance, in 1952 mass development of the alga *Stephanodiscus Binderanus* in the Maloye More was observed in February. Other Baikalian algae also begin to develop there early. In the sub-ice layers the young of *Epischura* appear en masse in January—February, reaching maturity soon after the ice break-up. In the southern part of the Maloye More the development of the aestival generation of *Epischura* begin in May, and at the end of June the biomass of zooplankton there may reach the annual maximum. In the centre of the Maloye More the annual maximum of the zooplankton biomass is observed from the second half of July until August, and in the northern part, in August and September. As in the gulfs, this maximum, resulting chiefly from the development of *Epischura*, shifts from internal shallows to external, according to the temperature of 12—14° C on the surface and 8—10° at a depth of 10 metres.

These relatively open shallow regions not more than 200—250 metres deep and the neigbouring deep-water regions are particularly rich in food for the pelagic fish of Baikal. Together with the littoral-sor zone they occupy an area of approximately 600,000 hectares.

The seasonal changes of the temperature and plankton strongly influence the horizontal and vertical migrations of pelagic fish.

The feeding period of plankton-eating fish in Baikal lasts the year round, but is least intensive in winter. In late autumn (November—December), after the temperature of the upper layers drops to 4—5°, the plankton-feeders omul, *Cottocomephorus* and *Comephorus* descend to deeper layers. Shoals of the omul and *Cottocomephorus* winter, as a rule, near the shallows which serve as spring convergence grounds, at a depth of 150 to 300 metres. In winter species of *Comephorus* are often found together with the omul, and in deep-water regions they may descend to still deeper layers.

The shoreward migrations of the omul from the wintering grounds begin in March, being accompanied by the movement of *Cottocomephorus* towards the shallows. In April—May the latter come close to the shores to spawn. The passage of the omul toward the shores in the sub-ice period takes place chiefly in the near-bottom layers. In June shoals of the omul converge near the shores, making possible beach-seine and gill-net fishing.

Deep-water regions with poorly developed near-shore shallows are avoided by the omul in spring and the first half of summer. In general, this is a littoral fish in the broad sense of the word, which prefers regions with relatively small depths not exceeding 200—300 metres. These are feeding and fishing grounds where the omul occurs throughout the year leaving them partially only in August—September.

Beginning with mid-July the omul rises from the near-bottom to upper layers, spreading to more open parts of shallows, where a rich crustacean plankton develops by this time. Thus, towards mid-summer the omul's feeding grounds expand greatly, coinciding with regions of the maximum density of crustacean plankton.

The fry of *Cottocomephorus grewingki* hatch in the second half of July and at the beginning of August and accumulate in vast numbers along the shores, feeding vigorously upon the littoral plankton. The omul, too, appears near the shores, attracted by them. Adult *Cottocomephorus* feed upon planktonic crustaceans, but also consume their own fry until the latter grow bigger and spread over extensive areas of the littoral.

The period between the end of July and beginning of September sees the formation of adult omul spawning shoals which migrate extensively in the open waters, tending in the direction of the pre-estuarine regions of spawning rivers. After returning from the rivers in autumn these shoals move towards the places of .future spring convergence.

In the taxonomic composition and vertical distribution of plankton Baikal somewhat resembles the deep Balkan Lake Ohrid. As in Baikal, the trophogenous stratum there extends approximately from 0 to 50 metres, Chlorophyceae and Diatomeae predominate in phytoplankton, the open waters have very few Cladocera, and *Keratella cochlearis, Notholca longispina* and *Filinia longiseta* prevail among the rotifers. But the specific composition of plankton in Lake Ohrid is much poorer than in Baikal (STANKOVIČ, 1960).

2. Annual Changes in the Plankton of Open Waters.

Shown in fig. 95—100 are changes in the quantity and biomass of the most important species and groups of plankton in the open waters of Baikal in different years. Especially extensive crop fluctuations are observed in the phytoplankton (ANTIPOVA & KOZHOV, 1953). A decisive part in these fluctuation is played, as already noted, by the diatom *Melosira baicalensis* and partially by *M. islandica* subsp. *helvetica*, and in the shallows by *Stephanodiscus Binderanus*.

Since 1941 exceptionally high crops of *M. baicalensis* have been recorded in 1943, 1946, 1950, 1953, 1957, 1960 (fig. 95).

In 1961 thick accumulations of *M. baicalensis* and *M. islandica* var. *helvetica* were recorded only in South Baikal. To the north of the Selenga shallow these diatoms were rare in 1961.

In *Melosira*-rich years the raw biomass of diatoms reaches in March—April an average of 4—6 g/m³ in the 0—25 m layer. In the years of poor development of *Melosira* the maximum biomass of vernal phytoplankton usually does not exceed a few hundred milligrams per m³. In such years *Cyclotella baicalensis* and sometimes *Synedra* species predominate in the open waters in spring, and the peridinean *Gymnodinium* in early spring.

The peridineans of the genus *Gymnodinium* annually appear in the open waters, occurring chiefly in the uppermost sub-ice layer of water. But their numerical density varies widely with years. In high-crop years their biomass in the maximum development period reaches an average of 0.5—1 g/m³ in the 0—50 layer, while in the 0—5 = 0—10 layers much denser concentrations are observed.

In years fruitful for *Melosira* species their mass development spreads practically over the whole of Baikal or its southern and middle part. For instance, in 1950 *M. baicalensis* populated the waters of the southern and middle parts of the lake very densely but was much less abundant in the northern part, although even there it predominated among all other algae. In 1960 the picture, on the whole, repeated itself, but the distribution of *M. baicalensis* was uneven. In its poor years *M. baicalensis* is either absent or develops extremely poorly in all parts of the lake, yielding predominance to other vernal algae, such as *Cyclotella baicalensis* and species of *Synedra* and *Gymnodinium*. It is only in extensive shallows of the region adjacent to the Selenga delta, the southern part of the Maloye More and other such similar sections that fluctuations in the crop of *Melosira* species, *M. baicalensis* included, are much less pronounced than in deep-water regions.

The biomass of aestival algae and the area occupied by them also vary widely with years, even in shallows. In deep-water regions, aestival algae appear in considerable amounts only in the immediate vicinity of shallows or in the belt of deep currents, and even there not every year by far.

The biomass of zooplankton and the qualitative composition of predominant species likewise differ from year to year, but the amplitude of annual fluctuations of the biomass is much less than is the case with phytoplankton (fig. 100).

As regards the zooplankton density we can distinguish in Baikal regions of 1. high aestival biomass reaching in August 400 kg per hectare, or an average of about 1 gram or more per m^3 in the 0—50 layer; 2. medium biomass of 200—400 kg/ha, or 0.4—0.8 g/m^3; and 3. low biomass of less than 200 kg/ha, or 0.4 g/m^3.

Regions of high biomass (400 and more kg/ha) even in zooplankton-rich years do not take up more than 1/5 to 1/10 of the area of Baikal, whereas regions with medium biomass occupy in such years up to half the total area of the lake.

In low-crop years regions with high and medium biomass of zooplankton are chiefly shallows, gulfs, bays, etc., but even there the biomass rarely exceeds 200—300 kg/ha.

Not much has been done yet in the study of plankton production in Baikal. Phytoplankton production in conjunction with the dynamics of biogenous elements was studied in the area of the Baikal biological station (Bolshiye Koty) in the course of several years by K. K. VOTINTSEV (1948b, 1952a, b, 1953a, b, c, 1955a, b, 1956, 1961) and A. V. SAMARINA (1960), who employed G. G. VINBERG's dark and light bottles procedure. S. I. KUZNETSOV (1955), who investigated photosynthetic production in Baikal with the help of radioactive C^{14}, found out that in August 1953 daily net carbon output (respiration consumption excluded) ranged in different regions of Baikal between 133 and 570 mg/m^2, with 1 to 1.5% falling to the share of haemosynthesising bacteria. In August the average daily carbon production in Baikal as a whole comprised 307 mg/m^2, and the monthly production 9—10 g/m^2 (18—20 grams of dry dry organic matter). S. I. KUZNETSOV notes that the values of net primary production found by him in Baikal are of the same order with the values observed by STIEMAN-NIELSEN in the Indian Ocean.

According to K. K. VOTINTSEV (1952b, c, 1953a, 1956, 1961) and A. V. SAMARINA (1960), in 1948 and 1955 (poor years in the development of *Melosira*) the total annual carbon production (expenditure on respiration included) in the area of Bolshiye Koty was 200 and 243 g/m^2 respectively.

It can be inferred from the materials available that *Melosira*-rich years in Baikal are characterised not only by an exceptionally high vernal biomass of algae, but also by a higher annual production, although the amplitude of annual fluctuations in production is less than in the case of biomass. This can be explained by the ability of the diatom *Cyclotella*, the peridinean *Gymnodinium* and also the tiny flagellates and protococcoid and blue-green algae appearing in summer to produce abundantly in a very low biomass. In deep-water regions the amplitude of annual fluctuations in production and biomass is evidently many times greater than in the shallows, since the vernal maximum of the development of algae there is, as a rule, the only one of the year.

In estimating the overall productive capacity of Baikal it is necessary to bear in mind that the area of photosynthesis studied by the method of dark and light bottles (Bolshiye Koty) is under a strong influence of currents from the plankton-rich Selenga shallow. The regions of extreme depths, which are free from such influences and which occupy four-fifths of the total area of the lake, are always very poor in aestival phytoplankton.

According to G. G. VINBERG (1956) the total annual phytoplankton production in the lakes of central regions of the European part of the Soviet Union comprises approximately 250—750 g/m^2 of dry organic matter. Assumably the annual fluctuations in phytoplankton production in Baikal may keep within the same limits. For Baikal as a whole the average amount will be of the order of 3—8 million tons of dry organic matter.

Comformably with the amount of algae production, a considerable number of bacteria has been recorded in the water mass of Baikal. For instance, according to S. I. KUZNETSOV, in October 1949 the biomass of bacteria in Baikal was established at 713,000 tons. If the exceptionally high propagation ability of bacteria is taken into account, it will become clear what a tremendous supply of organic matter is needed for the subsistence of these numbers of bacteria.

Zooplankton production in Baikal depends chiefly on *Epischura baicalensis*, and in shallows also on *Cyclops kolensis baicalensis* and Cladocera. The production of the amphipod *Macrohectopus* is evidently much smaller than that of the copepod plankton. There are no other data on the production of smaller forms of zooplankton, such as rotifers and infusorians. For a very rough estimate of total crustacean plankton production (with natural dying-off excluded) we double its maximum aestival biomass and obtain, for high-crop years, the annual production value of 1—1.4 g/m^3 for the 0—50 metres layer (500—700 kg per hectare), and 1,600,000 tons of raw mass for the whole of Baikal. In low-crop years it can range from one-half to one-third of the above. The annual production indicated above is probably the minimum value.

The fluctuations in the crop of zooplankton, along with fluctuations in the thermal regime, also influence its consumers, the omul and *Cottocomephorus*. In the years of high zooplankton production their fattiness, weight, size and fertility have been found to increase.

Let us try to make a brief analysis of the factors of water environment on which fluctuations in the crop of plankton may depend.

It is known that an important regulating part in the development of phytoplankton is played by the quantity and composition of the biogenous compounds contained in the body of water (GUSEVA, 1947, 1952). As has been established by the research conducted by K. VOTINTSEV (1955a, b, 1956, 1961), the amount of N, P, Si and Fe compounds in the photosynthesis zone of Baikal changes in the course of the year, as in all bodies of water, depending on the seasonal cycle of phytoplankton development and water mass circulation (fig. 24). But the annual differences in these fluctuations obviously do not correspond to the annual fluctuation in the crop of phytoplankton. High-crop years differ from low-crop ones merely in the increased amplitude of seasonal fluctuation of biogenous elements in the photosynthesis zone. For instance, in the years of high *Melosira* crop, its maximum development is accompanied by the almost complete disappearance of nitrates from the photosynthesis zone and a sharp decrease in the phosphates and silicon content. Hence, probably, the variation in the development of *Melosira* in such years, with vigorous explosive propagation followed by a temporary depression after which, thanks to a new supply of biogenous elements from the depths, the reproduction of the alga continues with renewed force.

It is commonly believed that fluctuations in the crop of lacustrine phytoplankton depend on the annual supply of biogenous compounds by the rivers. We have compared the periodicity of high crops of vernal phytoplankton in Baikal with the activity of its affluents, the amount of precipitation and the level of the waters of the lake, which depends on the water yield of the rivers, but found no direct connection between these factors and the biomass of phytoplankton. Indeed, this connection can hardly be expected in a lake the volume of which is 450 times bigger than the amount of water it receives in the course of a year. It can be supposed that in the deep-water regions of Baikal the allochthonous supply of biogenous compounds from rivers determines only their centennial and not annual budget. The amount of biogenous compounds which the algae have at their disposal in the photosynthesis zone in every given year depends mainly on their supply from the lower layers, which, in turn, is determined by the intensity of the vertical water circulation.

K. K. VOTINTSEV (1955a) pointed out, as one of the reasons for the fluctuations in the crop of diatoms in Baikal, the extraction by them, in high-crop years, of colossal amounts of silicon, which results in a sharp drop in the content of this element, essential for diatoms, in the water of Baikal and which inhibits the development of diatoms for the next two or three years.

This author considers that explosive reproduction of the diatoms without disturbing the total silicon budget in Baikal may occur throughout the lake once in about two or three years or take place annually, but embracing only some sections of the lake. But this explanation leaves the question open why in some years *Melosira* species extract so much silicon from the water that this prevents their normal development for several subsequent years. In addition, the silicon content in the photosynthesis zone returns to the normal the following year, but not the *Melosira* population.

Light is known to be essential for the development of algae, especially in the sub-ice period. In some years heavy snowfalls begin very early, covering the ice with a thick blanket over large areas. This blanket remains throughout the winter and becomes thicker with new snowfalls. But there are also winters when snow is scarce. These differences must influence light intensity in the water and photosynthetic activity under the ice. Still, no direct relationship is detected between the annual fluctuations in the vernal phytoplankton crop and the thickness of the ice and snow cover.

Among the factors of the water environment that may have a powerful influence on the development of plankton, the best-studied is the thermal regime of the Baikal waters.

The influence of the thermal factor can well be seen on the intensity of summer development and distribution of algae in deep-water regions. As the water gets warmer, aestival forms, which first appear in shallows, spread with currents to neighbouring more open areas and even, under favourable conditions, deep-water regions. But in these regions the intensity of their development is always sharply reduced and clearly depends on thermal conditions. In years with a cold summer only individual specimens of blue-green algae occur in deep-water regions, although they are not poorer in biogenous compounds than the shallows. But, given favourable conditions, they can form extensive and fairly stable fields of algal bloom in these regions as well.

For instance, in the second half of August 1956 a great development of the blue-green alga *Anabaena flos-aquae* was observed in the deep-water regions bounding the Selenga shallow. It was preceded by prolonged cloudless calm weather. After the water temperature in the superficial layer there reached 15—16° C, currents from the area of the shallow enabled this alga to spread wide in the neighbouring deep-water regions. But as soon as the water grew colder after a storm, it disappeared, its reign there lasting only 5 or 6 days. Analogous phenomena were also observed in other regions of Baikal.

The vernal development of diatoms over large depths far from the shallows is usually prolonged till July, when aestival forms already predominate in the shallows. In exceptionally cold years the areas taken up by aestival algal bloom sharply contract even in shallow regions.

Many authors have shown that a very important regulating influence is exerted on the density of algae by such a biological factor as their numerous consumers. The results of investigations in Baikal give reason to suppose that this factor may have not so much direct as indirect impact. Possibly, the very type of the dynamics of the algal density is developed as an adaptation protecting the species from being eaten up. The uneven development of *Melosira* species, i.e., alteration of high-crop years with years of their almost complete absence can serve as an example of this type of density dynamics.

Indeed, in high-crop years the biomass of *Melosira* species is hundreds of times greater than that of their consumers, and the latter can only reduce the numbers by an insignificant degree. As regards the duration of the depression period (2 or 3 years), it may depend on the thermal conditions of the ripening of the alga's resting stages, which occur in this period in deep layers or on the bottom, out of the consumers' reach. As distinct from *Melosira* species, such diatom species as *Cyclotella baicalensis* and *C. minuta* never form any dense accumulations in the photosynthesis zone. With progressive multiplication their cells go down without lingering in the upper layers, which leads to the dispersal of the population in the mass of water, although the energy of photosynthesis of these algae seems to be very great. It can be assumed that this type of density dynamics also can, to a certain extent, protect the species from being eaten up.

It is also a noteworthy fact that in *Melosira*-rich years the other vernal mass species of diatoms (*Cyclotella* and *Synedra*) develop very poorly and comprise only a minute fraction of the total algal biomass, which can probably be explained by the fact that in such years *Melosira* species inhibit the development of other vernal species. This proposition is advanced by Š. STANKOVIČ (1960) in explaining the fluctuations in the crop of algae in Lake Ohrid.

The thermal regime is undoubtedly one of the most important factors in the development of zooplankton. In years when the spring and summer warming of water is very slow, the total crop of zooplankton dwindles everywhere. In such years zooplankton is dominated by *Epischura*, but even the biomass of *Epischura* during maximum development is much lower than the perennial mean. This is caused by the slackening of the pace of the multiplication and growth of the crustaceans and also probably by the scarcity of food provided by the poorly developing phytoplankton. Another common Baikalian crustacean, *Cyclops kolensis* var. *baicalensis*, in cold years does not spread beyond gulfs, bays and pre-estuarine shallows, but in warm years its swarms are found outside the shallows, in the neighbouring deep-water regions. In *Melosira*-rich years considerable amounts of *C. k.* var. *baicalensis* appear in deep-water regions in spring but their maximum density is attained there in August–September, when the open waters of Baikal are the warmest. In all

250

probability, the swarmings of *Melosira*, which first appears in large numbers in shoals and then spreads with currents to deep-water regions, carry with them also *C. kolensis* var. *baicalensis* and this neritic complex of forms, on the one hand, feeds directly or indirectly (through bacteria) upon the diatom *Melosira* and, on the other, itself serves as food for the carnivorous species of *Cyclops*. But in *Melosira*-rich years as well the density of *C. k.* var. *baicalensis* in deep-water regions in the summer period depends on the temperature of the water; the warmer the summer, the more abundantly it is represented and the farther it spreads to deep-water regions.

For *Epischura*, as distinct from *C. kolensis* var. *baicalensis*, excessively high temperatures in summer are a negative factor. In such

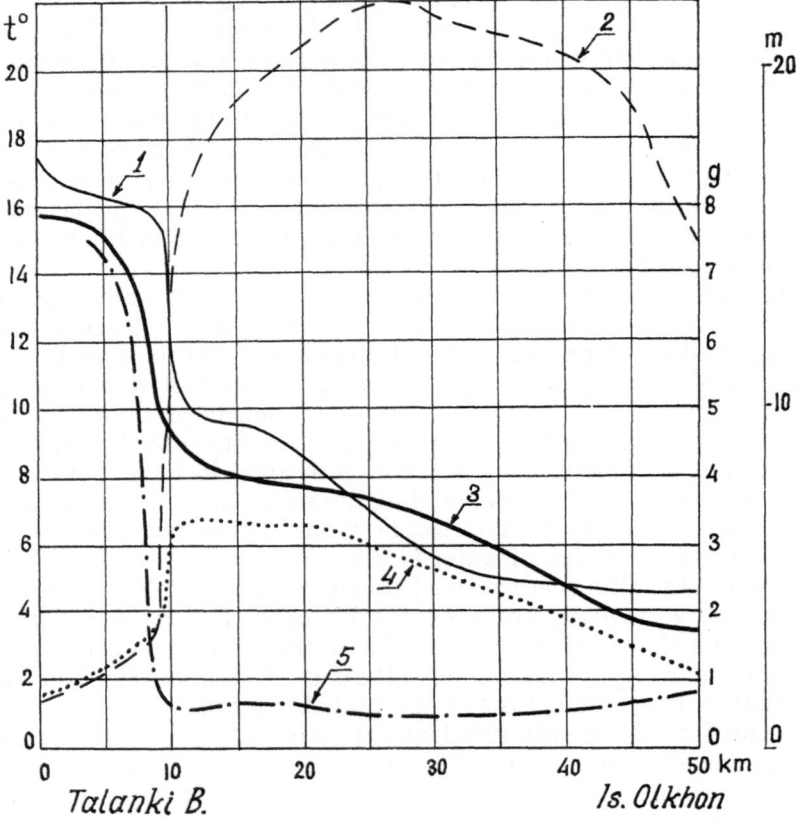

Fig. 102. Distribution of temperature and zooplankton in the Talanka Bay - Olkhon Island section on July 16, 1950. 1 - water temperature; 2 - Secchi disk transparency at daytime; 3 - total raw biomass of zooplankton in grams under m² in the 0—50 m and down to the bottom in shallow water; 4 - biomass of *Epischura baicalensis*; 5 - biomass of *Cyclops kolensis* var. *baicalensis*. After KOZHOV, 1957.

years they often become infested with saprolegnia and perish even in deep-water regions. The highest summer biomass of *Epischura* corresponds to 12—14° C in the upper 10-metre layer and 8 to 12° at a depth of 25 metres, while that of *Cyclops kolensis* var. *baicalensis* is observed in the zone where the temperature of about 14° embraces the entire 0—25 metres layer. Thus, in accordance with the distribution of temperatures, *C. k.* var. *baicalensis* usually predominate in the 4 to 5 km belt along the shores and *Epischura*, further on to the middle of Baikal (fig. 102).

It has been noticed also that in extreme-depth regions, where the water temperature is lower and phytoplankton relatively poor even in summer, the density of *Epischura* and the total mass of zooplankton are always lesser than in the neighbouring regions with a more favourable thermal regime.

It should be noted, however, that fluctuations in *Epischura* density depend not only on the thermal regime of the water. For instance, in *Melosira*-rich years, during the initial stage of their development, the density of *Epischura* in the photosynthesis zone grows parallel with the increase in the density of *Melosira* cells. But when the latter densely populate the upper layer, the bulk of *Epischura* is found underneath the photosynthesis zone and the number of juveniles drops sharply, probably owing to consumption by *Cyclops*. After the breaking up of the ice, adult *Epischura* specimens descend following the subsidence of the swarmings of *Melosira* and other vernal diatoms, intensely feeding upon them, which evidently favours their maturing and the hatching of a numerous generation of juveniles in autumn and winter. This is probably why *Epischura* is especially abundant the year following a rich crop of *Melosira*.

All these suppositions regarding the problem of the fluctuations in plankton crops are of a preliminary nature. Only a systematic comprehensive study in the course of many successive years will help to elucidate the causes of sharp plankton crop fluctuations. The solution of this problem for Baikal may shed light on many important aspects of the biological productivity of bodies of water.

3. Trophic Relationships, Interspecific Contacts and Vertical Migrations of Mass Pelagic Dwellers.

Need for food and food relationships are known to play an exceptionally important part in the biology of species. The main links in the food relationships in the open waters of Baikal are algae (producers), the crustacean *Epischura baicalensis* and the young of the amphipod *Macrohectopus* (the chief consumers of phytoplankton), then, among the fish, the omul and *Cottocomephorus* and *Comephorus* species (the chief consumers of zooplankton), and, finally, the seal,

252

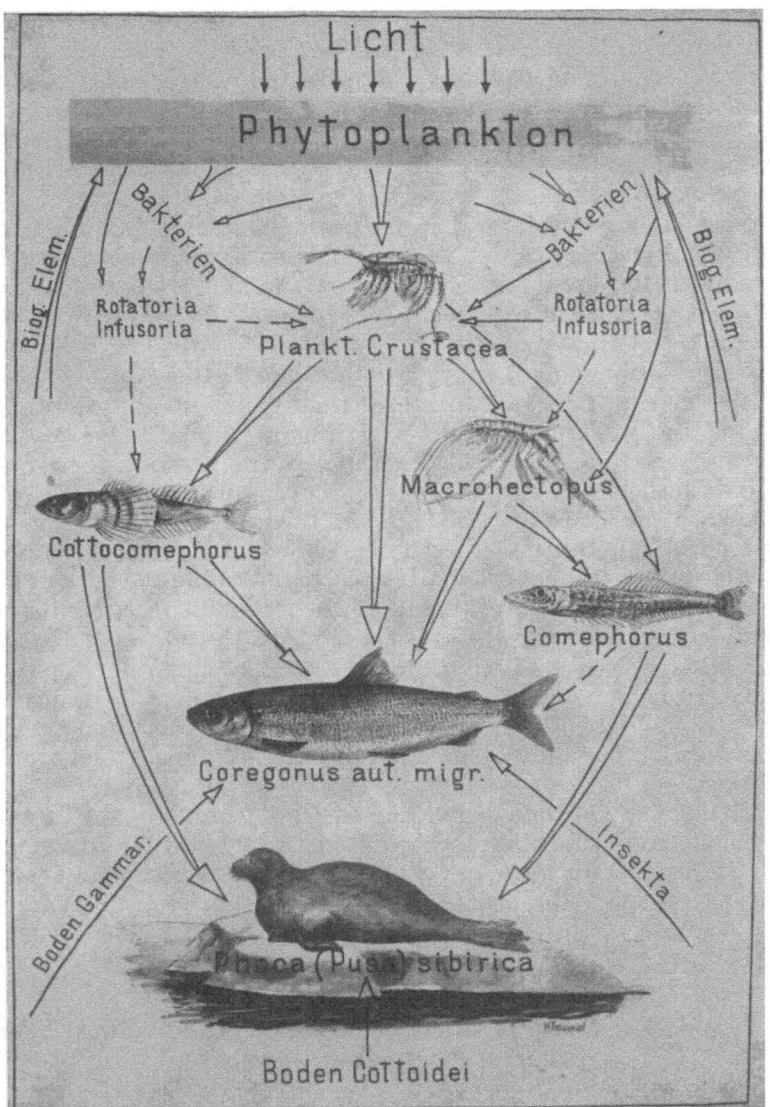

Fig. 103. Food relationships in the mass of water of open Baikal. Double lines denote main likes, single lines, secondary ones. After KOZHOV, 1959.

which feeds upon pelagic fish. All other links are of relatively secondary importance, with the exception of bacteria, which, on the one hand, participate in the processes of the disintegration of dying organisms and, on the other, serve as food for the masses of small filter-feeders from the zooplankton—in summertime, at least (KOZHOV, 1954; POTAKUYEV, 1954, 1956).

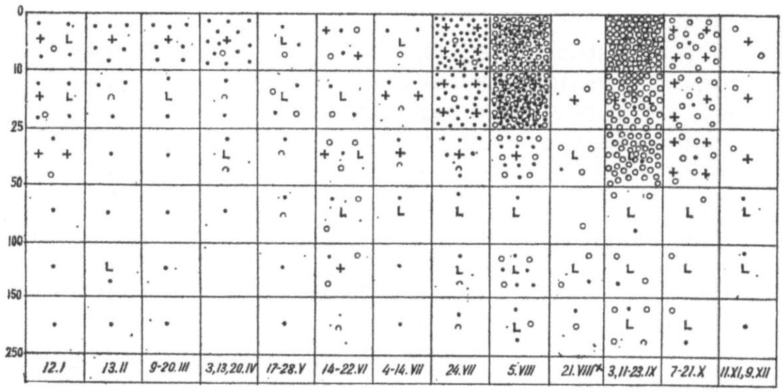

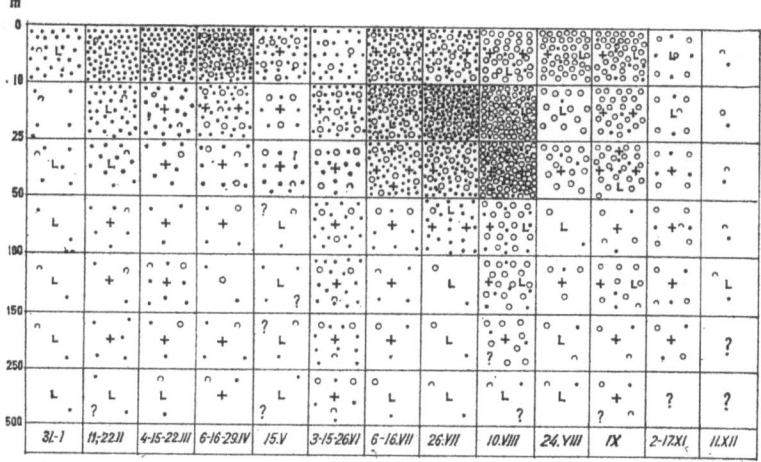

Fig. 104. Vertical distribution of *Cyclops kolensis* var. *baicalensis* in the mass of water of open Baikal at daytime in 1950; : : : - nauplii; ° °° - copepodits; ‡ + - adults. Each sign corresponds to 1,000 specimens per m³, on the average, in the layer of water studied. ∟ ∟ ⌒ - copepodits and adults in amounts less than 500 per m³. Original.

---

* There was a strong north-western storm on August 19 and 20.

Fig. 105. Vertical distribution of *Epischura baicalensis* in the mass of water of open Baikal at daytime in 1954. Symbols the same as in Fig. 104. Original.

254

Fig. 103 gives a simplified chart of food relationships among the most important components of the population of the mass of Baikal's open waters. Without a study of contacts between them stemming from food relationships it is impossible to find a satisfactory explanation of many essential peculiarities of their behaviour, including vertical seasonal and diurnal migrations. A one-sided interpretation, based only on taxes, direct effect of light, temperature, etc., produces a rather primitive concept of a complex biological phenomenon.

Many years' study of seasonal and diurnal migrations of mass planktonic species in Baikal has led us to the conclusion that the intensity, amplitude, period and time of migrations are developed by the migrating species on the basis of food relationships. Direct contacts are realised during these migrations when the consumer finds food, the predator meets its prey or, conversely, the prey finds a shelter protecting it from destruction. Among abiotic conditions, light and temperature are of particular importance. But since the manifestation of these factors is subject to a definite seasonal and diurnal rhythm, periodicity and rhythm also develop

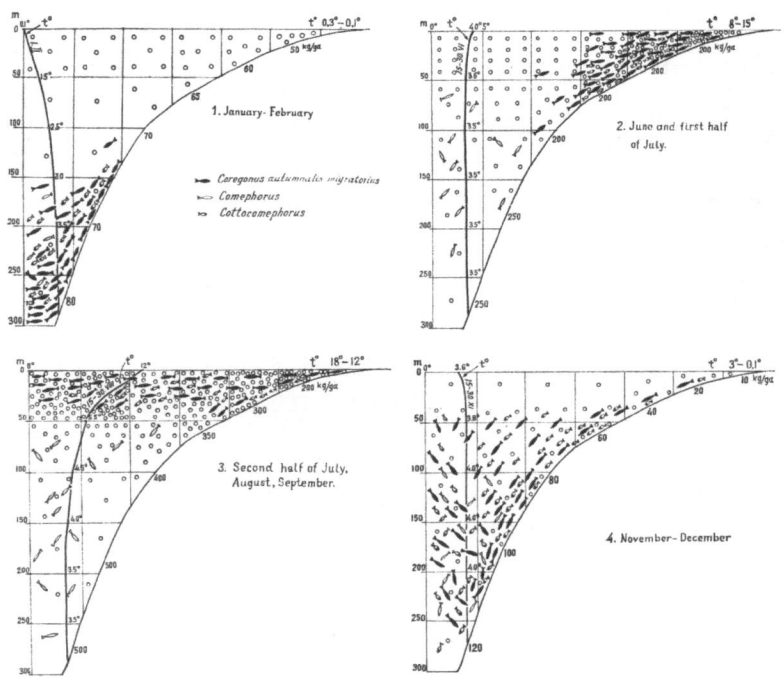

Fig. 106. Seasonal changes in the vertical distribution of water temperature, zooplankton and pelagic fish in lake Baikal. One circle denotes 10 kg/ha of crustacean plankton in the 0—250 m layer. After Kozhov, 1958.

in the vertical migrations of hydrobionts and gradually turn into inherited, physiologically essential reactions in the realisation of which periodical changes in environmental factors (light, temperature, etc.) play, in the main, only a part (KOZHOV, 1947, 1959; NIKOLAYEV, 1950; MANTEIFEL, 1959, 1961).

Stated below are the results of observations of the vertical migrations of the most important species of Baikalian plankton in different periods of the year and under different thermal, light and biotic conditions.

January—February (First Half of the Sub-ice Period)

In January the ice-cover forms all over the lake. Snow on the ice, as a rule, is very scarce, being blown off by winds and accumulating in snowdrifts only among hummocks. As a result, extensive openings are formed, allowing free passage of sunlight through the transparent ice. Table XXXI gives data on the most important factors of the water environment in the area of Bolshiye Koty (South Baikal) in the sub-ice period.

As already noted, in January-February begins the vegetation reproduction of vernal algae, chiefly peridineans and diatoms. They concentrate primarily in the 0—5 = 0—10 metres layer, attaining maximum of density under the very ice cover. In zooplankton, *Epischura* nauplii of the winter generation predominate, concentrating, for the most part, in the photosynthesis zone (fig. 105). Baikalian forms of rotifers and infusorians are also found there.

In January—February the consumers of the crustacean plankton are not in evidence in the upper layers yet. In this period the omul and *Cottocomephorus* winter on the slopes of shallows at depths of about 150—300 metres. The amphipod *Macrohectopus* lives in deep layers down to near-bottom ones.

The bulk (60 to 80%)* of the nauplii of *Epischura* keeps constantly in the 0—10 metres layer and in the 0—5 metres layer if the snow cover on the ice is considerable. In the second half of the day crustaceans form thick swarms under the ice (MAZEPOVA, 1952b); at night their concentration there diminishes. Not more than 15 to 20% of the total number of crustaceans take part in these periodic accumulations under the ice (in the 0—2 metres layer). The average amplitude of their diurnal migrations does not exceed 10 metres.

The copepodit stages of *Epischura* keep throughout the 24 hours slightly below the nauplii. Their partial rise to the 0—2 metres layer is observed after midday, at about 3 p.m., i.e., still at daytime, 2 hours before sunset. Increased concentrations remain there till 8—9 p.m., after which the crustaceans disperse slightly. At about

---

* Here and elsewhere below the number of migrating *Epischura* and *Cyclops* is given in % of their total numerical density in the 0—50 metres layer.

Table XXXI.

The most important factors of the water environment in the sub-ice period as observed 1.5 km from the shore over a depth of about 1,000 m in the area of Bolshiye Koty (1955-1956).

Factors	Middle of January	Middle of February	Middle of March	First 10 days of April	Last 10 days of April
Ice thickness in cm Snow thickness in cm	20—40 Very insignificant	50—80 Insignificant	80—110 5—10	80—60 Snow fused with ice	60—40 Snow fused with ice
Sunrise at observation point*) Sunset	About 9:00 About 17:00	About 8:00 About 18:00	About 7:00 18:30	6:30 19:10	About 6:00 19:30
Duration of the day	8 h	About 10 h	11 h 30 m	12 h 40 m	13 h 30 m
Duration of light period (light intensity more than 1 lux)	9 h 40 m	About 11 h	13 h 40 m	14 h 20 m	15 h
Maximum light intensity on ice at noon in 1,000 luxes	12	17	35—44	55	58
Secchi disc transparency at daytime in metres	20—30	15—25	7—8	8—14	10—12
Water t° (Centigrade) 0 m	0.1	0.1	0.3	0.8—1.0	1.4
5 m	0.3	0.3	0.5—0.8	1.0—1.4	1.4
20 m	0.4	0.4	0.8	1.0	1.2
50 m	1.3	1.0	1.3	1.2—1.3	1.2
250 m	3.4	3.6	3.6	3.6	3.3
Layer of maximum concentration of algae	0—2=0—5	0—2=0—5	0—2=0—5	0—2=0—5	0—2=0—5

*) Clouds or fog sometimes prevented the timing of sunrise and sunset accurate within minutes.

8 a.m. (before sunrise), a new short-lived concentration in observed in the upper 2 m layer. About 20—25% of the total number of crustaceans take part in vertical migrations (in both directions), the average amplitude being 10 to 15 metres.

In January—February the rotifers are scattered in the 0—30 m layer, and the extensiveness of their diurnal migrations is very insignificant.

The amphipod *Macrohectopus* also accomplishes vertical diurnal migrations in January-February, but they can hardly be ascertained by means of ordinary quantitative sampling nets. In this period it prefers to live in the layers between 150 and 250 metres.

Only a part of the crustaceans rise to the sub-ice layer and remain there till dawn.

No clearly defined vertical diurnal movements in January—February have been observed among the fish species *Cottocomephorus* and *Comephorus*.

March—April (Second Half of the Sub-Ice Period)

In March the temperature of water in the 0—50 m layer changes little compared with the previous period. But beginning with the middle of April the upper layer of up to 5—10 m, with the exception of the sub-ice surface layer, is warmed to 1—1.4° and a relative homothermy is established at this level in the upper 50 m layer.

The ice cover reaches 80—120 cm thickness in March, beginning to diminish since the middle of April. The thickness of the snow cover increases to 5—10 cm. In April the snow begins to thaw and fuses with the ice. This reduces light penetration through the ice cover considerably, although the daylight period grows longer and light intensity over the ice increases.

In March—April the vernal algae, especially peridineans and diatoms, continue to develop intensely and their density in the photosynthesis zone becomes greater (fig. 95—98). But the greatest concentration of algae is observed only in the uppermost 2-metre layer.

Among *Epischura*, nauplii of older ages and various stages of copepodits of the winter generation predominate in March—April. In March the maximum of density of the crustaceans is observed around the clock in the 10—25 = 10—50 m layer, i.e., somewhat deeper than in February, which coincides with the extension of the mass habitation zone of algae (fig. 105).

Increased concentrations of *Epischura* in the 0—2 m layer are observed also after midday. The crustaceans rise to the upper layer, right up to the ice, chiefly from the 2—10 m layer, whereas their concentration in the 15—25 m layer remains almost unchanged throughout the 24 hours. Not more than 10 to 15% of the total number of nauplii and a still lesser percentage of copepodits take part in migrations. About half the total number of the latter keep approximately in the 50 m layer around the clock, and we can speak of a weak tendency only towards their migration to the sub-ice layer in the lighter part of the day. Most of the adult females remain at about 150—200 metres and deeper. No distinct migrations have been detected among them.

The concentration of *Cyclops kolensis* var. *baicalensis* under the ice changes according to light intensity (fig. 104). If there is a snow cover on the ice they accumulate under the ice cover at day-time and scatter in deeper layers at night. Under patches of transparent ice free from snow the number of *C. k.* var. *baicalensis* drops sharply at daytime and grows towards midnight. In conditions of bright light the number of crustaceans under such patches of ice decreases (MAZEPOVA, 1957).

An increase in the vertical migrations of *Epischura* is observed at the very end of the sub-ice period, in April, when a considerable part of the crustaceans of the winter generation mature, which evidently results in greater demand for food. The bulk of the crustaceans of all the age groups of *Epischura* remain day and night in the 15—50 m layer. After midday up to 20% of the total number of nauplii rise to the 0—2 = 0—5 m layer. Increased concentrations of copepodits in the 0—2 m layer are observed not only during the day, but also in the evening and morning.

The bulk of *Macrohectopus* remain in the 150—250 m layers around the clock. In March—April their vertical migrations to the upper layers seem to be somewhat more intense than in February. In the 0—50 m layers their young appear in the dark period, forming increased swarms in the 0—10 m layer.

The rotifers remain, as before, in a more or less dispersed state in the layer between 0 and 30 metres, with a certain increase in concentration in the upper 5 m layer. In the daylight period their concentration grows somewhat in the 0—2 m layer.

Among the pelagic fish, adult *Cottocomephorus* rise from the depths and at the end of the sub-ice period they move towards the shores for spawning. The omul shifts slowly from its wintering places to the shores of extensive shallows, but this movement proceeds primarily in the near-bottom layers. Its feeding during the approach to the shores is very slight.

May—June (Transitional Period)

The breaking up of the ice brings about abrupt changes in the life conditions of the water mass dwellers. The most important factors in this period are: intense warming through of the upper layer of water, a sharp increase in convection and wind-induced mixing, and greater illumination of the upper layers. In May the temperature

of water in the 0—50 m layer ranges between about 2 and 3°, and
on June 20—25 a homothermy at a level of 3.5—3.8° sets in in
the southern part of Baikal. Thanks to the high transparency of the
Baikal waters (visibility limit in June: 30 metres and more), sunlight
penetrates deep into the mass of water, and the duration of the day
and light intensity increase.

This period constitutes the sharp dividing line between spring
and summer in the life of plankton. After the temperature of water
in the superficial layer exceeds 4—5°, vernal forms of algae go down
to deep layers and disperse in the mass of water. This descent proceeds very slowly, and even in July thick swarms of dying-off
vernal forms and especially *Melosira* still occur at a depth of 200—
300 metres (fig. 107).

Zooplankton (*Epischura*, Rotatoria, Infusoria) also disperses in
May—June. More than half the total number of individuals of the
most common species occur now below the 50 m isobath. In all
probability, tiny forms of zooplankton scatter in the mass of water
passively, due to intensive convection and wind-induced water
mixing. As regards the adult specimens of *Epischura* and older
copepodit stages, it appears that they actively descend following
the subsiding swarmings of algae and vigorously feed upon them
and the bacteria developing there due to the accumulation of algae.
This part of the zooplankton, of course, cannot participate in diurnal vertical migrations altogether.

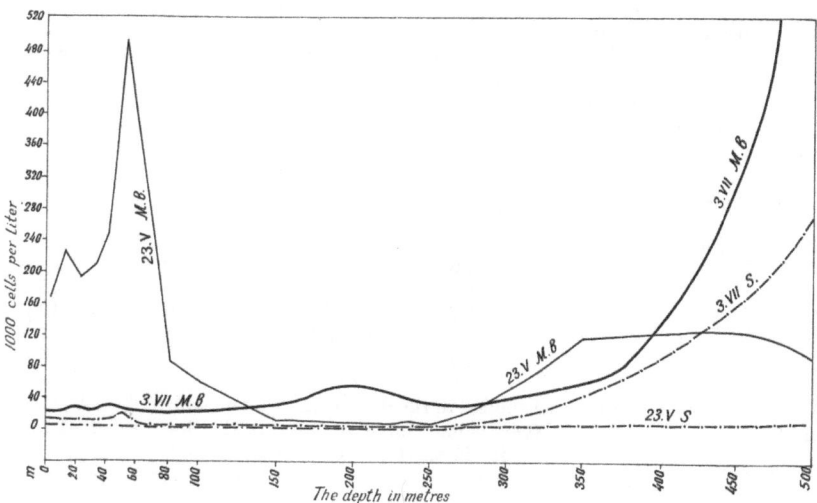

Fig. 107. Vertical distribution of mass algae species in May and July of 1950,
when *Melosira* was in abundance, from the surface to a depth of 500 m in the Bolshiye
Koty area, in thousands of cells per litre. M.b. *Melosira baicalensis*. S - species of the
genus *Synedra*. Original.

In this period most of the adult females of *Epischura* prefer deep layers, where they feed intensely and hatch the young of the new vernal-aestival generation.

The biggest concentration of the copepodit stages of *Epischura* is observed day and night in the 10—50 m layers. Towards 9 p.m. their density increases somewhat in the 0—2 m layer but decreases

Table XXXII.

The Most Important Factors of the Water Environment in the Area of Bolshiye Koty at 1.5 km from the Shore in the Spring-Summer Period of 1956.

Factors	May 23—26	June 25—26	July Last 10 days	August 15—25
Sunrise	5:30	about 5:00	about 5:20	about 6:30
Sunset	20:20	20:54	about 20:00	about 19:00
Duration of the day	14 h 50 m	about 16 h	about 14 h 40 m	about 12 h 30 m
Duration of light period (light intensity more than 1 lux)	17 h 20 m	18 h 30 m	about 17 h 00 m	15 h 00 m
Maximum light intensity over water in thousands of luxes at noon	—	103	90	80
Secchi disc transparency at daytime in m	15—20	15—25	9	8—6
Secchi disc transparency at midnight (moonless night	—	3—4	—	—
At full moon	—	8	—	—
Water t° (Centigrade)				
0 m	1.6	3.7	6.9	14—16
5 m	1.6	3.6—3.5	5.0	13.0
10 m	1.6	3.5	4.6	10.6—13.0
20 m	1.6	3.5	4.5	11.5
50 m	1.7	3.5	4.2	5.4
Maximum concentration of algae	Dispersal in the 0—200 m layer, some increase in density in the 0—2, 0—5 m layer		Dispersal in the 0—50 m layer with some increase in density in the 0—2 m layer	0—2 m 0—10 m

in the 2—5 m layer. Not more than 25% of the total number of the crustaceans occurring in the 0—50 m layer (12% of the number in the 0—250 m layer) seem to take part in the upward movement. After a certain dispersal at midnight a very insignificant, barely detectable increase in concentration is observed in the 0—2 m layer in the morning hours.

In June, in conditions of complete spring homothermy (according to the data for June 25—26, 1956), the share of nauplii migrating upward is quite insignificant, not exceeding 3—5% of their total number in the 0—250 m layer. During the course of the 24 hours the crustaceans are more or less evenly distributed in the mass of water, but mostly in the 5—25 m layer, where up to 40—50% of their total number is observed. Only a slight increase in concentration in the 0—2 m layer takes place in the evening hours (6—8 p.m.), at midnight and early in the morning (5—6 a.m.).

The diurnal vertical migrations of rotifers and the amphipod *Macrohectopus* in May—June are also insignificant.

June is the period of the mass approach of the omul *(Coregonus autumnalis migratorius)* to the shores of extensive shallows (fig. 106). Summer begins earlier there, and zooplankton develops faster. The summer type of vertical migrations, which will be dealt with below, should also start earlier there. Thick shoals of the omul move along the shores of shallows in search of sections with plankton-rich water heated to 8—12 = 8—14°. But in open deep-water regions, where water is cold even near the shores, the omul does not appear in June.

In May—June adult *Cottocomephorus* continue to spawn on rocky banks along open shores.

Summer (Second Half of July, August, September)

The most characteristic factors for the summer are the great light intensity, relatively high temperatures of the upper layer of water and clearly defined direct thermal stratification, discontinuation of convection mixing, and a sharp intensification of horizontal currents with the emergence of hypolimnetic waters to the surface, especially after north-westerly gales and storms.

At the end of July the temperature of the upper layers of water (0—5 m) in open regions reaches 8—10°. In this period an abrupt temperature drop often occurs at a depth of about 5—8 m (for instance, from 10° to 6—5°), while at a depth of 50 m the temperature remains at a level of about 4—4.5°. In August the temperature of the surface layer even in deep-water regions reaches 12—13° and in some years 15—16°, and at a depth of 10 m, 8—10°. In September the upper layers begin to cool, whereas the temperature of deep layers continues to grow. Towards the end of the summer light intensity in the upper layers gradually decreases along with the shortening of the light period and the reduction of the visibility limit from 20—30 m (in June) to 6—10 m (August—September).

In July the dying off of vernal algae in deep-water regions and their slow descent continue. Their place is taken by aestival forms, blue-green algae among them. They concentrate chiefly in the heated upper layers (0—2, 0—5 m). The density of living algae decreases with depth, but it is still fairly high in the 10—20 m layers. Considerable numbers of descending diatoms are also found in deeper layers, down to 150 m and more (fig. 107).

In summer a considerable density is attained in the open waters by bacteria, whose greatest concentration is observed in the 10—25 m. In August the layer of their maximum density extends from 0 to 50 m (fig. 99).

Among *Epischura*, July is the month of the highest numerical density of the nauplii of the summer generation. In summer these young grow rapidly and therefore need large amounts of food.

In July the nauplii and copepodits attain the maximum of density (up to 50—60% of the total number) at 10—25 m depths. At about 5 p.m. they begin to rise from there, reaching the greatest concentration in the 0—2 m layer in the evening (7—9 p.m.) and early in the morning (about 4 a.m.). In the darkest period of the night the concentration of the crustaceans in the uppermost layer

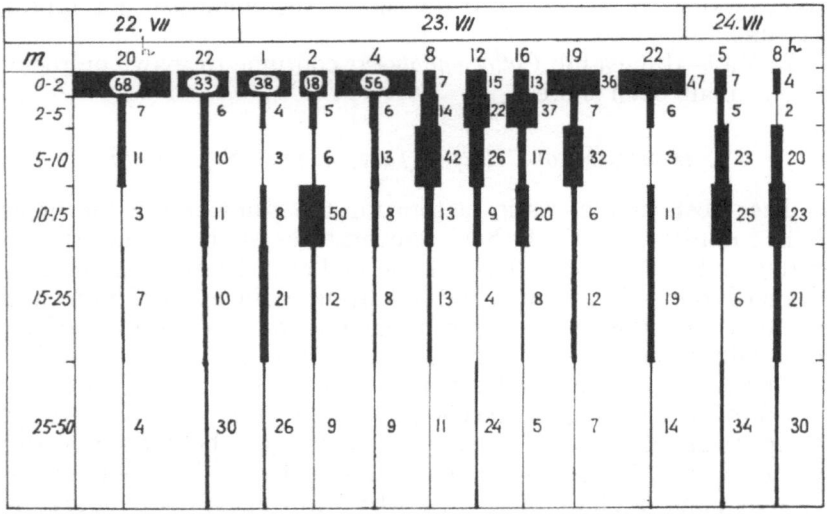

Fig. 108. Vertical distribution of copepodit stages of *Epischura baicalensis* in the course of July 22—24, 1955, in the open waters of Baikal in per cent of their total number in the 0—50 m layer. The areas of shaded sections are proportional to this value. After Kozhov, 1959.

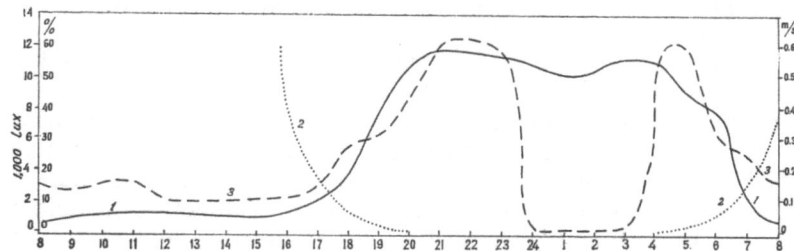

Fig. 109. Diurnal changes of: 1 - number copepodit stages of *Epischura* in the 0—5 m layer in per cent of their total number; 2 - illumination in 1000 luxes above the water surface; 3 - speed of the movement of the young omul in m/sec. August 1960. Original.

diminishes. Up to 20—25% of the nauplii and up to 40—70% of the copepodits take part in migrations. Thick concentrations of crustaceans are maintained in the 0—2 m layer till 4—5 a.m., i.e., during the course of about 7—8 hours.

In August, as in July, the zone of the mass habitation of nauplii and copepodits lies at depths of 0—15 = 0—25 m. The rise of nauplii to the upper layers begins at sunset, at about 7 p.m., and of copepodits at about 5—6 p.m., i.e., an hour or an hour and a half before sunset. Towards 9—10 p.m., two or three hours after sunset, the concentration of crustaceans in the 0—2 m layer reaches its maximum and is expressed in hundreds of thousands per m³ (up to 80% of the total number of crustaceans). Increased densities are sustained in the 0—2 m layer throughout the dark period (fig. 108, 109).

On overcast and foggy days the migrations of crustaceans are much less pronounced than on clear days.

At dawn the copepodits and nauplii leave the upper layer and towards 8—9 a.m. disperse in the mass of water. At daytime not more than 2—5% of the total number of crustaceans are found in the upper layer of 0—2 metres.

In September the diurnal migrations of crustaceans are very much like those observed in July and August.

In the summer of a *Cyclops*-rich year, *Cyclops kolensis* var. *baicalensis* keeps permanently to higher layers as compared with *Epischura*. Similar to the latter, *C. k.* var. *baicalensis* has intense periodical diurnal migrations, moving up in the evening and down in the morning. Also observed in the darkest hours is a certain decrease in concentration in the upper 2-metre layer, with the restoration of density before dawn.

As regards the rotifers, a more distinct trend towards diurnal migrations in summer is observed among *Filinia longiseta* and *Notholca longispina*. In the dark hours their densest swarms are observed in the 0—2 = 0—5 m layer and at daytime, at a depth of 5—15 = 5—25 metres.

The summer vertical diurnal migrations of *Macrohectopus* take place fairly intensely.

In the first half of the night *Macrohectopus* swarms in the surface layers are so thick that sometimes they can be observed with the naked eye. Evidently not less than 50—60% of the total number of *Macrohectopus* in the layer between 0 and 500 metres participate in upward movements. The rest remain in deep layers.

Summer is the period of mass consumption of crustacean plankton by plankton-eating fish, the omul and *Cottocomephorus*. *Epischura* is preyed upon also by *Macrohectopus* and the young of *Comephorus*, while *Macrohectopus* and the young of *Cottocomephorus* are themselves eaten by the omul and adult Cottoidei.

From the middle of July until October the horizontal feeding migrations of the omul and pelagic Cottoidei take place exclusively in the upper warm and life-rich layers of water (fig. 106). The routes of the omul's migrations coincide with the increased concentrations of crustacean plankton. In July the mass hatching of *Cottocomephorus* takes place. Vast schools of fry move first along the shores and then in more open regions, consuming tremendous amounts of zooplankton. In summer the sections of Baikal where pelagic fish feed for a long time become practically clear of crustacean plankton. Thus the highest intensity of the vertical migrations of the species comprising the basis of nutrition for the pelagic fish coincides with the period of intense feeding migrations and active feeding of the latter.

Summing up, the following propositions and conclusions can be made.

In the sub-ice period the development of phytoplankton and the increase in the number of its main consumers (*Epischura*, some rotifers and infusorians) proceed simultaneously, and contact between them can be accomplished throughout the photosynthesis zone, which expands gradually as the algae develop. But the greatest phytoplankton density is achieved in the uppermost sub-ice layer, and, in consequence, the concentrations of phytoplankton-feeders, the copepods and rotifers, are observed there, but only for a short time, chiefly after midday, when twilight sets in under the ice. But in view of a very low temperature, which remains almost the same in the whole of the 0—50 m layer during the sub-ice period, periodic upward movements of crustaceans are inconspicuous, while their consumers winter during this period in the near-bottom layers at a depth of 150—250 metres and feed much less actively.

In the transitory period of May—June, after the breaking up of the ice, general dispersal of phyto- and zooplankton takes place in the layers lying at 100—150 m from the surface and deeper, making possible contact between algae and their consumers throughout this depth. In this period most of the copepods, rotifers and infusorians scatter in the mass of water and cannot accomplish diurnal movements to the upper layers. The intensity of vertical diurnal migrations of crustaceans and other planktonic components is found to be extremely low (fig. 110). In this period the main consumers of crustaceans keep to the near-bottom layers of water near shallows and feed inactively. The intensity of migrations grows sharply in summer, when up to 70—80% of the total number of planktonic organisms take part in them. This increase proceeds simultaneously with the increase in water temperature and its stratification, whereas light intensity in the water and its transparency diminish noticeably as compared with May—June (fig. 110).

At this time the greatest density of algae is observed in the

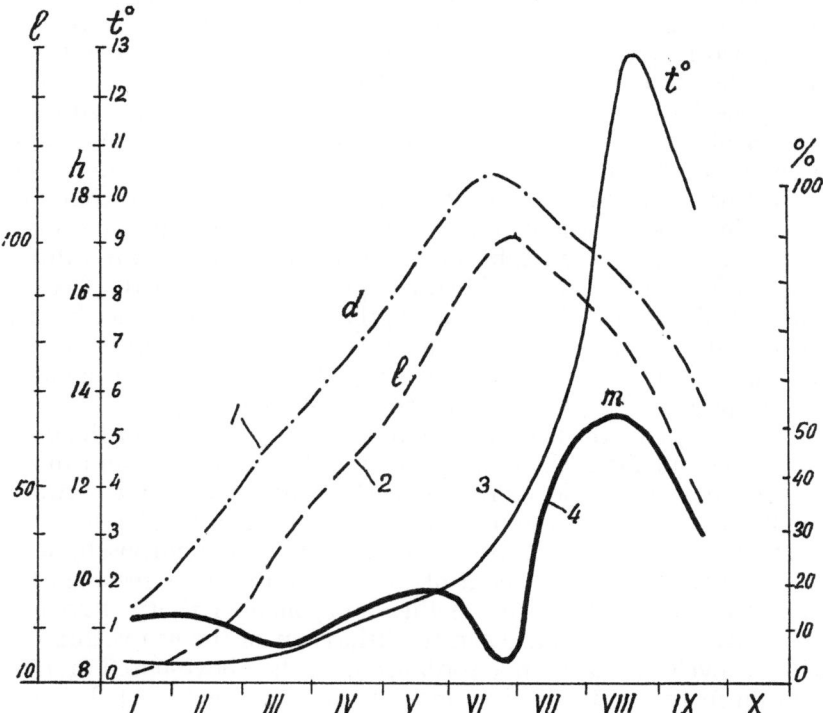

Fig. 110. Seasonal changes in water temperature, illumination and the intensity of diurnal vertical migrations of copepodit stages of *Epischura baicalensis* in open Baikal, as observed in 1955 and 1956. 1 - duration of light period in hours (light intensity more than 1 lux above the water at the point of observation); 2 - maximum light intensity in 1,000 luxes above the water at midday; 3 - average water temperature in the 0—20 m layer; 4 - intensity of migrations of *Epischura baicalensis* (migrating species in per cent of their total number in the 0—50 m layer) After KOZHOV, 1959.

0—2 = 0—5 m layer. This is where contact between phyto- and zooplankton is chiefly realised in the evening before sunset.

The rhythm of the diurnal vertical migrations of zooplankton described above decisively influences the diurnal rhythm of the behaviour of plankton-eating fish, which, too, have rhythmic changes in feeding and movement activity in the course of the 24 hours, particularly clearly defined in the summer feeding period (July—September). We have observed this in both natural and artificial conditions, (in round tanks of cement in which we studied the behaviour of young omuls in all seasons of the year) (fig. 109).

In conditions of low temperature (1 to 3° C in winter) the activity of the fish in the tanks is as low as in nature. They eat little even when offered abundant food in the form of live zooplankton. When the temperature rises (in the summer period), their need for food sharply increases and they become more active. It has been established that the fish search for and catch food with the aid of eyesight and perform purposeful movements. Older omuls filter plankton-rich water through their gills, but even in this case eyesight remains indispensable in search of food. These observations agree with the results of the study of other pelagic fish (BABURINA, 1955, and others). Blinded fish cannot catch food in sufficient amounts and sooner or later starve to death. The highest feeding activity of omuls in tanks in summer is observed, as in natural conditions, from 6 till 9 p.m., when light intensity on the surface of water drops below 1,500—1,000 luxes. When complete dark sets in the feeding activity of the fish decreases abruptly. It increases somewhat at sunset but remains low throughout the day. The rhythm of the feeding activity of the fish coincides with the rhythm of the changes in the speed and nature of their movements (fig. 109).

In the daytime the fish take shelter in shaded sections of the tank and move but little. At twilight they gather in a shoal and move rapidly (0.6 m/sec and more) along the walls of the tank, always counter-clockwise. In the darkest period the shoal breaks up, with the movements of the fish becoming slow and losing purposefulness. At dawn the rapid movement of the fish in a shoal is resumed for some time, but before sunrise their movement activity decreases markedly. All these changes in the behaviour of the fish within the 24-hour cycle is clearly in accordance with the above-noted diurnal rhythm in the vertical migrations of the crustaceans. At daytime the crustaceans are scattered in the mass of water, and therefore the search for them does not justify the expenditure of energy. In the darkest period the crustaceans are not visible, although they amass thickly in the upper layer. The conditions for hunting are most favourable in the period of poor (twilight) illumination in the upper layer, where the crustaceans accumulate densely in the evening, becoming relatively visible and accessible.

On the basis of the above observations we consider it possible to give the following explanation of the origin of the phenomena of diurnal changes in the behaviour of animals inter-connected by the food chain (KOZHOV, 1947, 1955, 1960). The main feeding ground of phytoplankton-eating crustaceans lies in the photosynthesis zone, with the highest concentration of algae taking place, as a rule, in the uppermost layer (0—2 = 0—5 m in Baikal). But this layer is brightly lit, and if there are carnivores in it which use eyesight in search for food, at daytime the crustaceans would be in danger of complete destruction there. This danger is averted by the development in them of the habit of periodically changing their zone of habitation. At daytime, when they are well seen by the enemies, they descend and disperse without forming thick swarms. In the dark period, when they are less visible, they stay on the feeding ground, i.e., in the upper layer, but remain there not longer than 6 to 8 hours (in the summer), which also prevents them from being eaten out. Repeating themselves in the course of countless generations, these rhythms in the behaviour of the crustaceans have acquired the force of instinct. These rhythms are controlled by abiotic factors (light, temperature, etc.).

However, the tendency of the plankton-eating crustaceans to remain in conditions of twilight or complete darkness throughout the 24 hours is not absolute. As we have seen, in some periods of the year the majority of the crustaceans remain in the "survival zone" day and night and do not migrate. Even in summer, in the period of intense migrations, a certain number of them remain in the depths. Evidently an important impulse to migrate is provided by the physiological condition of the animal, such as the degree of fattiness, changes in the demand for food in the pre-spawning period, etc. Even in summer certain numbers of crustaceans are frequently observed staying in the brightly illumined superficial layer at daytime. Referring to such phenomena among marine plankton, B. P. MANTEIFEL (1959, p. 108) writes that in some cases organisms are compelled to stay in the upper illumined layers of water, when they themselves or their offspring need constant ample nutrition. For *Calanus finmarchicus* it is the period of multiplication.

The keenness of sight of plankton-feeders enabling them to find plankton accumulations in a very dim light does not disprove the above-stated propositions regarding the adaptive value of the vertical migrations of the crustaceans consumed by them. It is to be assumed that this keenness of sight itself is developed in plankton-eating fish in view of the need to see their prey in dim light. What is taking place here is a two-way process of accommodation to the biotic conditions of life. The prey develops means of escaping destruction, while the consumer perfects the means of finding food.

VI

THE HISTORY OF LAKE BAIKAL AND ITS FAUNA

1. The Geological History of Baikal.

All geologists and palaeontologists agree that the Baikal and trans-Baikal areas entered the continental period of development as early as the Palaeozoic. It was only in the Jurassic period of the Mesozoic (Lower and Middle Jura) that the sea extended into the trans-Baikal area from the region of sea basins in the eastern part of Asia. In the Jurassic period, a long sea gulf stretched where the modern valleys of the Shilka and Onon (the Amur drainage) lie today, reaching 115° E. Long., 450 kilometres to the east of present-day Baikal (SOKOLOV, 1936; PRESNYAKOV, 1940).

It is assumed that in the Jurassic the region of the Eastern Sayan Range and the southern part of Baikal was occupied by a mountainous country from which rivers flowed into an extensive submontane depression known in the literature as the Irkut coal basin. This basin consisted of a ramified system of shallow lakes and marshes with a varying regime favourable for the accumulation of vegetable, mostly ligneous, matter which provided material for the formation of the coal found in the Angara region today. Thick deposits of coarse Jurassic conglomerates at the outflow of the Angara and to the north of it along the shores of Baikal, lying in some places below the surface level of the lake, are an indication of a prolonged period of intense activity of the Mesozoic affluents of the lake.

The foundations of the modern relief of the trans-Baikal area were laid back in the Mesozoic era, when long mountain ranges began to form, with deep troughs running between them for hundreds of kilometres in a north-easterly direction. The troughs were filled with extensive sedimentation which resulted in the formation of numerous coal deposits.

Stretching to the north-east and east of the Baikal and trans-Baikal mountain systems for thousands of kilometres, up to the Arctic Ocean, was the so-called Siberian platform which presented, as today, a relatively poorly undulating country slightly slanting to the N. and N.E. and abounding in shallow lakes and marshes.

V. A. OBRUCHEV (1929, 1932, 1938, 1948) and, with him, many other geologists, consider that the old relief of the Baikal mountainous regions and the trans-Baikal area, characterised by alteration of mountain ranges and deep intermontane depressions, is a result of fractures in the bulging rigid block of the "Baikal shield", followed by deep subsidence, along the lines of fracture, of huge blocks of the earth's crust with the simultaneous uplifting of adjacent sections. Thus, in the opinion of V. A. OBRUCHEV, the trans-Baikal Mesozoic depressions are grabens, while the ranges present horsts. Other geologists (FLORENSOV, 1948, 1954) hold that old depressions in the trans-Baikal area are chiefly a result of folding. "The mountain ranges of the trans-Baikal area are anticlinal by nature, while the depressions are gentle synclinal downwarps," writes N. A. FLORENSOV (1948). In his opinion, the basic elements of the modern relief of the trans-Baikal area date back to the Cretaceous period, but it could also inherit structural elements of an earlier age, the Lower Mesozoic and even perhaps the Palaeozoic.

"After a certain tectonic calm," Y. V. PAVLOVSKY & N. A. FLO-RENSOV (1956 p.p.12—13) write, "early in the second half of the Tertiary period, in the Neogene, East Siberia was reached by a wave of vigorous tectonic movements occurring along the periphery of the Pacific and in South Asia. There appeared extensive slowly deepening troughs in the Baikal area, approximately in the same way as has been outlined in the Mesozoic period. This process was accompanied by the resurgence and rapid formation of anticlines developing into mountain ranges. An alpine landscape was formed. This complicated process led to the development of colossal, sharply defined, intermountain areas of the Baikal type. Lake Baikal is the most typical and graphic expression of the gigantic tectonic process which concentrated all its tremendous force in the mountain belt stretching along the southern fringe of the ancient Siberian platform.

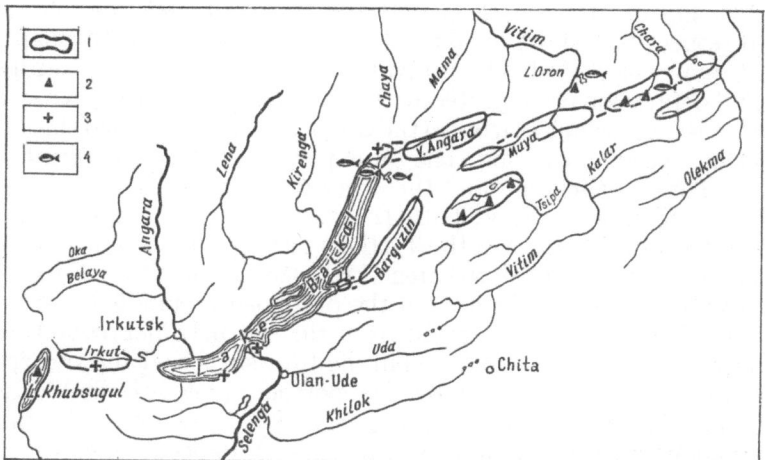

Fig. 111. Tectonic depressions of the Baikal system. After Y. V. PAVLOVSKY, 1941, with additions, 1 - tectonic depressions; 2 - habitation of living Baikalian fauna in lakes in the area of depressions; 3 - fossil Baikalian fauna in the deposits of the same depressions; 4 - habitation of *Salvelinus alpinus erithrinus*.

It is a significant fact that at the same time depressions similar to that of Lake Baikal were also developing along the eastern margins of the ancient African shield: those of Tanganyika, Nyasa, Rudolf, Albert and other lakes . . ."

The Baikal system of depressions, besides Baikal proper, includes depressions situated to the south-east, east and north-east of it, among them those of Tunka, Tora, possibly also Khubsugul in Mongolia, Barguzin, Upper Angara, Tsipa, Muya, Chara, Kalar (PAVLOVSKY & TSVETKOV, 1936; PAVLOVSKY, 1937, 1941, 1948b; fig. 111).

The present-day bottoms of these depressions lie at different levels: the Barguzin, which is closest to Baikal, is 550—600 metres above ocean level (90—140 metres above Baikal), the Tsipa 1,070 metres (more than 600 metres above Baikal), the Muya 700—800 metres, the Chara 700—1,000 metres, the Kalar 1,120 metres, the Tunka 800 metres above ocean level. However, the original bottom of many of these depressions lies much lower than the Baikal level and is covered with sediments thousands of metres thick. The oldest in the Baikal system, according to Y. V. PAVLOVSKY (1937, 1941), is the basin of South Baikal, which could have been formed as a deep depression back in the Mesozoic era, whereas the other depressions appeared only in the later half of the Tertiary period, when the southern depression was radically transformed and changed its direction to the south-eastern, i.e., that of present-day Baikal.

N. V. DUMITRASHKO (1948, 1949, 1952a, b) also points out that an intermountain depression existed in the southern part of present-day Baikal in the Jurassic period and that it stretched diagonally with respect to the depression existing today. It lasted throughout the Cretaceous period. The present-day orientation of the southern Baikal depression is connected with the Tertiary rise of the Khamar-Daban Range, which started in the Oligocene and continued throughout the subsequent period. The same period also saw the formation of the Tunka depression and those neighbouring on it.

In the opinion of N. V. DUMITRASHKO, the Barguzin depression was not yet in existence in the Tertiary period, but there could have been an extensive lake in the area of the delta of the River Barguzin. Lakes also existed in the area of the present-day middle depression of Baikal. They were separated from the southern depression by a mountain mass which had arisen in the Jurassic period and traversed the present-day lake diagonally. The northern part of Baikal formed only in the Pliocene, at first as a shallow depression. The other depressions of the Baikal area are also of Pliocene age. In that period they could be linked by rivers.

The emergence of Baikal as a deep lake with borders approximating to their present shape is dated by DUMITRASHKO to the end of the Pliocene and the beginning of the Quaternary, when the ridges crossing Baikal sunk below its surface level and a single water plane formed in the Baikal trough.

S. G. SARKISYAN (1955, 1958) holds that the Baikal trough originated in the Miocene or Pliocene and acquired its present shape in the Quaternary period. It is interesting that boreholes sunk on the southeast shore of Baikal and also in the Tunka and Barguzin depressions went through Quaternary and Tertiary sedimentary formations and reached Pre-Cambrian rocks but found no Mesozoic deposits. This means that in the Jurassic and Cretaceous periods dry land still lay in the southern part of Baikal, from which shingle, sand and other debris was washed down into the trough of the Irkut basin and other local depressions. It was only in the Tertiary period (possibly in the Oligocene) that the first lacustrine basins appeared in the area now covered by Baikal (SARKISYAN, 1955).

In the opinion of V. V. LAMAKIN (1952, 1955), who paid particular attention to the latest movements of the earth's crust, the whole of Lake Baikal formed in the Tertiary period.

Interesting research has been conducted in recent years on the Ushkanyi Islands situated in the middle part of Baikal. In the opinion of G. Y. VERESHCHAGIN (1949), N. V. DUMITRASHKO (1952a, b) and other scientists, the Ushkanyi Islands are summits of the now submerged Akademichesky Range, once an element of the Jurassic relief of the Baikal area. Back in 1878 CHERSKY pointed out the existence on the Ushkanyi Islands of well-defined terraces with clear traces of surf erosion. V. V. LAMAKIN counted eleven such terraces on the Greater Ushkany Island, the uppermost terrace extending to the highest point of the island, which rises to 211 metres above the surface level of the lake.

The presence of terraces at different sections of the banks of Baikal with marks from 100 to 200 metres above the level of the lake and even higher was noted by all students of the morphology of the Baikal trough. I. D. CHERSKY (1878), V. A. OBRUCHEV (1932), N. V. DUMITRASHKO (1952a, b) and G. Y. VERESHCHAGIN (1949) associated this terracing of the shores of Baikal with a Quaternary fall of its water level and also with the differentiated raising of the shores. An important part in the subsidence of the water level was played by the Angara, which is vigorously deepening its bed, as is indicated clearly by high terraces at various sections of its course.

Some authors associate the high level of the Baikal water in the past with the thawing of the glaciers which covered mountain ranges in the interglacial and postglacial periods. DUMITRASHKO (1952a, b) holds that this rise of the level could not exceed 50—100 metres. V. V. LAMAKIN (1952) attaches much greater importance to the thawing of glaciers as a factor for the rise of the water level of Baikal, especially in the interglacial period, when, in his opinion, the Baikal waters could penetrate into the Lena watershed and fill the Barguzin, Tsipa and other tectonic depressions of the Baikal system.

The present-day views on the geological history of the Baikal trough stated here could be summarised as follows:

1. It is assumed that as early as the Mesozoic era (Jura) there was an intermountain area on the site of South Baikal. It may, however, not have been the embryo of the Baikal of today. The lakes situated in it were neither deep nor big, as is seen from the fact that no Mesozoic deposits common in the Mesozoic depressions of the trans-Baikal area have been found within the confines of the southern Baikal basin.

2. The Baikal basins (two or three) began to expand and grow relatively deep only around the middle of the Tertiary period or in its earlier half. The subsidence of the bottom of the depression lasted throughout the Neogene and was especially rapid at the threshold of the Quaternary period. It has not ended yet.

3. On the threshold between the Tertiary and Quaternary periods the Baikal basins, which were more or less isolated, began to merge. The present tremendous depth of Baikal developed at the end of the Tertiary period and in the Quaternary period, but it is possible that they were relatively large and deep already in the initial stage of the formation of Baikal, in the middle of the Tertiary period.

2. The Most Important Results of Palaeontological Research.

Of exceptional importance for determining the age of Baikal and the stages of its colonisation with organic life is the exact definition of organic remains preserved in ancient deposits both within the morphological boundaries of the lake and the territory around it.

As noted above, in the Upper Jurassic and Lower Cretaceous periods there were big lakes in the trans-Baikal area and Mongolia, which filled intermountain depressions. They were populated, as organic remains show, by a rather homogenous fauna, the study of which was started back in the middle of the last century by MIDDEN-DORFF (1867). O. REIS (1910) also studied it. A more detailed study of this fauna was made by E. S. RAMMELMEIER (1931, 1935, 1940) and in recent years by G. G. MARTINSON (1940, 1948a, 1949b, 1951, 1952, 1957a, etc.) and NOVOZHILOV (1954). In the so-called Turga formation of the trans-Baikal Mesozoic basins remains of peculiar fishes from the family Lycopteridae and the genus *Lycoptera* were found. The latter, L. S. BERG writes (1949b), stands at the outset of the Teleostei stem and is one of the oldest bony fishes and close to the cyprinoids. Gastropod remnants from these deposits outwardly resemble the modern Baicaliidae.

E. S. RAMMELMEIER (1935), G. Y. VERESHCHAGIN (1940b) and

G. G. MARTINSON (1940) admitted that some turreted fossil gastropod shells from the Mesozoic deposits of the trans-Baikal area can be regarded as direct ancestors of endemic Baikalian mollusks. But this opinion is questionable. The remains of gastropods which E. S. RAMMELMEIER attributes to the Baikalian genus *Benedictia* are so strongly deformed as to make exact definition impossible, while the species *"Baicalia"* (*Probaicalia* after MARTINSON) from these deposits are not closely related to the modern Baicaliidae. Recently G. G. MARTINSON (1956) has conducted a detailed study of the Mesozoic trans-Baikalian mollusks and has now associated the fossil *"Probaicalia"* with the family Micromelaniidae and not with the Baicaliidae.

The finds in the deposits of the Upper Jurassic-Lower Cretaceous lakes of the trans-Baikal area include mollusks representing the *Viviparus*, *Limnaea*, *Planorbis*, *Cyrena*, Physidae and Bithyniinae, also big Ostracoda, larvae of the Odonata, Plecoptera and Trichoptera, aquatic and semi-aquatic Reptilia and the above-mentioned peculiar fish *Lycoptera*. An analogous fauna was found in the Mesozoic deposits outside the trans-Baikal area, in Mongolia and in regions adjoining China. This Mesozoic fauna has no relation to the modern and Tertiary fossil fauna of Baikal.

In recent decades an interesting fossil fauna of mollusks has been found in some depressions in Central Asia. A study of it gives reason to suppose that in Upper Cretaceous time Mongolia, the northwestern regions of China, and also the Far East had fairly extensive continental basins inhabited, along with a typically fresh-water fauna, by highly original costate lamellibranchiate mollusks from the genera *Trigonoides* and *Sainschandia*, which formed, as it were, an intermediate link between brackish and fresh-water mollusks (MARTINSON, 1955a, b). In G. G. MARTINSON's opinion (1955a), these costate mollusks are indicative of the existence, in Upper Cretaceous time, of a whole system of vast bodies of water extending from the Pacific through Manchuria to the Gobi area of Mongolia and on to Central Asia (Fergana, Kara-Tau).

But no remains of a fauna comparable to the modern or fossil Baikalian fauna have so far been found in any of the late-Mesozoic deposits of Asia either, although some gastropods with the turreted shell, in the opinion of G. G. MARTINSON, somewhat resemble the *Baicalia* and may have been the ancestors of the family Baicaliidae (MARTINSON, 1955a).

The results obtained by the palaeontological expeditions of the U.S.S.R. Academy of Sciences in Mongolia, which discovered abundant remains of turtles, crocodiles, dinosaurs, etc., also point to an abundance of waters in Central Asia in Upper Cretaceous time. Where there is desert in Mongolia now, in the Cretaceous period there were extensive boggy lowlands in the direct vicinity of the sea. Remains of swamp forests were found almost everywhere (YEFREMOV, 1954).

It was only towards the end of the Cretaceous period, as G. G. MARTINSON (1955a) and other authors point out, that the great basins of Central Asia disappeared and the fauna inhabiting them either disappeared or changed its habit.

The first half of the Tertiary period did not leave clear traces of any large basins in Central Asia and the trans-Baikal area. The fossil fauna of that period is practically unknown. But in recent years an interesting fossil fauna has been discovered in the Middle and late Tertiary continental deposits of north-western China (the area of Sinkiang), where remains of a varied typically fresh-water fauna were found, including gastropods which, indeed, strongly resemble modern and fossil Baicaliidae. Thick sedimentary formations and an abundant fresh-water fauna in Sinkiang, in the opinion of MARTINSON (1955b), again prove the existence of numerous continental basins in the Oligocene and Miocene. There was, he writes, a whole system of lakes which covered the territory of Dzungaria and Kashgaria and extended into Mongolia and West Siberia and also into the adjoining regions of Central Asia. In one of the Tertiary deposit formations of Sinkiang, namely the upper green rock dated by G. G. MARTINSON to the Middle-Upper Miocene, gastropod forms were found which he directly attributes to the modern Baikalian genus *Baicalia*, pointing out that the Chinese *Baicalia* lived in deeper zones of big Tertiary lakes, separately from such lacustrine mollusks as the Unionidae, Planorbidae or Limnaeidae, just as is the case with the mollusks of present-day Baikal. MARTINSON (1955b) holds that the affinity of the Sinkiang and Baikalian forms points to extensive communication between the lakes of China and Siberia, which were connected by all kinds of lake dykes and by river drainage. It was only in the Pliocene, when violent uplift took place, that the lake systems of Central Asia began to shrink, turning into small isolated bogged lakes, with the deep- and open-water fauna in them being replaced by the fauna of shallow lakes and marshes akin to the present-day fauna.

In their recent works MARTINSON (1960) and MARTINSON & POPOVA (1959) reported on the results of the study of fossil mollusks from the fresh-water Tertiary deposits of West Siberia in the south of the Omsk Region, almost on the border with North Kazakhstan. Continental deposits of the Pliocene, Miocene and Upper Oligocene age were reached there, underlain by a marine Palaeogene and early Neogene suit containing foraminifers and ostracods of the Lower and Middle Oligocene age. The gastropod remains found in this area were classed by the authors as belonging to the palaearctic genera *Hydrobia*, *Valvata* and *Bithynia* and also the Baikalian genera *Baicalia* and *Kobeltocochlea*. On the basis of these finds MARTINSON assumes now that the endemic Baikalian gastropod fauna could originate in the marine and semi-marine basins of the Turgai depression and their freshening waters, where this fauna lived in the Oligocene epoch and whence it spread by various ways eastward, to Central Asia, including West China (Sinkiang), and later on to Baikal, as well as westward, to the ancient Pontocaspian basin and the Balkan Peninsula.

But, judging by the specimens the author saw in G. G. MARTINSON's collection and the photographs that have been published, the inclusion of the small turreted gastropods of West Siberia in the family Baicaliidae is not indisputable. They may represent a western branch of the Rissoida, which is closely related to the modern and fossil Tertiary Hydrobiidae, Micromelaniidae and Pyrgulinae characteristic of the Pontocaspian basin. As regards the "*Kobeltocochlea*", there are equal grounds to ascribe it to the genus *Lithoglyphus*, which occurs in Pontocaspian reservoirs now as well.

Of great importance for the chronology of the initial stages of the formation of the Baikalian fauna and certainly of the lake itself is the study of fossils within the modern morphological boundaries of the Baikal depression itself and in other depressions of the Baikal system.

Highly interesting in this respect are faunal remains in the deposits of Tertiary terraces on the south-eastern shores of Baikal, which were known back in the 19th century. Among fossil mollusks E. S. RAMMELMEIER (1931, 1940) came across a great number of large lamellibranchiates from the family Unionidae (species of the genera *Unio*, *Nodularia*, *Lepidodesma*, *Acuticosta*, and others). These mollusks were found to be closely related to the modern Chinese species from the same genera. Besides, remains of large *Viviparus* were found there. But of special interest was the discovery of remains of gastropod shells of the family Baicaliidae occurring in present-day Baikal. Some of them proved to be so close to the modern species that the same specific names were given to them.

On the basis of finds of vegetable remains in the same deposits (and also in the analogous deposits of the Tunka depression) E. S. RAMMELMEIER (1940) agreed with I. V. PALIBIN (1936) in dating them to the Oligocene.

In recent years a detailed palaeontological study of these deposits has been conducted by G. G. MARTINSON (1938, 1940, 1948a, 1951, etc.). The 29 species of the Bivalvia found there contained not a single form close to the modern species from Baikal's open waters. In the Viviparidae family, new species were found of the genera *Viviparus* and *Tulotoma*, which do not exist in Baikal today. Remains of gastropod shells of the genus *Lithoglyphus* related to the modern Chinese genera were found for the first time. It is to be noted that the genus *Lithoglyphus* is close to the Baikalian genera

Kobeltocochlea and *Benedictia*. Remains of shells which G. G. MAR-
TINSON attributed to the genus *Benedictia* were also discovered.
The number of fossil species of the genus *Baicalia* reached 15. The
first recorded finds were made of remains of shells of the Baikalian
genera *Liobaicalia* and *Choanomphalus*, common in Baikal today.

The same Tertiary deposits of the south-western shore of Baikal
contain siliceous spicules of the Baikalian sponges Lubomirskiidae
and also of the Spongillidae. Also noteworthy is the discovery there
of tests of foraminifers of the genera *Discorbis* L.? or *Anomalina*
ORB.?

MARTINSON disagrees with RAMMELMEIER as to the age of the
deposits of the Baikalian terraces containing remains of this fauna
and considers that they are younger and date from the Miocene and
Lower Pliocene.

Analysing the fauna of the Baikalian terraces as a whole, G. G.
MARTINSON (1951) differentiates it into two main genetic complexes:
Chinese or Chinese-Mongolian, and Balkanian. In the Chinese com-
plex he lists almost all Unionidae and Viviparidae and in the
Balkanian, the Baicaliidae, *Lithoglyphus* and *Benedictia*, as well as
Viviparus and certain *Unio*. Representatives of the genera *Pisidium*
and *Planorbis* are believed by him to be remnants of the widespread
fauna of the continental waters of Eurasia.

MARTINSON ascribes the Baicaliidae, *Lithoglypus* and *Benedictia*
to the "Balkanian" group on the basis of the long-recorded outward
similarity between some fossil gastropod shells from the Pliocene
deposits of South-East Europe, the Balkan Peninsula in particular,
and the fossil and even modern turreted shells of the Baicaliidae
species. Many palaeontologists have noted this similarity. But this
rather superficial similarity is evidently explained by the convergent development of two independent centres of speciation in
Tertiary time: the Ponto-Aralo-Caspian ("Balkanian") and the
Baikalian (Baicaliidae). For this reason the term "Balkanian" as
applied to the Baikalian gastropods cannot be regarded as appropri-
ate. As to the numerous fossil Unionidae species, it is true that they
are closely related to the modern and fossil Chinese species.

It is a noteworthy fact that sedimentary strata from the Tertiary
terraces of Baikal with remnants of Baicaliidae contain almost no
Unionidae and that, conversely, there are no Baicaliidae where
Unionidae prevail. This corroborates the assumption earlier expressed by the author (KOZHOV, 1936, 1947), that sediments in the
terraces on the south-eastern shore of Baikal were deposited in the
littoral zone in conditions similar to those still found in the littoral-
sor region adjacent to the Selenga delta. Evidently these terraces
were alternately inundated by the open waters of the lake and
sediments were deposited in them containing remains of the typical
indigenous Baikalian fauna (mollusks Baicallidae, sponges Lubo-

mirskiidae), or turned into sors or even shallow marginal lakelets and were colonised by ordinary fauna common in the waters of South Siberia.

Spicules of Baikalian sponges of the family Lubomirskiidae and the widespread family Spongillidae were found in the Tertiary deposits of the Tunka tectonic depression, Baikal's neighbour (DUMITRASHKO & MARTINSON, 1940).

Of great importance for specifying the time of the isolation of the Baikalian fauna from the neighbouring fauna of Siberia is the study of fossil remains contained in the deposits of old shallow lakes in areas of Siberia outside the Baikal mountain region. In recent years, a fossil Tertiary fauna has been found in the Western Sayan Range, in the vicinity of the Tannu-Ola Range (N. ZAITSEV, 1947) and in the area of the Irkut coal basin (MARTINSON, 1949a, 1954a). The latter was found to contain remains of Planorbidae, such as *Planorbis; Gyraulus, Hippeutis, Spiralina* and *Segmentina*, as well as *Radix, Stagnicola*, Physidae, Bithyniinae, then Melaniidae *(Melania)* and, finally, representatives of the genus *Hydrobia*. G. G. MARTINSON stresses the relationship between them and the Pliocene mollusks of Mongolia and China. It should be emphasised that this fresh-water Pliocene lacustrine fauna of South Siberia has no relation to the modern or fossil Baikalian fauna. Nor does the fauna from the Pliocene lacustrine deposits of the Tannu-Ola bear any trace of a fauna related to the fauna of Baikal. It consist of *Unio, Valvata piscinalis, Viviparus*, etc. Tertiary (Miocene?) deposits of the island Olkhon (Baikal) contain remains of heat-loving fresh-water turtles of the genus *Clemmis* and a tortoise of the genus *Testudo*, as well as Planorbidae *(Gyraulus, Hippeutis)*, the fish *Amia* and some silurids and salmonids (KITAINIK & IVANYEV, 1958), The pollen of coniferous and broad-leaved trees was also found there. No traces of the Baikalian fauna were found in the Olkhon deposits, which were sedimented under a shallow body of water, or in the Tertiary deposits of trans-Baikal depressions (Lake Gusinoye and others).

All the data of palaeontological research cited here show that already in the Pliocene and perhaps even in the Miocene the Baikalian fauna as a whole was not only formed, but also ecologically distinct from the fauna of the shallow Siberian lakes surrounding Baikal, which means that at that time Baikal already existed as a relatively large and deep lake or a system of such lakes.

Of great importance for the history of organic life in Baikal is the study of the fossil terrestrial flora of South Siberia and the Baikal area, which point to possible changes in the climate and relief of the country (KRISHTOFOVICH, 1928, 1930, 1932). This study proves conclusively that throughout the Tertiary period the climate of South Siberia was much warmer and more humid than now. In

the middle of the Tertiary period the so-called Turgai flora characterised by the presence of heat-loving broad-leaved forests were
dominant there. At the same time, in the middle of the Tertiary
period, there was a distinct zonal (vertical) differentiation of the
flora of South Siberia's mountain regions: along with broad-leaved
forests in valleys there were coniferous forests in the mountains.

But beginning with the middle of the Tertiary period the climate
of South Siberia gradually began to grow colder and drier. Signs of a
general cooling appeared already in the Palaeogene, and the later
half of the Oligocene was decisive in the history of the covering of
vegetation (BARANOV, 1942). In the Pliocene the cooling continued,
and towards the end of it the broad-leaved forests gradually disappeared; their place in the mountain regions was taken by the
present-day taiga, while steppes and forest-steppes began to predominate in the valleys and on the southern slopes of mountain
ranges.

At the end of the Pliocene, ice began to accumulate on the ridges
bounding Baikal, and the glacial period which ensued has left vivid
traces both in the relief of the ridges around Baikal and in the flora
and fauna of Baikal itself. Most geologists hold that there are traces
of two glaciations in the Baikal area. The first and more extensive
glaciation was in the nature of an complete ice-sheet, while in the
second glaciation only the tops of mountain systems were covered by
glaciers.

N. V. DUMITRASHKO (1952a, b) rejects the idea of repeated
glaciations in the Baikal area, but sees four phases of one glaciation.
In the maximum phase the glaciers reached 80 kilometres in length
and were 300 to 350 metres thick. The snow line was at an absolute
height of 800 or 1,000 metres.

V. V. LAMAKIN (1950, 1952, 1953, 1959, etc.) points out that in
their downhill creep during the maximum glaciation the glaciers
reached Baikal and in some places even slid below its water level.

Owing to the dissected relief and relatively humid climate rich in
precipitation the Tertiary tributaries undoubtedly carried more
water than now. In the Tertiary period the powerful water channels
of the Baikal area, the ancient Selenga included, probably drained
a much greater territory than now, with their sources extending
deep into Central Asia. Likewise, in view of a more humid and
warmer climate, the water temperatures of Baikal in the Tertiary
period were much milder. In connection with the progressing cooling
in the Pliocene the Baikal waters grew increasingly colder, with
their temperatures becoming most rigorous in the glacial time.

But the importance of glacial phenomena as a factor for changing
the organic life that populated the shores and waters of Baikal must
not be over-estimated. The botanists N. A. YEPOVA (1955, 1956)
and M. G. POPOV (1954, 1957) point to the existence of relics of the

Tertiary broad-leaved forests in the modern mountain forests of the Khamar-Daban Range, which certainly could not have survived had the climate of the glacial period been entirely detrimental to the whole of the Tertiary mountain flora. As will be shown further on, it was not detrimental to the population of Baikal proper, either.

3. Development of Concepts of the Origin and History of the Baikalian Fauna and Flora.

Views on the history of the Baikalian fauna and flora have been developing parallel with the growth of knowledge of the living world and geological history of the lake and its area.

In the early period of Baikal studies the greatest puzzle was the presence of the seal and other animals of marine habit in Baikal. A. v. HUMBOLDT (1843), the outstanding geographer of the last century, considered the presence of closely related seals in the Caspian Sea and Baikal to be an evidence of the fact that these lakes had communicated with each other in the past. O. PESCHEL (1878) believed Baikal to be a former gulf of the Arctic Ocean which in a geologically recent past could have extended to Baikal and perhaps even to the Caspian and Black Seas.

BENEDICT DYBOWSKY and the well-known malacologist WLADISLAW DYBOWSKY were inclined to believe that Baikal had been directly linked with the Arctic Ocean, from which it had received its marine forms (W. DYBOWSKY, 1884a).

A decisive blow shattering the hypotheses about Baikal's direct communication with the Arctic Ocean or any other sea was dealt by geological and palaeontological studies in South Siberia, which detected no traces of the presence of the sea in the Baikal area even in the most ancient periods of the history of the earth. As far back as 1877 I. D. CHERSKY came to the conclusion that neither the post-Tertiary Arctic Ocean nor the waters of any other sea had reached the latitude of Baikal in East Siberia. To explain the presence of the seal in Baikal, I. D. CHERSKY advanced the hypothesis of its penetration into Baikal from the Arctic Ocean through the Yenisei-Angara system.

R. CREDNER (1887—1888), basing his conclusions on CHERSKY's views, also maintained that Baikal could have received its marine population from the Arctic Ocean, but through migration rather than direct communication.

On the threshold between the 19th and 20th centuries many scientists still adhered to the hypothesis of the marine origin of some elements of the Baikalian fauna. A. A. KOROTNEV (1901, p. 35) wrote: "It is difficult . . . to judge of Baikal's association with the Arctic Ocean or the centre of the Central Asian basin which existed here in the past; nevertheless, I am personally inclined to support the former theory."

The same views were expressed by V. P. GARYAYEV (1901), a student of Baikalian gammarids, and the prominent German palaeontologist M. NEIMEIER (1886), both of whom considered that the fauna of Baikal as well as that of the Caspian was a relict of an originally marine fauna.

R. HOERNES (GERNES) (1897a, b, 1898), relying chiefly on palaeontological research and the outward similarity between fossil gastropod shells from the Sarmatian-Pontian deposits and modern Baikalian forms, suggested the hypothesis that the fauna of Baikal had originated in an inland sea in which Sarmatian, Maeotian, Pontian and Aralocaspian sediments were deposited. Baikal received its population from an Upper Tertiary inland sea and not from the Arctic Ocean.

In the initial period of his studies of the Baikalian fauna L. S. BERG arrived at a similar conclusion (1900).

In 1902 N. I. ANDRUSOV put forth the hypothesis that many Baikalian animals owed their marine habit to convergence induced by the unusual depth and size of the lake making life conditions in it similar to those of the sea. As to the roots of its peculiar fauna, Baikal received them from various basins and zoogeographical regions in different periods of its life. He did not specify from what regions the ancestors of the Baikalian fauna penetrated into the lake. ANDRUSOV (1902) attached particular importance to the long geological history of Baikal as a factor for evolution. The longer the life of a lake, the richer and more varied its fauna, Baikal owes the remarkable variety of its fauna, in ANDRUSOV's opinion, also to the quaternary shifts of climatic zones, which brought closer to Baikal organisms now living far away from it.

The study of a wealth of faunistic data obtained by the Baikal expedition of A. A. KOROTNEV and other researchers at the turn of this century made it possible to judge with greater certainty of the origin and history of some important groups of the animals of Baikal. Thus, a study of the oligochaetes of Baikal led V. MICHAELSEN (1901) to the conclusion that this fauna is characterised by the prevalence of phylogenetically old forms and is to be regarded as being of a geologically considerable age. In any case, he wrote, it is older than the fauna of the European waters. In MICHAELSEN's opinion, in its geologically long life Baikal could have retained forms of independent local origin as well as immigrants from other freshwater lakes, which disappeared long ago, and even from the seas. Similar view were held by the well-known malacologist V. A. LINDHOLM (1909) and V. K. SOVINSKY (1915), the author of a fundamental summary on the gammarids of Baikal.

The hypothesis of the indigenous fresh-water character of the fauna of Baikal was defended and developed most energetically and consistently by L. S. BERG (1910, 1922, 1925, 1928, 1934, 1937, 1949a). In his work "The Fauna of Baikal and Its Origin" (1910) L. S. BERG thus formulated his views:

The modern fauna of Baikal consists of two elements: 1. the forms which have developed in Baikal itself in the course of its long geological life (Oligochaeta, Comephoridae, Mollusca, etc.); 2. the vestiges of the Upper Tertiary heat-loving fresh-water fauna of Siberia and the adjacent areas of Central Asia.

L. S. BERG stresses in particular that in the Upper Tertiary period the aquatic and terrestrial fauna of China, Siberia and Europe differed far less than now, being, on the whole, of the Chinese type. It was a homogenous fresh-water fauna remnants of which have been preserved in Lake Ohrid, Baikal, Lake Tali-fu (China) and the Caspian Sea. Subsequently, in the glacial epoch, this fauna perished, and now we find insignificant remnants of it in Baikal, the Amur and some other places. Even the Baikalian and Caspian seal was regarded by BERG in his early works as a relict of the heat-loving fresh-water fauna that was widespread in Eurasia and North America in the Tertiary period. Later on, however, BERG admitted the possibility of the seal and omul having penetrated into Baikal from Arctic regions.

In a more distinct form the hypothesis of "the homogenous heat-loving fauna and flora of the Chinese type" which lived all over North Asia, Europe and North America at the end of the Tertiary period (the Pliocene) and became extinct in the glacial period, was developed by L. S. BERG in his work "The Fishes of the Amur drainage" (1909). These views found their most important substantiation in the discovery, in the second half of the 19th century and at the beginning of the 20th century, of a fossil late-Tertiary fauna of mollusks in regions bordering on the Caspian and Black Seas, the seas that washed the Balkans and the neighbouring regions of South-East Europe, and also in Asia to the east of the Caspian Sea. The numerous fossil gastropod species from the genera *Caspia*, *Micromelania*, *Goniochillus*, *Prososthenia*, *Lithoglyphus* and many others were believed to be related to the gastropods of Baikal, the Caspian Sea and the continental waters of China and North America (Pleuroceridae). Great importance was also attached to the establishment of discontinuity in the distribution in northern Eurasia of a considerable number of relatively heat-loving modern species of broad-leaved trees (oak, hornbeam, lime, etc.) as well as some animals (*Silurus*, *Cyprinus carpio*, *Abramis*, the crayfish *Potamobius*, and others), which are widespread in Europe, absent in East Siberia and reappear in the drainage basin of the River Amur.

L. S. BERG's hypothesis was accepted by almost all zoologists and biogeographers and provided a guide for biogeographical studies and for the development of conceptions on the history of the continental hydrofauna of Eurasia, the Baikalian fauna included.

The hypothesis of the exclusively fresh-water origin of the modern Baikalian fauna was also supported by V. C. DOROGOSTAISKY (1923b), the author of the proposition that in the present epoch the fauna of Baikal is in a state of intense speciation. He saw a substantiation of this view in the fact that we find a number of gradual transitions from one species to another in some groups of Baikalian animals, and no evidence of the gaps which are observed in the fully developed fauna of ancient origin. When still a shallow lake, this scientist believes, Baikal was populated by a sub-tropical fauna which began to disappear by the end of the Tertiary period, surviving only in bigger water sheets, especially Baikal. In the glacial period the fauna of Baikal, in all probablity, was poor both qualitatively and quantitatively. In post-glacial time, when the climatic conditions had changed for the better, the species that had survived there began to spread vigorously, invading almost unpopulated depths and the littoral regions, bays and gulfs freed from the ice cover. In the same period many aquatic dwellers of the faraway North, such as the seal, omul and others, penetrated into Baikal via the deep Angara.

In recent years the idea of the Quaternary origin of almost all of the modern Baikalian fauna has been developed by D. N. TALIYEV (1948).

Despite the popularity of L. S. BERG's hypothesis of the purely fresh-water roots of the Baikalian fauna, the idea of the marine origin of its more enigmatic elements persisted. In the "twenties" and "thirties" the polemics on this question flared up with renewed force.

With great persistance the hypothesis of the marine origin of the main elements of the Baikalian fauna was developed in numerous articles by G. Y. VERESHCHAGIN (1928, etc.). This author allowed the possibility of the existence of an extensive sea basin in Central Asia in the Mesozoic era and a sea basin in West Siberia in Lower Tertiary time. The latter was supposedly connected with the Caspian and the Black Seas and extended eastward, reaching the Yenisei. From these seas, VERESHCHAGIN believed, Baikal could have received progenitors of some elements of its fauna. But the idea of an early Tertiary sea extending to the Yenisei and covering large areas of West Siberia found no corroboration.

Developing his views on the history of the fauna and flora of Baikal, G. Y. VERESHCHAGIN (1930, 1940a, b) differentiated it into the following genetic elements:

1. general Siberian, which consists of forms currently widespread both in the territory of Siberia and in Baikal; 2. old fresh-water, comprising remnants of the Tertiary fresh-water fauna common in

the past but now occurring, besides Baikal, only in some big water bodies in India, China, the Balkan Peninsula and some other places; 3. marine, which includes forms having close relatives in the marine fauna and flora and occurring in fresh waters, besides Baikal, either as marine relics or immigrants from the sea; 4. elements whose origin has not been ascertained yet.

In 1940 G. Y. VERESHCHAGIN published a fundamental work entitled "The Origin of Baikal and Its Fauna and Flora," in which he endeavoured to prove that Baikal or the basins which formed it began to develop as early as the Mesozoic era. At that time it was inhabited, he writes, by old fresh-water elements the remnants of which, in the modern Baikalian fauna, we can consider, are the oligochaetes, some turbellarian worms and the mollusks Benedictii-nae. In the Tertiary period the area of present-day Baikal was populated by a fauna bearing some resemblance to that of North America. This fauna has become almost fully extinct in the fresh waters of Siberia, but some of its elements have survived in Baikal and in the extreme East of Asia. Among such relics VERESHCHAGIN lists the Baikalian *Asellus* from the subgenus *Mesoasellus*, some Harpacticoida crustaceans and *Epischura baicalensis* from the Calanoida. In Upper Tertiary times Baikal was invaded by a heat-loving fauna which has been preserved in South China, India and Indo-China, and in the glacial period, by a cold-loving fauna, such as the diatoms *Melosira baicalensis* and *M. islandica*, which was also pointed out by R. VERTEBNAYA (1929) and B. SKVORTSOV (1937). Finally, elements of the general Siberian fauna and flora still keep penetrating the morphological boundaries of Baikal.

As to the "marine element" to which VERESHCHAGIN refers the sponges Lubomirskiidae, the majority of gammarids, the polichaete *Manayunkia*, etc., his interpretation of the genesis of this faunal group has changed considerably. He considers now (1940a) that the problem of "marine elements in the fauna of Baikal can only be solved in connection with the general problem of the origin of young marine immigrants to continental waters, Baikal is only one of the few collectors of this fauna."

VERESHCHAGIN points to the existence of several main sources in Eurasia from which "young" immigrants from the sea invaded continental waters. The Aralo-Pontocaspian region served as such a source chiefly since Upper Tertiary times; the advance of sea in the Far East, in Lower Tertiary times; the advance of sea from the east and north-west to the Vilyui depression and the south-east trans-Baikal area, in the Mesozoic (the Lower Cretaceous); finally, South-East Asia could serve as such possibly in Tertiary times, and the Arctic basin in the Quaternary period. A source of the immigration of marine elements in the Quaternary period could also be provided by the freshening seas of the Far East. The main channel

for the immigration of marine forms to inland basins, in VERE-
SHCHAGIN's opinion, was provided by the freshening of basins becoming detached from the sea or of its littoral sections. As regards
active immigration via rivers, this channel could not be of any great
importance for it was open only to those marine forms which could
actively work their way upstream.

Of all these sources which contributed, beginning with the Mesozoic, to the formation of continental elements in the fauna of Eura-
sian basins, of special importance for Baikal could be, according to
VERESHCHAGIN, marine relict basins in the south-east of the trans-
Baikal area, which he suggests existed in the Mesozoic era. The system of lakes which subsequently developed in their place could
have served as an intermediate point on the way of marine elements
to Baikal. Baikal might or might not have been a component part
of this system and the recipient of its waters preserving marine
relicts in its fauna. In the latter instance some "marine elements",
already modified and adapted to typically fresh-water habitat, could
have reached Baikal via running waters. In this connection VERE-
SHCHAGIN advanced the hypothesis of water convergence, in which
he discussed the possibility of successive transgressions of the trans-
Baikal lake system in the direction of Baikal with the flowing down
of its waters into the gradually deepening trough of the lake.

Thus, in the opinion of VERESHCHAGIN (1940a, b), the initial
marine elements of the Baikalian fauna migrated into the system
of ancient basins of which Baikal is a member in three different
periods: at the end of the Mesozoic (from the relict basins of the
trans-Baikal area), in Tertiary time (from the seas of South-East
and East Asia), and in the Quaternary period from the Arctic Ocean
transgressions, with active migration by rivers (the seal, the omul).

Proceeding from the results of the study of fossil Mesozoic and
Tertiary mollusks of Asia, G. G. MARTINSON (1959b, 1960) calls
the fresh-water fauna which could have developed from marine
progenitors in Mesozoic and Tertiary time neolimnic, as distinct
from the palaeolimnic fauna which was formed in most remote
times and which is to be found the world over now. According to
MARTINSON, the Baikalian fauna (Baicaliidae, Lubomirskiidae etc.)
is to be referred to the neolimnic complex.

In his last years L. S. BERG (1949a) continued to defend his old
hypothesis of the Pliocenic, indigenously fresh-water roots of the
main elements of the Baikalian fauna. He wrote that there were no
marine elements in the fauna and flora of Baikal and that it was
useless, therefore, to look for the roots of their "marine elements"
in Mongolia, South-East Asia, or elsewhere. In his opinion, there is
not a shade of plausibility in the supposition that the hypothetical
marine elements in the Tertiary lacustrine basins of the trans-Baikal
area had their roots in the Tertiary seas of South-East Asia.

The present state of the problem of the origin and history of the Baikalian fauna will be dealt with further on.

4. The Main Genetic Groups in the Biota of Baikal and Their Origin.

Proceeding from the present-day knowledge of the composition and distribution of the Baikalian biota and also the latest biogeographical and palaeontological research, we can distinguish the population of Baikal into the following genetic groups (KOZHOV, 1958a, b, 1959b):

Representatives of the modern Siberian-European fauna

In this complex, heterogenous in its origin, ecological groups also have to be distinguished.

a. *Limnophiles*. The group of the modern population of Baikal comprises forms which are identical with or very closely related to the forms now dwelling in shallow lakes around Baikal and other eutrophic waters. Siberian limnophiles in Baikal inhabit almost exclusively littoral-sor sections: sors, sheltered bays, enclosed parts of gulfs, etc. On the whole they correspond to the "biogeographical" complex which G. Y. VERESHCHAGIN (1935) called the Siberian complex. It consists of the sponges Spongillidae, lacustrine Siberian-European species of Mollusca, Oligochaeta, Hirudinea, Rotatoria, Copepoda, Cladocera, Trichoptera, Chironomidae, Gammaridae *(Gammarus lacustris)*, etc. The fish include *Rutilus rutilus lacustris, Leuciscus idus, Phoxinus, Perca fluviatilis, Esox lucius*, and other lacustrine species which, according to G. V. NIKOLSKY (1947, 1955), from their origin can be classed, genetically, as belonging to the Siberian plainland complex. For reproduction they either enter marginal lakes or spawn in Baikal's sors and gulfs. But in summer and in some parts also in winter they spread along the shores of more or less open shallows of the lake. In summer Siberian species of planktonic Rotatoria, Copepoda, Infusoria and also algae are carried by currents from sors and sheltered gulfs into the open waters of Baikal. But there they are dispersed and quickly perish in deepwater regions.

b. *Limnorheophiles*. They unite a fairly large number of forms closely related to or even identical with species common in Siberia's big and relatively deep running-water lakes, rapid rivers and streams and also partially in eutrophic waters, but only in the cold season of the year. Immigrants from such waters inhabit in Baikal mainly the open littoral and some of them also the sublittoral zone, constituting part of the open-water biocoenoses along with indigenous Baikalians. In littoral-sor regions they occur only in the cold season. In other words, their habitat in Baikal resembles that of their relatives.

As a rule, those forms of this complex which dwell in the deeper zones of Baikal (the sublittoral or the lower littoral) have deviated to some degree or other from the parental forms and comprise endemic varieties, whereas the littoral and plankton forms differ very little from their ancestors. To this group should be referred, in all probability, certain species of chironomids and oligochaetes from the genera *Mesenchitraeus*, *Nais* and *Paranais*, possibly some species of littoral nematodes, fresh-water hydras, etc. The fish include *Thymallus arcticus baicalensis* and *T.a. brevipinnis*, *Lota lota* with its Baikalian deep-water variety, and also the Baikalian forms of bottom-living gwyniads *Coregonus lavaretus*, with their parasites. The dace *Leuciscus leuciscus baicalensis* should probably be included in this group as well. Some of the species of the group reviewed here belong to the submontane complex of G. V. NIKOLSKY *(Thymallus)*, others are migratory species but these have settled in Baikal (the sturgeon, gwyniads). Their closest relatives live in mountain and big valley rivers of Siberia (grayling, gwyniad, sturgeon, dace, burbot) and in the littoral of deep mountain lakes of Siberia.

The pelagic organisms belonging to this group apparently include some infusorians and flagellates which live in the open littoral (*Mallomonas*, etc.), the Baikalian forms of such widespread species of rotifers as *Notholca longispina*, *N. striata acuminata*, *Keratella quadrata*, *K. cochlearis* and *Filinia longiseta*, and also perhaps the copepod *Cyclops kolensis*, which sporadically occurs in the water bodies of North Europe and Siberia.

It was this group of Siberian limnorheophiles that G. Y. VERE-SHCHAGIN (1935) had in view when he established the Siberian-Baikalian biogeographical complex among the inhabitants of Baikal. But he also referred to it a number of elements which should be identified with other genetic groups, reviewed below.

G. Y. VERESHCHAGIN considered that the Siberian species in Baikal's open waters were genetically young, i.e., still in the process of incorporating themselves into the old fauna of Baikal. But this proposition cannot be accepted without serious reservations. Siberian limnorheophiles, fish included, undeniably lived in Baikal and the waters around it since the initial stages of its history, i.e., in the Tertiary and later periods. The reason for their insignificant deviation from the modern Siberian species lies in not their being "young" immigrants to Baikal, but in the fact that life conditions in the Baikal littoral could not and do not differ sharply from those of the oligotrophic Siberian lakes and particularly rapid rivers. As regards the fish (gwyniad, grayling, burbot), it should also be borne in mind that their Baikalian populations, which need rivers for spawning, can mingle there with ordinary fluviolacustrine populations, which should also serve as a certain obstacle to the process of their speciation.

Descendents of dwellers of the Tertiary Holarctic region

We refer to this group those inhabitants of the open waters of present-day Baikal whose closest relatives do not live in the bodies of water adjacent to it but sporadically occur in places very far removed one from another, chiefly within the confines of the Siberian-European sub-region of the Palaearctic and partly in North America.

In Baikal, this complex is represented, as a rule, by endemic varieties and species and sometimes genera. They inhabit all kinds of biotopes but chiefly the open littoral, sub-littoral and deeper zones, right up to extreme depths. Among them are many Baikalian oligochaetes (together with the endemic genus *Agriodrilus*), probably some species and genera of turbellarians, leeches from the family Piscicolidae, the crustacean *Epischura baicalensis* (Copepoda-Calanoida), which is one of the most numerous inhabitants of Baikal's pelagic zone, the isopods *Mesoasellus* and *Baicaloasellus*, some species of the Harpacticoida and Ostracoda, as well as Trichoptera of the group Baicalinini, Chironomids of the sub-genus *Baicalosergentia*, some endemic pelagic infusorians described by N. S. GAYEVSKAYA (1933), and so on. This complex probably includes also the endemic species of peridineans of the genus *Gymnodinium* and the diatoms *Cyclotella baicalensis, C. minuta* and *Melosira baicalensis*.

The old age of the roots of the complex reviewed here can be illustrated on the example of the Baikalian Trichoptera. Their endemism is very distinct; they have lost connection with their nearest relatives, and their separation from the stem common with limnophilids was dated by A. V. MARTYNOV from the middle of the Tertiary period. Since then they have broken up into several species and genera and acquired many distinct morphological and biological characters which have developed in the conditions of this lake.

It also took much time for the development, in the Baikal deeps, of such an endemic genus of Turbellaria as *Polycotylus*, a giant tubellarian with hundreds of lateral suckers (fig. 38).

Baikal and other lakes of its system were colonised by the group under review chiefly during the Pliocenic and Miocenic cooling of the climate of Siberia and the retreat of the fauna of the northern regions to the south. The influx of immigrants from the north increased gradually at the end of the Pliocene and the beginning of the Pleistocene. In that period even high-Arctic elements, such as, for instance, *Salvelinus alpinus erithrinus*, could work their way into Baikal and other water bodies of the Baikal mountain region.

Descendents of Dwellers of Ancient Central Asian Basins

In this group we include the more enigmatic elements of the Baikalian fauna whose endemism is expressed most vividly. It incorporates endemic species, genera and even families. They are remotely related to the modern fauna of the running waters of China (mollusks of the genus *Kobeltocochlea* related to *Lithoglyphus*) the fossil faunas of the Tertiary Central Asian basins (Baicaliidae mollusks), the modern dwellers of the big lakes of South Siberia and Mongolia (the sponges Lubomirskiidae in Lake Dzhegetai, the mollusks *Choanomphalus* and *Kobeltocochlea* in Lake Khubsugul), and, finally, some species inhabiting bodies of water in South Asia (the leeches Toricinae, the bryozoan *Hislopia*). It is possible that gammarids too may be attributed to the same group.

It was this complex of forms that was regarded by G. Y. VERESHCHAGIN (1940b, etc.) as being chiefly marine, and by L. S. BERG (1949a, etc.) as a survival of the heat-loving fresh-water fauna that was widespread in Eurasia and North America at the end of the Tertiary period. Basing himself on palaeontological data, G. G. MARTINSON in his early works called approximately the same group of mollusk forms "Balkanian." At present, as mentioned above, he prefers to call it neolimnic.

So far there is no possibility of clearly distinguishing the Central Asian complex from Tertiary Siberian Holarctic forms. In any case, it can be asserted that in the first half and in the middle of the Tertiary period North Siberia (the geologists' Siberian platform), on the one hand, and Central Asia with adjacent regions of South Siberia, on the other, presented two independent centres of the formation of both aquatic and terrestrial fauna.

The part of Siberia to the north and north-east of the South Siberian mountain systems, drained by the great Siberian rivers, abounded in shallow lakes. There, in a temperate climate, the hydrofauna which served as the foundation for the modern Holarctic fauna had been developing for millions of years.

At the end of the Mesozoic and in the Tertiary period Central Asia, as distinct from Siberia, was a tectonically active land of mountain structures and big lacustrine basins. Evidently some of them often approached extensive semi-marine basins which intruded deep into Asia from the east, or on remnants of seas of the Tethys Ocean in the west.

The dissected relief of the country, climatic contrasts and the nearness of sea basins favoured the development of a fauna differing from the North Siberian fauna of the Tertiary period. From its origin till the Pliocene cooling the region of the Baikal lake system fringing on the northern outskirts of Central Asia must have been under a strong influence of the Central Asian and also perhaps, to some degree, East Asian fauna, both the valley and heat-loving fauna and the more cold-loving fauna living in mountain basins, rivers, cave waters, etc.

The colonisation of the lakes of the Baikal mountain region by the Central Asian fauna must have taken place in the earliest stages of the formation of Baikal, in the first half or the middle of the Tertiary period and certainly before the Tertiary fauna began to penetrate there from North Siberia. This explains the particularly pronounced endemism of its descendants in Baikal.

The outward similarity between some modern species of Baikalian gammarids and gastropods and Caspian species (BAZIKALOVA, 1940, 1945; MARTINSON, 1945) is not a proof of their close kinship. Today we can consider it a sufficiently well established fact that in the Tertiary period the Pontocaspian region with adjoining areas as well as the Balkans presented independent centres of the development of hydrofauna, not connected with the Siberian and Central Asian centres characterised above. Of course, in different periods the faunae of these widely separated regions could fall under each other's influence, without losing, however, their distinctive features (KOMAREK, 1953; S. STANKOVIČ, 1932, 1955a, b, 1960).

As shown above, the Baikalian fauna has been found to contain a number of species whose closest relatives occur in underground waters and springs in Western Europe, the Balkans, as well as Japan, North America and other regions of the world far removed from each other. Among them are the crustaceans Bathynellidae (species of the genus *Bathynella*), water mites from the genus *Cerebrothrombidium*, and species of crustaceans from the genera *Acanthocyclops* and *Orthocyclops*, etc. Evidently this fauna is a remnant of a very ancient, partially perhaps even pre-Tertiary fauna that inhabited the underground waters of Asia and Europe in conditions of a warm climate. They and their descendants are extant in very few places, Baikal among them.

In recent years information has been accumulating about the presence of genera, related to Baikalian forms, among the turbellarians, oligochaetes, leeches, Toricinae, isopods *(Asellus)* and other groups of the fauna inhabiting the drainage basin of the Amur, the regions of China adjacent to it, and Japan. It can be supposed, therefore, that the old faunae of Central and East Asia exerted a considerable influence upon each other. A channel for this intercourse could be provided by the ancient Amur and the chain of extensive ancient basins referred to by G. G. MARTINSON (1955a), which stretched from Japan to Central Asia inclusive.

Immigrants from the Arctic Ocean and Basins on Its Coast

This group includes *Coregonus autumnalis migratorius* and the seal *Phoca sibirica*, together with the parasites they brought with them. They must have penetrated into Baikal during the great advance of the Arctic Ocean in the Quaternary period, which was pointed out by I. D. CHERSKY back in 1877. During this advance the waters of the ocean covered a vast territory of the modern Arctic coast, and the great Siberian rivers had to cover much shorter distances before emptying into it. Besides, during the retreat of the glaciers, they carried much more water than now. The omul and the seal, having penetrated into Baikal via these rivers, found there conditions resembling those of their distant birthplace.

The origin of the progenitors of the modern Baikalian Cottoidei, including the family Comephoridae, cannot be considered to be sufficiently elucidated. D. N. TALIYEV (1955) propounds the idea that their progenitors penetrated into Baikal along the Amur system from the Far-Eastern seas in Quaternary and even post-Quaternary time. These progenitors, according to D. N. TALIYEV, diverged from the Far-Eastern marine Cottoidae.

If the supposition concerning the Far-Eastern marine progenitors of the Baikalian Cottoidei is confirmed, a group of animals of marine extraction should also be distinguished in Baikal. But there are no grounds to accept the idea that they appeared in Baikal only in post-Pliocene time. They could not have penetrated there later than the progenitors of gammarids and other distinct endemic species of Baikal.

5. Evolution of Fauna in Baikal.

As shown in the foregoing pages, in its long history Baikal has communicated with different biogeographical regions and received immigrants from them. This type of colonisation of old lakes, islands and other isolated areas by immigrants is referred to in the literature as multiple colonisation, and some authors hold it chiefly responsible for the variety and abundance of the fauna in old lakes. For instance, E. MAYR (1944, 1947) considers that old fresh-water lakes are, for fresh-water fauna, very much what old islands are for terrestrial fauna. They permit the survival of old elements which have long since become extinct in the surrounding areas. In MAYR's opinion, such old lakes as Baikal, Tanganyika, Nyasa and others were peopled by means of multiple colonisation.

But all the variety of life in old lakes cannot be explained by colonisation alone. Students of their fauna come to the conclusion that the majority of species inhabiting them today have evolved in these lakes themselves. This opinion regarding the African lakes was expressed by WORTHINGTON (1954) and other authors who have studied the fauna of the great lakes of Africa. Identical views were voiced on the fauna of the old lakes of Central Celebes by WOLTE-RECK (1931), on the fish of Lake Lanao in the Philippines by HERRE (1933), on Lake Ohrid by STANKOVIČ (1932, 1955a, b, 1960), and, finally, on Lake Baikal by DOROGOSTAISKY (1923b) as well as L. S. BERG (1949a), G. Y. VERESHCHAGIN (1940b) and almost all modern students of the systematic composition of the Baikalian fauna.

The scope of this autochthonous process of the evolution of fauna in Baikal itself can be judged by the following data.

Table XXXIII.

The number of presumed parental species for some groups of the fauna of Baikal and the number of species and genera which have evolved from them in the lake itself.

Groups	Number of presumed ancestors (parental species)	Developed from them in Baikal		
		Species	Genera	Families and Subfamilies
Porifera, fam. Lubomirskiidae	1—2	6	3	1
Turbellaria, gen. *Sorocelis, Baicalobia, Bdellocephala*	3—4	up to 40	?	—
Ostracoda, gen. *Candona*	1—2	19	—	—
Isopoda, gen. *Asellus*	2	5	—	—
Amphipoda, fam. Gammaridae	4—5	239	34	—
Trichoptera, group Baicalinini	2—3	10—12	2	—
Gastropoda, fam. Baikaliidae	1—2	34	2	1
subfam. Benedictiinae	2	5	2	1
subfam. Choanomphalinae	2	7	2	1
Pisces, fam. Comephoridae, subfam. Abyssocottinae and Cottocomephorinae	2—3	23	7	3

Intense speciation is also observed among the benthic Cyclopoida, chironomids and many other groups of fauna.

At the same time, another significant peculiarity of the Baikalian fauna is conspicuous: missing from it are representatives of many important taxonomic groups. The open waters of Baikal have no big Unionidae; Ephemeroptera occur very rarely; nor are there the gastropod families Viviparidae and Bathyniidae, as well as Plecoptera, Decapoda, Branchiopoda and other groups common in the waters of Siberia and Europe. Therefore, the conditions of Baikal must have proved suitable for only a few representatives of the old hydrofauna that inhabited the fresh waters of adjacent regions. Also notable is the higher individual and group variability of Baikalian species, especially those belonging to families and genera with numerous species. For instance, D. N. TALIYEV (1948, 1955) emphasises the strong individual variability of the Baikalian species of Cottoidei, while A. Y. BAZIKALOVA points in a number of her works to the exceptionally high individual and group variability among gammarids. Among the Mollusca, too, we find numerous species which vary strongly and slightly differ one from another.

It should be noted also that numerous species from groups where differentiation has been particularly vigorous unite around a few basic types in what may be called series of very similar forms. For instance, the mollusks of the family Baicaliidae are represented by 34 species differing in size, the number of whorls and the kind of ornamentation on the shell. But as regards the structure of such vital anatomical characters as the nervous, genital and digestive systems, all of them present a single group which are very close to each other (KOZHOV, 1951, etc.) A. Y. BAZIKALOVA (1945) points out the same phenomenon among the Baikalian gammarids. D. N. TALIYEV (1940a), studying the Baikalian Cottoidei by serodiagnostic methods, established that differences between them are very insignificant and much lower than generic and specific serodiagnostic indices for representatives of the family Salmonidae.

These peculiarities of the Baikalian fauna bear a strong resemblance to analogous features of the faunae of other old lakes (Tanganyika, Nyasa, and others) and oceanic islands. Baikal can be called an "island" amidst the "ocean" of the Palaearctic, inhabited by a peculiar "insular" fauna with its own specific features, such as profound endemism, high variability of species and the presence of slight differences among them.

According to modern conceptions Baikal consisted of an extensive body of water or a system of such waters in the middle of the Tertiary period, and thus there has been sufficient time for the evolution of its fauna. To substantiate their hypotheses, V. C. DOROGOSTAISKY (1923b), who insisted on the Quaternary age of the whole of the Baikalian fauna, and D. N. TALIYEV (1948), referred to the once popular mutation theory. D. N. TALIYEV (1948) wrote: "Is it not that Baikal has, in view of its young tectonic age and deep position in the lithosphere, specific features conducive to the mutational variability of deep-water dwellers?" We think that it is perfectly possible to do without such assumptions in explaining the evolution of the Baikalian fauna.

Two main trends in the process of the divergent differentiation of species into large groups of new species are discussed in present-day literature.

1. Allopatric speciation (in E. MAYR's terminology, 1944), with geographic spatial isolation of the populations of the initial species as the main factor for the development of hereditarily stable subspecies and species.

2. Sympatric speciation without preliminary isolation of populations by any geographic (topographic and other) obstacles, i.e., in a non-dissected lake or any other region isolated from its neighbours.

E. MAYR (1944) contends that the formation of new species from old ones can be, in the main, only allopatric, i.e. induced by spatial geographic isolation. Sympatric speciation, in his opinion, is hardly possible. He is supported by J. BROOKS (1950), who endeavours to prove in his compendium on old lakes that in these lakes, Baikal included, species were formed chiefly thanks to geographic, spatial isolation of the populations of the parental species in separate sections of the lake. Every new fauna, in the opinion of these scientists, becomes hereditarily stable, i.e., physiologically distinct, only after prolonged full or partial spatial isolation from congeneric forms. Only this fauna can retain its originality when the conditions change, i.e., when the barriers break down and it again comes into contact with forms allied with it.

Discussing the question of speciation in Lakes Tanganyika and Nyasa, BROOKS points to the presence of spatially isolated biotopes along their shores, inhabited by populations of different species. Among such biotopes he mentions shallow firths, bays, gulfs separated by deeps and other insuperable obstacles. Special importance is attached to numerous marginal lakes into which the fauna of the main lake can penetrate and remain there in isolation for a long time, till their new merger with it. Resettling in the main lake, the new forms, which have now become hereditarily stable, find suitable biotopes and can diverge further on, but already on the basis of biological relationships. Exactly the same picture of allopatric fauna formation BROOKS tries to draw for Baikal.

At the present time we have sufficient data to affirm that the development of forms in Baikal on the basis of isolation did and does take place indeed.

As shown above, the modern zoogeographical zoning of the lake stems from the differences in the composition of the fauna inhabiting its various sections. We have seen that differences are especially marked between the faunae of the northern and southern parts of Baikal, which can only be explained if we assume that but recently it consisted of at least two parts not fully separated by an above-water elevation. There are both topographic and biological grounds for such an assumption. Local forms also exist in such areas of Baikal with a distinct habitat as the Ushkani Archipelago, the Maloye More, the Selenga shoal, some sors, etc. The great length of the Baikal shoreline itself (more than 2,000 km) is evidently conducive in some measure to the isolation of far separated populations of one and the same species occurring in one depth zone all along the shoreline. Here are a few examples.

A common species amongst the Baikalian mollusks is *Choanomphalus maacki* (fig. 66,8), which inhabits the littoral of the whole lake. A study of the populations of this species shows how the forms of its shell vary from place to place. In the south-western part of

Baikal, where the bottom of the littoral is predominantly rocky, *Ch. maacki* has a very tall, almost turreted shell and a wide funnel-shaped umbilicus with steep and flat walls (var. *andrussowianus* LDH.). Farther to the north along the south-western shore (the Goloustnaya section), the shell of *Ch. maacki* becomes shorter and correspondingly broader (f. *typica*). In the Maloye More and to the north of it, where sandy bottom predominates, the shell of this species is entirely flat, with a wide umbilicus (var. *elatior*). This variety is typical of this species in the northern part of Baikal and along its south-eastern shore (the North Baikal section), while to the south of the Selenga shoal we again find the tall South Baikalian var. *andrussowianus*. A similar picture is to be observed among *Ch. amauronius*. The differences in the form of the shells of these (and many other) species from various habitats are so conspicuous that the majority of local forms were described formerly as separate species. The mollusk *Baicalia herderiana* abounds on the rocky bottom of the littoral along the southern shore, where the cone-shaped shell of this species has well defined costae lying across the whorls (fig. 67,2).

In the littoral of the northern half of Baikal *B. herderiana* is replaced by a closely related species, *B. variesculpta* (fig. 67,1), whose shell is characterised by the conversion of the costae into rows of knolls. But at the meeting-point of these two species, along the western shore in the area of the Anga-Buguldeika, forms can be found that are clearly of a transitional, apparently hybrid nature. The species *B. angarensis*, which is very close to *B. herderiana*, lives in the Angara. Its shell is entirely devoid of costae and is much smaller than that of *B. herderiana*, and these characters are apparently explained by conditions of life in a rapidly flowing river. Several local varieties and even species are formed by the gammarids of the genera *Micruropus, Hyalellopsis, Pallasea, Eulimnogammarus, Odontogammarus*, and others.

Allopatric speciation must also have been favoured by the fact that at one time Baikal was the centre of the giant Baikal system of tectonic troughs situated to the west, east and north-east of it which, as has been noted above, contained extensive lakes. There may have been intercourse between the littoral faunae which were developing autochthonously in these lakes and a few remnants of the Baikalian fauna still survive in the big running-water lakes filling these troughs.

But a deeper analysis of the composition and distribution of the Baikalian fauna in the lake itself and outside of it leads one to believe that the main part in its evolution from relatively few parental forms was played by the gradual deepening of the lake, the resulting appearance of new biotopes, their colonisation and, as a consequence of it, the gradual splitting of species into littoral and deep-water forms without any marked participation of spatial geographic isolation of the diverging populations, i.e., by the sympatric way.

The subsidence of the bottom gradually changed the physical outlook of the lake, which was developing into a vast and deep reservoir. The thermal and chemical regimes of the lake, seasonal rhythmics, the composition and quality of its bottom soil, light-intensity, etc., were also changing. All this in itself served for the inhabitants of the lake, which were accomodating themselves to the new conditions, as a factor for their physiological isolation from the dwellers of ordinary lakes or rivers of the area around Baikal. Thus the deepening of the trough and the appearance of new bio-topes in the lake itself created conditions for the differentiation of littoral species into deep-water populations which subsequently evolved into varieties and species without any isolating geographic obstacles.

The proposition that spatial geographic isolation could play only a secondary part in the evolution of Baikal can be illustrated with the following facts.

As has been shown above, via the systems of the Angara and the Yenisei and ancient arms of the latter some Baikalian forms pene-trated far to the north and settled in such Arctic bodies as Lake Taimyr and the lakes of the Gyda watershed (between the Yenisei and the Ob). There they were fully isolated from the initial Baikalian and even the Angara-Yenisei population of the same species. Appar-ently they have lived in this isolation for many millenia and have remained very close to the Baikalian species, even to the point of remaining identical with them.

Despite the still longer period of isolation from the initial Baikal-ian species, those Baikalian species which have survived in the vestiges of old lakes of the Baikal system and in the drainage area of the Vitim and the Chara (the Lena watershed), differ from them only within the limits of very closely related species and varieties.

An essential fact, in our opinion, for a correct understanding of the processes of autochthonous speciation in Baikal is that the greatest number of endemic forms is observed, not in the shallow littoral area, but in the zone of depths of 8 to 200 metres, even though horizontally it is uniform for the entire lake. This zone is the habitat of all Baikalian species of sponges, 80% of mollusks and 75% of oligochaetes and turbellarians. Only 6 from the 28 forms of the fish Cottoidei live in the 0—5 metre zone, whereas the 5—100 metre zone is inhabited by 22 forms, the 100—300 metre zone by 25 forms, the 300—500 m zone by 21 forms; and 16 forms live deeper than 500 m (TALIYEV, 1948).

According to BAZIKALOVA & TALIYEV (1948), 49 species and varie-ties of gammarids (17.1%) live in the 0—5 metre zone; 147 (49.8%) in the 5—70—150 metre zone; 41 (14.4%) in the 70—150—300

zone; and 28 (18.7%) in the 300—to—500 metre zone. Two-thirds
of the gammarid genera (20 out of 31) are represented by numerous
species living almost exclusively below the 10—15 metre depths,
with about 15 of them being represented by species whose main
habitat lies below the 80—100 metre zone. The Bathynellidae, the
majority of the Cyclopoida, etc., also live beyond the littoral. It
should be borne in mind that the deep-water fauna of Baikal has
not been studied so carefully as the fauna of the littoral. In the
future the number of deep-water species known is likely to grow
considerably.

The deeper and more comprehensive our knowledge of the
Baikalian fauna becomes, the more we become convinced that the
zone of depths between 5—10 and 100—200 metres was and remains the original birthplace of the vast majority of Baikalian
endemics, a centre in which the fauna colonising Baikal, shallow-
water fauna in the broad sense, developed.

G. Y. VERESHCHAGIN (1935) considered that the fauna inhabiting
depths below 500 metres consisted solely of old endemic forms,
while the fauna of lesser depths presented a mixture of old endemic
and younger elements. But we cannot agree with him. Old parental
forms can only be an exception at great depths. As is shown by
detailed morphological studies of series of related species succeeding
one another vertically, the colonisation of the deep zones of Baikal
started from the shallows and was accompanied by the divergence
of littoral forms settling in new deep-water biotopes. Thus, the
majority of abyssal species of gammarids, as was shown by A. Y.
BAZIKALOVA (1945), are closely connected with modern littoral and
sublittoral species, as follows from this comparison (Table XXXIV).

Table XXXIV.

Primarily littoral and sublittoral genera of gammarids	Sublittoral-supra-abyssal and abyssal genera or species that diverged from them
Eulimnogammarus	Abyssogammarus, Pachyschesis, possibly Lobogammarus, Ommatogammarus, Polyacanthisca, as well as numerous sublittoral-supra-abyssal species of Eulimnogammarus
Crypturopus	Homocerisca
Acanthogammarus	Brachiuropus, Carinurus, Axelboeckia, Garjajewia, abyssal species of Acanthogammarus
Pallasea	Abyssal species of Pallasea, as well as Parapallasea, Hackonboeckia, Ceratogammarus, Cheirogammarus

It is interesting that the burrowing forms of abyssal gammarids
are closely related to the burrowing forms inhabiting the shoals.
For instance, according to BAZIKALOVA (1945), the abyssal species

Homocerisca caudata is closely related to the shallow-water *H. perla*; *Carinurus reissneri* to *C. ssolskii*, etc.

Analogous series of species are also found among the mollusks. For instance, the initial form for the gastropods of the genus *Benedictia*, which includes 4 species with several varieties, probably was *Benedictia baicalensis* (fig. 70,3), which abounds in the zone of depths between 2 to 3 and 15 to 20 metres, or *B. limnaeoides*, which is anatomically close to it and inhabits the zone between 10 and 30 metres. *B. fragilis* (fig. 70,1, 2) which is closely related to the latter, lives at 30 to 40 metres and deeper, and, finally, *B. maxima*, its more distant relative, is common for the 50—200 metre zone and greater depths. The abyssal fauna of oligochaetes, according to V. V. IZOSIMOV (1949), is also young, i.e., it is in the period of formation and is allied with the shallow-water fauna. Sponges occur chiefly in the littoral and sublittoral, but they, too, form a series of species succeeding one another vertically: *Lubomirskia baicalensis* (fig. 32) is a littoral species, while *Baicalospongia bacillifera* (fig. 33) and *B. intermedia* reach depths of 400 and 500 metres and more. Possibly, some abyssal species have already lost their initial roots. As regards the Bathynellidae found at great depths, A. Y. BAZIKALOVA (1954c) assumes that their ancestors colonised ever greater depths parallel with the progressive subsidence of the bottom of Baikal.

Nor is it accidental that it is precisely the groups of fauna with the greatest radiation of species that populate Baikal from the edge of water to the greatest depths (Gammaridae, Turbellaria, Oligochaeta, Cottoidei, Cyclopoida), or, at least, to very considerable depths.

With the gradual deepening of the lake and increase of its water mass a distinct pelagic fauna was autochthonously evolving from originally bottom and near-bottom forms of great depths. For instance, the genus *Poekilogammarus* (fig. 61,3) diverged from the benthonic genus *Pallasea* (fig. 59,2), while the nectobenthic but deep-water species of the latter gave rise to the genus *Macrohectopus* (fig. 60,1) with only one species, *M. branickii*, which assumed an exclusively pelagic way of life. First nectobenthic and later typically pelagic species comprising the endemic subfamilies Cottocomephorinae (fig. 74,4) and the endemic family Comephoridae (fig. 73,1) developed from the originally benthonic and then semi-pelagic species of Cottoidei.

How, then, in what spatial isolation, in what bodies of water or their parts could be formed the abyssal benthic and pelagic species, genera and families inhabiting Baikal today?

If it is assumed that the abyssal species could form somewhere outside Baikal in deep lakes of the Baikal area, then it remains a mystery how they could penetrate into its depths, being absolutely incapable of enduring the conditions of life in rivers and lake shallows.

It can be supposed that, as the depths grew and free biotopes appeared, populations of various littoral species were forced to move there from the densely populated littoral area. There, in the deep zone, they found themselves in a new environment, especially as regards the temperature, light intensity and composition of the bottom soil, which led to a change in the characters of the immigrants.

The new thermal and light conditions directly affected the physiology of the new settlers first and foremost. Their growth and development were inhibited in view of the lower temperatures and their insignificant seasonal changes; the seasonal rhythm in the life cycle and especially in the process of propagation began to disappear; the life span grew longer; the period of immaturity was lengthened, and so on. All this was bound to bring about the physiological deviation of the immigrants from the parental littoral species and, consequently, their evolution into hereditary settled forms.

As a result of millions of years of existence in these conditions, and under the influence of natural selection, the cycle of reproduction and growth in the abyssal descendants of littoral species has become greatly protracted, while some of them have acquired the ability to multiply all the year round. The colder water retarded the development of the embryo in the body of the mother, which led some species to becoming viviparous (Comephoridae, some turbellarians, etc.). As a result of the lower temperatures (especially in the glacial period), as is pointed out by A. Y. BAZIKALOVA (1948a, 1951a, 1954b) and M. Y. BEKMAN (1958), some species of the Baikalian gammarids have become neotenic, as it were, with respect to their progenitors i.e., after reaching maturity they retain juvenile features owing to the underdevelopment of certain morphological characters.

Also conspicuous is the "gigantism" of many abyssal species. For instance, among the mollusks of the genus *Benedictia* the biggest species, *B. fragilis*, lives deeper than others. The abyssal species of gammarids and oligochaetes are also, as a rule, bigger than their littoral relatives. The biggest turbellarian, *Polycotylus* sp., which reaches 30 centimetres in length, occurs only beyond the littoral. The biggest of the isopod species, *Asellus dybowskii*, lives at great depths, while the biggest Baikalian oligochaetes, *Limnodrilus inflatus, L. bythius* and *Rhynchelmis brachycephala*, are abyssal dwellers. Gigantism is also observed among the abyssal species of Cottoidei. V. N. YASNITSKY (1952) pointed out, as a specific feature of the Baikalian algae, the greater size of local forms as compared with related forms from ordinary lakes, with gigantism being especially pronounced among endemic open-water species.

It should be noted, however, that along with gigantism an opposite phenomenon, "nanism", is observed among the abyssal species, which include dwarf forms. For instance, the mollusk, *Baicalia nana*, which is the smallest of all the presently known Baicaliidae, lives at great depths. There are also small-size species among the deepwater burrowing gammarids, especially those of the genus *Micruropus*.

A clear picture of changes induced by the low light intensity is also to be observed. Practically complete darkness at depths of more than 200 metres reduced the value of the organs of sight. When studying the abyssal species of Gammaridae, Isopoda, Mollusca, Chironomidae, and Cottoidei we see that as they settled at ever greater depths their eyes grew more and more useless. The gammarids have retained only recesses on the sides of the head where the eyes used to be and orient themselves in space and find food with the help of antennae, which have become extraordinarily long. The eyes of the abyssal species of Cottoidei species have been reduced.

It is interesting that, parallel with the disappearance of the eyes' ability to perceive light, i.e., as the organs of sight grew increasingly useless, the eyes sockets of gammarids lost the regularity of shape, a feature which was observed by A. A. KOROTNEV back in 1901. The irregular variability of such "eyes" can only be explained by the fact that, having become useless, they were no longer subjected to the strict control of natural selection.

ᵡThe absence of light had a direct and adequate effect on the pigmentation of the body, which lost the colour pattern and variety of colours so characteristic of shallow-water dwellers and turned pale-pink or dirty-grey all over, as distinct from the bodies of species inhabiting the illuminated zone. In the latter, the vivid colouring and pattern in many cases are obviously of protective importance. The littoral species of gammarids and turbellarians can be unmistakably determined by the colour pattern and pigmentation of the body. In abyssal species, the development of characters needed for survival proceeded in obviously different directions, such as perfection and growth of tactile organs (antennae in the Gammaridae (fig. 58,2, 4) and organs of the lateral line in the Cottoidei), compression of the body and adaptation of the position of the feet for movement in tough substratum in some gammarids and *Asellus dybowskii* (fig. 57c), or extraordinary elongation of the ventral seta in the Oligochaeta, etc.

The influence of these abiotic conditions on the populations of species colonising deep-water biotopes was bound to lead to their physiological isolation from the parental populations and the impossibility of crossing with them, which was enough for the development of genetically stable forms.

This process of speciation and evolution of new forms proceeded in almost all groups of animals settling in deep-water zones and was especially intense among the Gammaridae, Cyclopoida, Turbellaria, Mollusca and Cottoidei. Thus new distinct biocoenoses formed in the depths of the lake, in which the species entered into close and contradictory relationships (struggle for food and for the preservation of the progeny, parasitism, commensalism, etc.). These contradictions were resolved through the development of new characters facilitating survival. λ

As regards the kind of food and method of feeding, the deep-water fauna differentiated into detritus-eaters and predators. Among the gammarids there developed bottom-dwelling and pelagic carnivores moving about briskly or lying passively under cover in wait for food. Some species of deep-water gammarids, attacking weak fish, get under the skin and eat away the intestines; the turbellarian *Polycotylus* (fig. 38), which has hundreds of lateral suckers, envelops its prey and sticks to it, and so on. D. N. TALIYEV (1947, 1948, 1955) points out that the feeding habits of deep-dwelling Cottoidei vary widely. For instance, *Cottinella boulengeri* and *Abyssocottus werestchagini* feed mainly upon vegetable and animal matter and burrow in the upper layer of the soil. The bathypelagic predatory forms *Limnocottus bergianus* and *Batrachocottus multiradiatus* as well as *Abyssocottus gibbosus* partially burrow in the soil and snap up at everything that swims by with their large mouths. *Limnocottus godlewskii* and *Abyssocottus korotneffi* actively hunt for prey but they also burrow in the silt with their duck-bill snouts and catch amphipods there. *Batrachocottus baicalensis pachytus* can literally walk on silty soil in search of prey. *Batrachocottus nikolskii* waits for prey in narrow slits on rocky patches. D. N. TALIYEV (1955) writes that this species is not only a bottom-dwelling but also bathypelagic form which has adapted itself to feeding upon species of *Comephorus*. Thanks to its extremely low specific weight (1.028) equalling that of *Comephorus*, it passively rises to above-bottom strata and floats there, snapping at everything that happens to pass by (TALIYEV & KORJAKOV, 1949).

In reply to this adaptation to the predatory way of life, the prey develop a wide diversity of protective faculties. Some species of gammarids bear long and sharp spines of different forms on their bodies; their ability to sense the approach of the enemy becomes keener; they begin to burrow better, and so on.

Mention should also be made of one more interesting form of adaptation, developed by caddis-flies (Trichoptera) in the conditions of Baikal.

The modern 14 to 16 species of endemic Baikalian caddis-flies have doubtless descended from winged members of the ancient family Limnophilidae. But in such a deep and extensive lake as Baikal its progeny have lost the ability to fly, while in some species *(Baicalina reducta, Thamastes dipterus)* the wings have turned into "paddles" used for swimming (fig. 63). Why and how has it happened? It is known that in the ordinary species of caddis-flies the wings are needed by the imaginal forms for flying during the mating period, when thick swarms of these insects fly along the shores of ordinary reservoirs above the water, into which the females deposit fertilised eggs. In shallow waters winds can facilitate the dispersal of the species and thus play a positive role. But in Baikal, where the shallow zone is very narrow, if the insect flies or is driven by the wind farther than 50 to 100 metres from the shore its young perish, for they cannot get out of the deep-water zone. Consequently, it is to be supposed that, as was the case with the wingless insects of oceanic islands, the reduction of the wings or their muscles proceeded by the selection of individuals with an impaired faculty of flight.

The further divergent evolution of the Trichoptera species must have been influenced to a certain extent by the temperatures. Some species deposit clutches at the edge of water, with their larvae developing close to the shores; others, in deeper zones. There are species (*Radema* sp.) which mate early in spring, when the ice has not melted away yet. Their pupae squeeze themselves to the surface through cracks in the ice, where the mating takes place and eggs are deposited in cracks and openings in the ice.

Of course, it was not temperatures alone that influenced the physiological isolation of the diverging populations of species. It obviously may have been induced, at least among the fish, also by differences in the location of spawning grounds.

For instance, the omul *(Coregonus autumnalis migratorius)*, which penetrated into Baikal comparatively recently, has already split into a number of types or races, depending on the place and period of spawning; the bottom-dwelling Cottoidei also differ in spawning habits; different periods of spawning divide the pelagic fish *Comephorus* into two distinct species.

These examples do not exhaust all possible factors which induce and determine the trend of species formation in Baikal. But they show that sympatric speciation in the deep zones of Baikal not dissected horizontally played a very important part role in the evolution after fauna of this lake.

As has been noted, the formation of the Baikalian fauna began in a climate which was warmer than today. It is possible therefore that some of its species which developed in the Tertiary period could not endure the cooling and particularly the rigorous conditions of the glacial period. The fossil fauna of the Tertiary terraces of Baikal offers many examples of the modification of the composition of the fauna in the period since the Miocene to our days. Such species of mollusks as *Baicalia duthiersioides* MART., *Baicalia kožovi* MART. and *Choanomphalus fossilus* MART., remains of which have been preserved in the Tertiary deposits of the Baikal terraces, no longer exist.

But the importance of the influence of the glacial cold on the living world of Baikal should not be over-estimated, for even before it the temperature of Baikal's deep waters could not have been appreciably higher. It is known that the temperature of the bottom strata of water in the Balkan Lake Ohrid does not exceed 6° C the year round (STANKOVIČ, 1960), and that at present this lake is approximately in the same climatic conditions which were found in South Siberia in the later half of the Tertiary period.

Under the impact of these factors there has developed from genetically different roots a distinct world of Baikalian organisms sharply divided into two main ecological community complexes: littoral-sor and Baikalian proper.

The littoral-sor complex community, as has been shown, inhabits the littoral-sor region of the lake. Genetically, it consists of Siberian limnophiles but it also includes a few native Baikalians: about a dozen species of gammarids, the oligochaete *Limnodrilus arenarius* MICH., and partly the polychaete *Manayunkia baicalensis*. The Baikalian complex proper lives in the open waters of the lake. This complex, in turn, should be differentiated into the following ecological groups (communities):

The littoral group community, which inhabits the zone from the edge of water to the depths of 15 to 20 metres. It was described in detail in the chapter on the distribution of benthos. The vast majority of the species of this littoral group are descendants of dwellers of Siberian Tertiary waters (Tertiary Holarctic forms) and immigrants from Central Asian basins (the Central Asian genetic complex). But it also incorporates a multitude of species of the modern Siberian limnorheophile fauna most of which live in the littoral. Amongst them are the fish *Coregonus lavaretus*, *Thymallus arcticus* with its varieties, *Lota lota*, *Leuciscus leuciscus baicalensis*, and open-water Baikalian forms of other groups of the Siberian Limnorheophile fauna described in the chapter devoted to benthos.

Bathyal group community, which occupies the sublittoral and supra-abyssal zones and in its specific composition forms a kind of intermediate link between the littoral and abyssal groups. Almost all species of this group are descendants of Tertiary northern Holarctic forms and immigrants from Central Asia, and all of them are characteristically endemic, with only a few Siberian limnorheophiles.

Abyssal group community, which inhabits depths below 250—300 metres. It is represented exclusively by descendants of the ancient Holarctic and Central Asian genetic complexes exhibiting clear traces of the effect of abyssal habitation.

Pelagic group community, which lives in the open waters of Baikal. Its specific composition, which is rather scanty, includes Siberian limnorheophiles with weakly expressed endemism, immigrants from big running-water Siberian and Siberian-European lakes, especially mountain lakes, (Rotatoria, *Cyclops*, *Epischura*), and recent immigrants from the Arctic regions (*Coregonus autumnalis migratorius* and the seal).

CONCLUSION

In this book we have tried to sum up the results of the most important studies of the living world of Baikal. Appraising these results as a whole, we feel entitled to maintain that at present Baikal is one of the most thoroughly studied great lakes of the world. All these achievements, however, are merely a stage on the way to a most comprehensive knowledge of the lake. The results of recent studies make possible the delineation of the basic problems and directions to be followed in further work, some of which are discussed below.

Many researches have laid a firm foundation of the knowledge of the hydrological regime of the lake. But there still exist gaps in this field, such as an insufficient study of wind-induced and thermal circulation of the waters; the depth of the layer reached by the influence of strong storms and gales; the horizontal currents and their intensity and direction. All these are essential factors in the distribution of life in Baikal, and their year-round study is an important task for future researchers.

Baikal provides exceptionally favourable conditions for the study of such an important phenomenon as diurnal vertical migrations of pelagic organisms. But success in this study requires the knowledge of light intensity in different layers and the depth of penetration of various rays of the solar spectrum into the mass of water with and without the ice cover. This research must proceed parallel with the study of the migrations themselves. It should also be borne in mind that Baikal, being as it is the world's greatest fresh-water body, can serve as a site of serious research into the penetration of cosmic rays into the mass of water and their impact on biological phenomena.

Recently it has been found that years with abundant crops of algae and years with very poor crops tend to alternate in Baikal. Sharp fluctuations have also been found to occur in the crop of zooplankton. Such fluctuations take place in many continental and sea basins, but their real causes are far from being known. In this respect, too, Baikal presents an exceptionally convenient "laboratory" owing to the relative simplicity of interspecific contacts in the mass of water and the possibility of stationary studies of the biology of species over many years. It is necessary to continue and develop, with the use of the latest methods, experimental and field investigations into the biology of planktonic species, their life cycle, their requirements as regards life conditions, and the conditions of their productivity. There is pressing need for the continuation of systematic observations of the seasonal and annual fluctuations in the amount of the most important biogenous compounds, their vertical and horizontal distribution and the pace of regeneration in different seasons of the year.

The data on the distribution of life in the deep-water region, which occupies four-fifths of the area of the lake, are still scant and fragmentary. A systematic study of the benthos of the deep-water region of Baikal should be launched.

Only a modest start has been made on the study of the production of the main groups of the Baikalian benthos. To continue with this work is one of the immediate tasks of the hydrobiologists studying life in Baikal.

No qualitative studies have been made of phytobenthos, which plays an important part in the life of the littoral of the lake. It is necessary to study the phytobenthos, with its seasonal phenomena and production, etc.

Only the first steps have been undertaken in the study of the microbenthos, a highly important nutrient of many invertebrates which, in turn, serve as food for commercially valuable animals. Work on this problem must be organised in different zones of the bottom, with the use of qualitative methods. Studies in this direction can also enrich the knowledge of the specific composition of some poorly known groups of animals, such as small benthonic crustaceans, rotifers, tardigrades, mites, etc. The data on the microbiology of the water mass of the lake also cannot be considered sufficient. Systematic year-round studies should be started of the micro-population of Baikal in connection with biotic and abiotic factors.

Data are insufficient or entirely lacking on the systematic composition of turbellarians, free nematodes and some other groups.

The exceptional variety and wealth of endemic species in the Baikalian fauna, their increased variability, the presence of numerous closely related groups inhabiting different depths, the development of protective colouring and other adaptive characters, the peculiarity of biocoenotic contacts—all this makes it possible to raise and solve in Baikal important problems of species-formation and evolution. Of particular interest in this respect is the study of the abyssal fauna and the morphological and physiological changes in species succeeding each other with change of depth of habitation (organs of sight, tactile organs, etc.). A study of all these phenomena must become a most important task for many years to come.

The study of the living world of Baikal cannot be detached from the other aspects of its life bearing upon the solution of problems connected with the history of its fauna and flora. Indispensable for the success of biological research in these directions is the study of the relief of the lake, more detailed than ever before. Many problems still remain open, such as to what degree, and when, Baikal was connected with other tectonic depressions of the Baikal system.

A primary task is to study bottom deposits by modern methods making it possible to obtain cores dozens of metres in depth. It is extremely important to study ancient sediments in other depressions of the Baikal system, especially in the Barguzin, Tsipa and Muya-Chara depressions, for the big running-water lakes in the region of these depressions are even now inhabited by some species of the Baikalian fauna. In this connection it is essential that a study be made of the history of the modern outflow of water from Baikal into the Angara and the Yenisei, and that of an older one, into the Vitim and the Lena.

In view of the discovery of living and fossil fauna of the Baikalian type in some localities of Central Asia, research should be undertaken jointly with the scientific organisations of Mongolia and China into the fauna and flora of big water bodies of Mongolia and the adjacent regions of Central Asia. The further development of the studies of the fossil fauna of these regions is also of utmost importance.

This enumeration does not exhaust all the problems connected with Baikal. As the knowledge of the nature of this unique lake grows, ever new problems will arise. An integrated study of Baikal calls for the participation of many sciences. Therefore a uniform programme should be drawn up in the nearest future embracing the entirety of the diverse life of Baikal and serving as a guide in its studies over many years.

Such are the tasks facing the scientific establishments systematically studying the nature of the lake in all its aspects.

BIBLIOGRAPHY

ABRIKOSOV, G. G., 1924. To the knowledge of the bryozoan fauna of Lake Baikal. *Russ. Hydrob. J., Saratov,* **11—12:** *260—265*.*

ABRIKOSOV, G. G., 1927. On the freshwater bryozoans of the USSR. *Doklady Acad. Sci. USSR,* **19:** *307—312*.*

ABRIKOSOV, G. G., 1959. On the generic divisions and geographical distribution of the Gymnolemata bryozoans of continental waters. *Doklady Acad. Sci. USSR,* **126,** 6: *1378—1380*.*

AFANASYEV, A. N., 1960. The water budget of Lake Baikal. *Trudy Baik. limnol. St. Acad. Sci. USSR,* **XVIII:** *155—241*.*

AKATOVA, N. A., 1949. The zooplankton of the River Kolyam and its drainage. *Trans. Leningr. Univ.,* 126, biol. ser., **21:** *341*.*

AKHMEROV, A. K. & DOMBROVSKAYA-AKHMEROVA, O. S., 1941. New Acanthocephala from the fishes of the River Amur. *Doklady Acad. Sci. USSR,* **XXXI,** 5: *517—520*.*

ALPATOV, V. V., 1923. A water-bug new to Baikal and a comparison of the Asellus species of Baikal with those of Europe and America. *Russ. hydrobiol. J., Saratov,* **3—4:** *64—66*.*

ANDRUSOV, N. I., 1902. On two new gastropod species from the Apsheron stage. *Trudy SPB Nat. Soc., geol. and mineral. dept.,* **XXXI,** 5: *55—75*.*

ANNANDLE, N., 1911 (1912). Systematic notes on the Stenostomus Polyzoa oi freshwater. *Rec. Indian Mus. Calcutta,* **IV,** 4.

ANNANDLE, N., 1913. Notes on some sponges from Lake Baikal in the collection of the Imp. Academy of Science. St. Petersburg. *Yearb. Zool. Mus. Acad. Sci.,* **XVIII:** *96—101.*

ANNENKOVA, N. P., 1930. Fresh and brackish water Polychaeta of the USSR. In Determinative tables of freshwater organisms of the USSR. Freshwater fauna, **2,** Leningrad. Pub. USSR V.I. Lenin Acad. of Agric. Sci*.

ANPILOVA, V. N., 1956 a. The Baunt gwyniad in Leningrad Region reservoirs. *Sci. and tech. Bull. USSR Res. Inst. Lake and River Fish.,* **3—4:** *54*.*

ANPILOVA, V. N., 1956 b. On the systematic position of the Baunt vendace. *Trudy Acad. Sci. USSR,* **111,** 4: *898—902*.*

ANTIPOVA, N. L. & KOZHOV, M. M., 1953. Materials on the seasonal and annual fluctuations in the crop of some widespread forms of phytoplankton in Lake Baikal. *Trudy Irk. Univ.,* **7,** 1—2: *63—68*.*

ANTIPOVA, N. L., 1955. New species of the genus *Gymnodinium* Stein (Gymnodinicea) from Lake Baikal. *Doklady Acad. Sc. USSR,* **103,** 2: *325—328*.*

ANTIPOVA, N. L., 1956 a. On the formation of auxospores in *Cyclotella baicalensis. Bot. mat. cryptog. plants dept.,* **XI:** *39—42*.*

ANTIPOVA, N. L., 1956 b. About a new species of the genus *Cyclotella* from Lake Baikal. *Bot. mat. cryptog. plants dept.,* **XI:** *35—39*.*

ANUDARIN DASHI-DORZH, 1953. To the knowledge of the water bodies and hydrofauna of East and North Mongolia. Synopsis of the Master's theses, Irkutsk: *3—20*.*

ARNDT, W., 1937. Ochridaspongia rotunda, ein neuer Süsswasserschwamm aus d. Ochrida-see. *Arch. Hydrobiol.,* **31:** *636—677.*

---

* denotes works published in Russian

Ass, M., 1935. Ectoparasites of the Baikalian seal. *Trudy Baik. limnol. St. Acad. Sci. USSR*, **VI**: *23—29**.

Baburina, E. A., 1955. The eye and retina of the Caspina shad. *Doklady Acad. Sci. USSR*, **100**, 6: *1167—1170**.

Baranov, V. I., 1942. Development of the vegetative landscapes of the USSR in Tertiary time. In *Priroda*, **1—2**: *47—70**.

Bauer, O. N., 1948 a. Parasites of the fishes of the River Yenisei. *Izvestiya USSR Res. Inst. of Lake Fish.* **XXVIII**: *97—153**.

Bauer, O. N., 1948 b. Parasites of the fishes of the River Lena. *Izvestiya USSR Res. Inst. of Lake Fish.* **XXVIII**: *157—171*.

Bauer, O. N., 1948. On the systematic position of *Ankyrocotyle baicalensis* Wlasenka. *Doklady Acad. Sci. USSR.*, **69**: *383—386*.

Bazikalova, A. Y., 1935. Concerning the systematics of the Baikalian amphipods. *Trudy Baik. limnol. St. Acad. Sci. USSR*, **VI**: *31—52** (French summary).

Bazikalova, A. Y., 1940. Caspian elements in the fauna of Baikal. *Trudy Baik. limnol. St. Acad. Sci. USSR*, **X**: *358—367** (French summary).

Bazikalova, A. Y., 1941. Materials on the study of the amphipods of Baikal. Consumption of oxygen. *Izvestiya Acad. Sci. USSR. biol. ser.*, **I**: *67—81**.

Bazikalova, A. Y., 1945. Amphipods of Lake Baikal. *Trudy Baik. limnol. St. Acad. Sci. USSR*, **XI**: *5—440** (French summary).

Bazikalova, A. Y., 1948 a. The adaptive value of the size of the Baikalian amphipods. *Doklady Acad. Sci. USSR*, **61**, 3: *569—572**.

Bazikalova, A. Y., 1948 b. Notes on the systematics of the Baikalian amphipods. *Trudy Baik. limnol. St. Acad. Sci. USSR*, **XII**: *20—32**.

Bazikalova, A. Y., 1951 a. Morphological peculiarities of the young stages of the Baikalian amphipods. *Trudy Baik. limnol. St. Acad. Sci. USSR*, **XII**: *120—205**.

Bazikalova, A. Y., 1951 b. Concerning the growth of some amphipods from Lake Baikal and the River Angara. *Trudy Baik. limnol. St. Acad. Sci. USSR*, **XIII**: *206—216**.

Bazikalova, A. Y., 1954 a. Some data on the biology of *Acanthogammarus (Brachyuropus) grewingki* (Dyb.). *Trudy Baik. limnol. St. Acad. Sci. USSR*, **XIV**: *312—326**.

Bazikalova, A. Y., 1954 b. Age changes in some species of the genus *Acanthogammarus*. *Trudy Baik. limnol. St. Acad. Sci. USSR*, **XIV**: *327—354**.

Bazikalova, A. Y., 1954 c. New species of the genus *Bathynella* from Lake Baikal. *Trudy Baik. limnol. St. Acad. Sci. USSR*, **XIV**: *355—368**.

Bazikalova, A. Y., 1957. On the amphipods of the River Angara. *Trudy Baik. limnol. St. Acad. Sci. USSR*, **XV**: *377—387**.

Bazikalova, A. Y., 1959. New amphipod species from the Maloye More. *Trudy Baik. limnol. St. Acad. Sci. USSR*, **XVII**: *512—519**.

Bazikalova, A. Y., Birstein, Y. A. & Taliyev, D. N., 1946 a. Osmotic pressure of the cavity liquid in the amphipods of Lake Baikal. *Doklady Acad. Sci. USSR*, **53**, 3: *293—295**.

Bazikalova, A. Y., Birsrein, Y. A. & Taliyev, D. N., 1946 b. Osmoregulating capabilities of the Baikalian amphipods. *Doklady Acad. Sci. USSR*, **53**, 4: *381—384**.

Bazikalova, A. Y. & Vilisova, I. K., 1959. The feeding habits of benthophagous fish in the Maloye More. *Trudy Baik. limnol. St. Acad. Sci. USSR*, **XVII**: *382—497**.

Bazikalova, A. Y. & Taliyev, D. N., 1948. Some peculiarities of the divergent evolution of Amphipoda and Cottoidea in Lake Baikal. *Doklady Acad. Sci. USSR*, **59**, 3: *565—568**.

Bebutova, I. M., 1941. Biology and systematics of the Baikalian caddis-flies. *Izvestiya Acad. Sci. USSR, biol. ser.*, **1**: *82—104**.

Beklemishev, V. N., 1923. Some questions of the geographical distribution of freshwater triclads. *Russ. hydrobiol. J.*, **II**: *167—173** (French summary).

308

BEKMAN, M. Y. & BAZIKALOVA, A. Y., 1951. The biology and productive capabilities of the Baikalian and Siberian amphipods. *Trudy Conf. on fundamental and topical problems*, **1**: *61—66**.

BEKMAN, M. Y., 1952. Concerning the possibility of specific influence of the Baikal water on the organism. *Doklady Acad. Sci. USSR*, **87**, 2: *293—296**.

BEKMAN, M. Y., 1954. The biology of *Gammarus lacustris* Sars in Baikal area lakes. *Trudy Baik. limnol. St. Acad. Sci. USSR*, **XIV**: *263—311**.

BEKMAN, M. Y., 1958. Dwarf males among the Baikalian endemics. *Doklady Acad. Sci. USSR*, **120**, 1: *208—211**.

BEKMAN, M. Y., 1959. Distribution and production of zoobenthos species in the Maloye More. *Trudy Baik. limnol. St. Acad. Sci. USSR*, **XVII**: *342—381**.

BELYSHEV, B. F., 1956. The southern dragon-fly species from the hot springs of the north Trans-Baikal area. *Zool. J.*, **XXXV**, 2: *1735—1736*.

BERG, L. S., 1900. The fishes of Baikal. *Yearb. Zool. Mus. Acad. Sci.*, **V**: *326—372**.

BERG, L. S., 1903. Notes on the systematics of the Baikalian Cottidae. *Yearb. Zool. Mus. Acad. Sci.*, **VIII**: *99—114**.

BERG, L. S., 1907. Die Cataphracti des Baikalsees. Ergebn. Zool. Exp. nach d. Baikalsee, **III**, Kiew u. Berlin: *1—75*.

BERG, L. S., 1909. The fishes of the Amur drainage. *Trans. Acad. Sci. USSR, phys.-math. dept.*, **24**, 9: *1—269**.

BERG, L. S., 1910. The fauna of Baikal and its origin. *Biol. J.*, **1**, 1, Moscow: *10—45**.

BERG, L. S., 1922. The fauna of Baikal and its origin. In Klimat i zhizn (Climate and life), Moscow: *28—53**.

BERG, L. S., 1925. Die Fauna des Baikalsees und ihre Herkunft. *Arch. Hydrobiol. Suppl.*, **IV**: *479—526*.

BERG., L. S., 1928. New data on the origin of the Baikalian fauna. *Doklady Acad. Sci. USSR*, A, **22**: *459—464**.

BERG, L. S., 1934. Concerning the supposedly marine elements in the fauna and flora of Baikal. *Izvestiya Acad. Sci. USSR*, **IV**, 2—3: *459—464**.

BERG, L. S., 1937. Southern elements in the Baikalian fauna. *Trans. Leningr. Univ.*, **IV**, 17: *249—254**.

BERG, L. S., 1948—1949. The fishes of the fresh waters of the USSR and adjacent countries, **1, 2**: Leningrad*.

BERG, L. S., 1949 a. Essays on physical geography. *Publ. Acad. Sci. USSR: 280—338*.

BERG, L. S., 1949 b. About the Lower Cretaceous fish *Lycoptera* (fam. Lycopteridae). *Trudy Zool. Inst. Acad. Sci. USSR*, **VII**, 3: *58—75**.

BERG, L. S. 1955. The nature of the USSR. The Baikal and trans-Baikal areas. *Geografizdat, Moscow: 375—388**.

BEREZOVSKY, A. I., 1957. Concerning the study of the Baikalian omul. *Doklady Acad. Sci. USSR*, **21**: *353—358**.

Bibliography of Irkutsk region, biology, 1956. *Trudy sci. libr. Irk. State Univ., Irkutsk**.

BIRSTEIN, Y. A., 1939. A zoogeographical characteristic of the water-bugs of Baikal. *Doklady Acad. Sci. USSR*, **25**, 3: *248—251**.

BIRSTEIN, Y. A., 1951. Aselotta. In the fauna of the USSR. Crustaceans, **VII**, 5: *3—143**.

BIRSTEIN, Y. A. & SPASSKY, N. N., 1952. The bottom fauna of the Caspian Sea before and after the invasion of nereides. Acclimatisation of nereides in the Caspian Sea. Coll. of works ed. by V. N. NIKITIN. Materials to the knowledge of fauna and flora. New ser., Zool. Dept., publ. Mos. Nat. Soc., **33/68**: *36—114**.

BOCHKARYOV, P. F., 1935. Hydrochemical studies conducted in the Chivyrkui Gulf of Baikal in the summer of 1932. *Izvestiya Biol.-Geogr. Inst. Irk. Univ.*, **6**, 2—4: *134—151**.

BOCHKARYOV, P. F., 1959. The Hydrochemistry of the rivers of East Siberia. Book Publ. House, Irkutsk: *1—155**.

BOGOLEPOVA, I. I., 1950. Monogenetic flukes of the endemic fishes of Baikal. *Doklady Acad. Sci. USSR*, **72**, 1: *229—232**.

309

BORUTSKY (BORUTZKY), Y., 1931. Materialen zu d. Harpacticiodenfauna d. Baikal-sees. *Zool. Anz.*, **93**, 7/10: *203—273*. **94**, 9/10: *281—287*.

BORUTSKY (BORUTCKIJ), Y. V., 1932. A description of new species of the Harpacticoida of Lake Baikal. *Trudy Baik. limnol. St. Acad. Sci. USSR*, **II**: *15—27**.

BORUTSKY (BORUTZKY), Y. V., 1947 a. About the new species of the genus *Epischura* from te Amur drainage—*E. udulensis sp.n*, Copepoda-Calanoida. *Doklady Acad. Sci. USSR*, **58**, 7: *1513—1515**.

BORUTSKY (BORUTZKY), Y. V., 1947 b. Materials on the Copepoda- Harpacticoida fauna of Baikal. The genus *Bryocamptus*. *Doklady Acad. Sci. USSR*, **59**. 9: *1669—1672**.

BORUTSKY (BORUTZKY), Y. V., 1947 c. Materials on the Copepoda-Harpacticoida fauna of Baikal. The genus *Canthocamptus*. *Doklady Acad. Sci. USSR*, **58**, 8: *1825—1827**.

BORUTSKY (BORUTZKY), Y. V., 1948. Materials on the Copepoda-Harpacticoida fauna of Baikal. The fauna of the littoral-sor region. *Doklady Acad. Sci. USSR*, **60**, 9: *1593—1596**.

BORUTSKY, (BORUTZKY), Y. V., 1949. Materials on the Copepoda-Harpacticoida fauna of Baikal. The genus *Moraria*. *Doklady Acad. Sci. USSR*, **64**, 6: *873—876**.

BORUTSKY, (BORUTZKY) Y. V., 1952. Fresh-water Harpacticoida. In the fauna of the USSR. Crustaceans. III, 4: *3—424**.

BRESLAU, 1933. Turbellaria. Handbuch d. Zoologie. Kükenthal u. Krumbach, **2**, 6.

BRESSON, JANIKE, 1955. Asellus de sources et de grottes d'Eurasie et d'Amérique du Nord. *Arch. Zool.*, **92**, 2.

BRONSTEIN, Z. S., 1930. To the knowledge of the ostracod fauna of Baikal. *Trudy Comm. for the Study of Baikal*, **III**: *117—158**.

BRONSTEIN, Z. S., 1939. On the origin of the ostracod fauna of Baikal. *Doklady Acad. Sci. USSR*, **25**, 4: *333—337*.

BRONSTEIN, Z. S., 1947. Freshwater ostracods. In the fauna of the USSR. Crustaceans, **II**, 1: *5—339*.

BROOKS, J. L., 1950. Speciation in Ancient Lakes. *Quart. Rev. Biol.* **25**: *30—60, 131—176*.

BURMAKIN, E. V., 1956. Prospects for the improvement of the Balkhash ichtyofauna. *Izvestiya USSR Res. Inst. of Lake and River Fish.*, **XXXVII**: *91—128*: Pishchepromizdat*.

BUROV, V. S., 1931. The oligochaetes of the Baikal area. Three new species of *Styloscolex* from Lake Baikal. *Izvestiya, Biol.-Geogr. Inst. Irk. Univ.*, **V**, 4: *79—86**.

BUROV, V. S., 1935. Productivity of the bottom of the Chivyrkui Gulf. *Izvestiya Biol.-Geogr. Inst. Irk. Univ.*, **VI**, 2—4: *168—181**.

BUROV, V. & KOZHOV, M., 1932. Concerning the distribution of bottom fauna in the Maloye More of Baikal. *Trudy Irk. Univ.*, **I**, *60—82**.

BUYANTUYEV, B. P., 1955. The Baikal area. Buryat-Mong. Book Publ. House, Ulan-Ude: *1—116**.

CHAPPUIS, P. A., DELAMARE-DEBOUTTEVILLE, CL., BALASUE, I. & RUFFO, S., 1954. Recherches sur les crustacées souterrains (Première série). *Arch. Zool.*, **91**, *1*.

CHEKANOVSKAYA, O. V., 1956. Concerning the oligochaete fauna of the Yenisei watershed, part 1. *Zool. J.*, **XXXV**, 5: *657—667**.

CHERNOVSKY A. A., 1949. Determinative tables for the larvae of gnats of the family Tendipedidae. USSR Acad. Sci., Moscow.

CHERSKY, I. D., 1877. Opinions on the extensive post-Tertiary spread of the Arctic Ocean waters in Siberia. *Izvestiya Sib. br, Russ. Geogr. Soc.*, **VIII**, 1—2: *70—72**.

CHERSKY, I. D., 1878. Preliminary report on the geological study of the coastline of Lake Baikal (second year, 1878). *Izvestiya East-Sib. br, Russ. Geogr. Soc.*, **IX**, 5—6, Irkutsk: *119—165**.

310

CREDNER, R. I., 1887—1888. Die Reliktenseen. Eine physische-geographische Monographie, Ergänzungsheft zu *Peterm. Mitteil.* **86**: *1—110*: **89**: *1—51*.

DOGEL, V. A., BOGOLEPOVA, I. I. & SMIRNOVA, K. V., 1949. The parasitofauna of Baikalian fishes and its zoogeographical significance.*Bull. Leningr. Univ.*, **7**: *13—34**.

DOGEL, V. A., & BOGOLEPOVA, I. I., 1957. The parasitofauna of the Baikalian fishes. *Trudy Baik. limnol. St. Acad. Sci. USSR*, **XV**: *427—464**.

DOROGOSTAISKY, V. Č., 1917. On the crustacean fauna of the River Angara. *Yearb. Zool. Mus., Acad. Sci.*, **XXI**: *302—322**.

DOROGOSTAISKY, V. Č., 1922. Materials on the carcinological fauna of Lake Baikal. *Trudy Comm. for the Study of Baikal*, **I**, 3: *105—153**.

DOROGOSTAISKY, V. Č., 1923 a. Concerning the systematics of the graylings of the Baikal basin. *Trudy Irk. nat. Soc.*, **I**: *75—80** (German summary).

DOROGOSTAISKY, V. Č., 1923 b. The vertical and horizontal distribution of fauna in Lake Baikal. *Coll. of works of Irk. Univ. prof. and lect.*, **V**: *103—131**.

DOROGOSTAISKY, V. Č., 1930. New materials for the carcinological fauna of Lake Baikal. *Trudy Comm. for the Study of Baikal*, **III**: *49—76**.

DOROGOSTAISKY, V. Č., 1936. The gammarids of the Barguzin Gulf. *Izvestiya Biol-Geogr. Inst. Irk. Univ.*, **VII**, 1—2: *42—51**.

DRIZHENKO, F., 1908. Sailing directions for Lake Baikal. St. Petersburg: *1—172**.

DUMITRASHKO, N. V. & MARTINSON, G. G., 1940. Results of the study of the sponge fauna of terraces in the Baikal area. *Izvestiya Acad. Sci. USSR, geol. ser.*, **5**: *114—125**.

DUMITRASHKO, N. V., 1949. Young and old age of the relief of South-East Siberia. *Trudy Geogr. Inst. Acad. Sci. USSR*, **XXXIX**: *21—40**.

DUMITRASHKO, N. V., 1949. The relief of the shores of Baikal, to the history of its formation. In G. Y. VERESHCHAGIN's "Baikal", Moscow: *205—227**.

DUMITRASHKO, N. V., 1952 a. The geomorphology and palaeography of the Baikal mountain region. *Trudy Geogr. Inst. Acad. Sci. USSR*, **55**, 9: *3—189**.

DUMITRASHKO, N. V., 1952 b. The history of the Baikal depression and its development in the Quaternary period. In Materials on the Quaternary period of the USSR, **3**: *196—203**.

DYBOWSKY, B., 1873 a. Über die *Comephorus baicalensis* Pall. *Verh. zool.-bot. Ges.*, *Wien*, **XXIII**: *475—484*.

DYBOWSKY, B., 1873 b. Über die Baikal Robbe *Phoca baicalensis*. *Arch. Anat., Physiol. wiss. Med. 109—125*.

DYBOWSKY, B., 1874. Beiträge zur näheren Kenntniss der im Baikalsee vorkommenden niederen Krebse aus der Gruppe der Gammariden.*Horae Soc. ent. Rossica, St. Ptb.*, **X**: *1—190*.

DYBOWSKY, B., 1875. The Gammarids of Lake Baikal. *Izvestiya Sib. Russ. Geogr. Soc.*, **VI**, 1, 2: *10—80**.

DYBOWSKY, B., 1876. The fishes of the Baikal system. *Izvestiya Sib. Russ. Geogr. Soc.*, **VII**, 1—2: *1—25**.

DYBOWSKY, B., 1884. Neue Beiträge zur Kenntnis der Crustaceen-Fauna des Baikalsees. *Bull. Soc. Nat., Moscou.* **LX**: *17—57*.

DYBOWSKY, W., 1875. Die Gasteropoden Fauna des Baikal-Sees anatomisch und systematisch bearbeitet. *Mem. Acad. Imp. Sci. St. Petersburg*, **VII (XXII)**, 8: *1—73*.

DYBOWSKY, W., 1882. Studien über die Spongien des Russischen Reiches. *Mem. Acad. Imp. Sci. St. Petersburg*, **VII: (XXVI)**, 6: *1—71*.

DYBOWSKY, W., 1884 a. Notiz über eine die Enstehung des Baikal-Sees betreffende Hypothese. *Bull. Soc. Nat., Moscou*, **LIX**: *175—181*.

DYBOWSKY, W., 1884 b. Ein Beitrag zur Kenntniss der im Baikal-See lebenden *Ancylus*-Arten. *Bull. Soc. Nat., Moscou*, **LX**: *145—195*.

DYBOWSKY, W., 1886 a. Mitteilungen über einen neuen Fundort des Schwammes *Lubomirskia baicalensis*. *SB. naturforsch. Ges., Dorpat*, **VII**: *44—45*.

DYBOWSKY, W., 1886 b. Über zwei neue Sibirische Valvata Arten. *Jb. dtsch. malacol. Ges.*, **13**: *107—121.*

DYBOWSKY, W., 1901. Diagnosen neuer *Choanomphalus* Arten. *Nachrichtsbl. dtsch. malacol. Ges.*, **33**: *119—125.*

DYBOWSKY, W., 1902. Die Cycladiden des Baikalsees. *Nachrichtsbl. dtsch. malacol. Ges.*, **34**: *81—97.*

DYBOWSKY, W., 1903. Zur Synonimik d. *Choanomphalus* Arten. *Trudy Zool. Mus. Acad. Sci.*, **VIII**: *254—266.*

DYBOWSKY, W., 1912. Mollusken aus d. Uferregion des Baikal-sees. *Yearb. Zool. Mus. Acad. Sci.*, **XVII**: *123—143.*

EPSTEIN, V. M., 1959. On the systematical position, habits and origin of the Baikalian leech *Trachelobdella torquata* (Grube). *Doklady Acad. Sci. USSR.*, **125,** *935—937*.*

FEDYNSKY, V. A., 1951. A gravitational characteristic of submontane and intermountain depressions in geosynclines. Coll. of works in memory of D. D. ARKHANGELSKY. Publ. Acad. Sc. USSR*.

Fish and fisheries in the Baikal basin. 1958. Coll. of works ed. by M. M. KOZHOV and K. I. MISHARIN, Irkutsk: *1—746*.*

FITINGOF, A., 1865. A description of the Selenga mouth area which subsided during the earthquake of December 30 and 31, 1861. In *Gorny journal*, **9**: *95—101*.*

FLORENSOV, N. A., 1954. On the role of fractures and downwarps in the structure of the Baikal-type depressions. In Problems of the geology of Asia, **1,** publ. Acad. Sc. USSR: *670—685*.*

FRIDMAN, G., 1933. The anatomy of *Baicalarctia gulo. Trudy Baik. limnol. St. Acad. Sci. USSR*, **V**: *179—256*.*

GAIGALAS, K., 1958. To the knowledge of the rotifers of Lake Baikal. *Izvestiya Biol.-Geogr. Inst. Irk. Univ.*, **XVII**, 1—4: *103—143*.*

GALAZY, G. I., 1956. The botanical method of determining the time of occurrence of high historical levels of water in Baikal. *Bot. J.*, **41,** 7: *1006—1020*.*

GARBER, B. V., 1941. The postembryonal development of *Epischura baicalensis* Sars. *Izvestiya Acad. Sci. USSR*, **I**: *105—115*.* (English summary).

GARBER, B. V., 1946. To the knowledge of the plankton of Lake Baikal's depths. *Trudy Baik. limnol. St. Acad. Sci. USSR*, **XII**: *33—56*.*

GARYAYEV, V. P., 1901. The gammarids of Lake Baikal. *Trudy Kazan Univ. Nat. Soc.*, **XXXV**, 6: *1—63*.* (German summary).

GAVRILOV, G. B., 1949. Concerning the period of propagation of the amphipods and isopods of Lake Baikal. *Doklady Acad. Sci. USSR*, **64,** 5: *739—742*.*

GAVRILOV, G. B., 1950 a. The faunal wealth of the littoral zone of Baikal. In *Priroda*, **9**: *67*.*

GAVRILOV, G. B., 1950 b. The rocky littoral fauna of Lake Baikal. Synopsis of the Master's theses. *Baik. limnol. St. Acad. Sci. USSR*, *1—20*.*

GAYEVSVAYA N. S. (GAJEWSKAJA, N.), 1929. Über einige seltene Infusorien d. Baikalsees. *Izvestiya Acad. Sci. USSR*, **VII**: *845—854.*

GAYEVSKAYA, N. S., 1932. Marine elements in the infusorian fauna of Lake Baikal. *Trudy Baik. limnol. St. Acad. Sci. USSR*, **3**: *1—14*.*

GAYEVSKAYA, N. S. (GAJEWSKAJA, N.), 1933. Zur Oekologie, Morphologie und Systematik der Infusorien des Baikalsees. *Zoologica*, H-85, **VIII**: *1—298.* Stuttgart.

GENKEL, A. G., 1925. Some materials pertaining to the knowledge of the plankton of Lake Baikal. *Izvestiya Biol. Inst. Perm Univ.* **III**, 8: *285—290*.* (German summary).

GEORGI, J. G., 1775. Bemerkungen auf einer Reise im russischen Reich im Jahre 1772. St. Petersburg, **2**: *194—242.*

312

GERNES, R., (HOERNES, R.) 1898. The Fauna of Baikal and its relictary nature. *Fish. Ind. Bull.*, **XIII**, 4: *237—244**.

GERSTFELDT, G., 1859 a. Über einige zum Teile neue Arten von Platoden, Anneliden Myriopoden u. Crustaceen Sibiriens. *Mem. Acad. Imp. Sci. St. Petersburg*, **VIII**: *261—269*.

GERSTFELDT, G., 1859 b. Über Land und Süsswasser-Mollusken Sibiriens and d. Amur Gebietes. *Mem. Acad. Imp. Sci. St. Petersburg*, **IX**: *507—548*.

GMELIN, I. E., 1751—52. Reise durch Sibirien 1733—43, Göttingen.

GREZE, V. N., 1947. Lake Taimyr. *Izvestiya USSR geogr. Soc.*, **3**: *289—301**.

GREZE, V. N., 1951 a. Concerning the discovery of the parasitic copepod *Paraergasilus Rylovi* Mark. in Baikal. *Doklady Acad. Sci. USSR*, **79**, 2: *361—363**.

GREZE, V. N., 1951 b. Baikalian elements as an acclimatisation stock. *Trudy USSR hydrobiol. Soc.*, **III**: *221—226**.

GREZE, V. N., 1953. The lakes of the western fringe of the Mid-Siberian Highland. Coll. of works dedicated to V. A. OBRUCHEV. Tomsk br. Georg. Soc. In Problems of the Geography of Siberia: *201—216**.

GREZE, V. N., 1954. On the laws governing the distribution of bottom fauna in the River Yenisei. *Trudy Conf. on fundamental and topical problems, Zool. Inst. Acad. Sci. USSR*, **II**: *68—74**.

GREZE, V. N., 1956. The hydrofauna of the River Yenisei. Synopsis of the Master's theses, Leningrad: *3—22**.

GREZE, V. N. 1957. Basic features of the hydrobiology of Lake Taimyr. *Trudy USSR, hydrobiol. Soc.* **VIII**: *183—218**.

GRUBE, E., 1871 (1872). Die Fauna d. Baikalsees sowie einige Hirudineen u. Planarien anderer Faunen. *Schrift. schles. Ges. vaterl. Kultur*, Breslau, **49**: *53—57*.

GRUBE, E., 1872. Beschreibungen von Planarien des Baikalgebietes. *Arch. Naturgesch.*, **38**: *273—292*.

GRUBE, E., 1872 (1873). Einige bisher noch unbekannte Bewohner des Baikalsees. *Jber. schles. Ges. vaterl. Kultur.* **50**, Breslau: *66—68*.

GURYANOVA, N. F., 1929. Concerning the Crustacea-Malacostraca fauna of the River Yenisei. *Russ. hydrobiol. J.*, **5**, 10—12: *285—299**.

GUSEV, O. K., 1956. The caddis-flies of North-East Baikal. In *Priroda*, **12**, *105—106**.

GUSEVA, K. A., 1947. Causes of periodicity in the development of phytoplankton in the Ucha reservoir. *Bull. Mos. nat. Soc., Biol. Dept.*, **LII**, (6): *49—62**.

GUSEVA, K. A., 1952. Water bloom, its causes, prognosis and measures of combating it. *Trudy USSR hydrobiol. Soc.*, **IV**: *3—84**.

HERRE, W., 1933. The fishes of Lake Lanao. A problem in evolution. *Amer. Nat.*, **67**: *154—162*.

HOERNES, R., 1897 a. Die Fauna des Baikalsees und ihre Reliktennatur. *Biolog. Cbl.*, **XVII**: *657—664*.

HOERNES, R., 1897 b. Die Reliktennatur der Fauna des Baikalsees. *Jber. K. K. geol. Reichsanst., Wien*, **XLVII**: *89—94*.

HRABE, S., 1929. *Lamprodrilus michaelseni. Arch. Hydrobiol.*, **XX**: *163—179*.

HRABE, S., 1931. Die Oligochaeten aus den Ochrida und Prespa. *Zool. Jb., System.*, **61**: *1—62*.

HUMBOLDT, A. v., 1843. Asie centrale. Paris.

IOHANSEN, H., 1925. Der Baikalsee. *Mitt. geogr. Ges., München*, **18**: *1—202*.

IVANOV, T. M., 1938. The Baikalian seal, its biology and fishing. *Izvestiya Biol.-Geogr. Inst. Irk. Univ.*, **VIII**, 1—2: *5—116**.

IZOSIMOV, V. V., 1949. The Lumbriculidae of Lake Baikal. Synopsis of the Doctor's theses, Kazan; *3—11**.

IZOSIMOV, V. V., 1962. The oligochaetes of the Lumbriculidae fam. of Lake Baikal. *Trudy Baik. limnol. St. Acad. Sci. USSR*, **XX***.

313

JAMAGUCHI, H., 1937, Studies on the aquatic oligochaeta of Japan. *Annot. zool.*, **6**, *3*.

KARANTONIS, F. E., KIRILLOV, F. K. & MUKHOMEDIAROV, F. B., 1956. The fishes of the middle course of the River Lena. *Trudy Inst. Biol., Yakutian br. Acad. Sci. USSR*, **2**: *3—212**.

KARNAUKHOV, A. S., 1936. On the hydrochemistry of the waters of the Barguzin Gulf. *Izvestiya Biol. - Geogr. Inst. Irk. Univ.*, **VII**, 1—2: *70—92**.

KHEISIN, E. M., 1930 a. On the biology of the parasitic infusorians of various invertebrates of Lake Baikal. *Doklady Acad. Sci. USSR*, **5**: *121—126**.

KHEISIN, E. M., 1930 b. New marine infusorians commensal on mollusks from Lake Baikal. *Doklady Acad. Sci. USSR*, **24**: *659—661**.

KHEISIN, E. M. (CHEJSSIN, E.), 1931. Infusorien Ancystridae u. Boveridae aus dem Baikalsee. *Arch. Protistenk.* **73**, 2: *280—304*.

KHEISIN, E. M., 1932. Concerning the morphology and systematics of the Baikalian parasitic infusorians of the fam. Ptychostomidae *Trudy Baik. limnol. St. Acad. Sci. USSR*, **2**: *29—53**.

KISELYOV, I. A. (KISSELEW, J. A.) & ZWETKOV, W. N., 1935. Zur Morphologie und Ökologie von *Peridinium baicalense* n. sp. Beihefte zu *Bot. Cbl.*, **53**: *518—524*.

KISELYOV, I. A., 1937. The phytoplankton of some mountain bodies of water in the Baikal Range. *Trudy Baik. limnol. St. Acad. Sci. USSR*, **VII**: *53—70**.

KITAINIK, A. F. & IVANYEV, L. N., 1958. A note on the Tertiary deposits of the Olkhon Island in Lake Baikal. *Trans. Irk. Region. Stud. Mus.* 1958: *55—60**.

KNYAZEVA, L. M., 1954. Sedimentation in the lakes of the humid zone of the USSR. South Baikal. In Sedimentation in modern bodies of water; publ. Acad. Sc. USSR. Moscow: *180—236**.

KOMAREK, J., 1953. Herkunft des Süsswasser-Endemiten der Dinarischen Gebirge. Revison der Arten, Artenentstehung bei Höhlentieren. *Arch. Hydrobiol.*, **48**: *269—349*.

KONDAKOV, N. N., 1960. On the problem of the systematic status of the Baikal seal. *Bull. Mos. nat. Soc.*, **LXV**, 4: *120—121*.

KOROTNEV, A. A., 1901. Report on the study of Lake Baikal. Coll. of works to commemorate the 50th anniv. of the Sib. br. Rus. Geogr. Soc., Fauna of Baikal, Kiev: *28—42**.

KOROTNEV (KOROTNEFF), A., 1905. Die Comephoriden des Baikal-Sees. Erg. zool. Exp. nach d. Baikalsee, **II**, Kiew u. Berlin: *1—39*.

KOROTNEV (KOROTNEFF), A. A., 1912. Die Planarien des Baikal-Sees. Erg. zool. Exp. nach d. Baikalsee, **V**, Kiew u. Berlin: *1—28*.

KORYAKOV, E. A., 1951 a. A new representative of parasitic copepods of the genus *Coregonicola* on Baikal cottids. *Doklady Acad. Sci. USSR*, **79**, 2: *365—368*.

KORYAKOV, E. A., 1951 b. On the discovery of the male of *Salmincola cottidarum* Messjatz. *Doklady Acad. Sci. USSR*, **79**, 5: *907—908**.

KORYAKOV, E. A., 1952. Distribution of the parasitic copepod *Salmincola cottidarum* Messjatz. in the depths of Baikal. *Doklady Acad. Sci. USSR*, **83**, 2: *325—327**.

KORYAKOV, E. A., 1954. New finds of Copepoda parasitica on Baikalian fishes. *Doklady Acad. Sci. USSR*, **99**, 4: *657—659**.

KORYAKOV, E. A., 1955. On the fertility and type of the spawning population of *Comephorus*. *Doklady Acad. Sci. USSR*, **101**, 5: *965—967**.

KORYAKOV, E. A., 1956. Some ecological adaptive features in the propagation of *Comephorus*. *Doklady Acad. Sci. USSR*, **111**, 5: *1111—1114**.

KORYAKOV, E. A., 1958. Cottoid fishes of Baikal. Coll. of art. on fish and fishing in the Baikal basin. Irkutsk: *389—419**.

KORYAKOV, E. A., 1959 a. On one of the causes of the immiscibility of the Baikalian fauna in connection with the problems of its reconstruction. In Biological principles of fishing, Tomsk: *345—350*.

KORYAKOV, E. A., 1959 b. A parasitic baicalian neoendemic in the Lena drainage. *Trudy Conf. of the Icht. Comm. Acad. Sci. USSR*, **9**: *168—173**.

314

KOZHOV, M. M., 1928. Observations of *Benedictia baicalensis* and other representatives of the fam. Benedictiidae. *Izvestiya Biol.-Geogr. Inst. Irk. Univ.*, **4**, 1: *81—96**.

KOZHOV (KOSHOW), M., 1930. Material zur Spongilidenfauna Ostsibiriens. *Zool. Anz.*, **90:** *155—163*.

KOZHOV, (KOSHOW), M. M., 1931 a. To the knowledge of the fauna of Baikal, its distribution and life conditions. *Izvestiya Biol.-Geogr. Inst. Irk. Univ.*, **V**, 1: *1—171**. (German summary).

KOZHOV, M. M., 1931 b. Materials on the fauna of the River Angara. *Izvestiya Biol.-Geogr. Inst. Irk. Univ.*, **V**, 4: *1—9**.

KOZHOV, M. M., 1934 a. Materials on the distribution of soils and fauna in North Baikal. *Izvestiya Biol.-Geogr. Inst. Irk. Univ.*, **VI**, 1: *154—175**.

KOZHOV, M. M. 1934 b. Hydrological and hydrobiological investigations in the Barguzin Gulf in 1932. *Izvestiya Biol.-Geogr. Inst. Irk. Univ.*, **VI**, 1: *9—83**.

KOZHOV, M. M., 1936 a. Materials on the hydrobiology of the Maloye More in Baikal and the migrations of the omul. *Izvestiya Biol.-Geogr. Inst. Irk. Univ.*, **VII**, 1—2: *93—129**.

KOZHOV, (Kožov) M. M., 1936 b. The mollusks of Lake Baikal. *Trudy Baik. limnol. St. Acad. Sci. USSR*, **VIII**: *1—352** (German summary).

KOZHOV, M. M., 1942. On the presence of the Baikalian polychaete *Manayunkia baicalensis* Nusb. in the lakes of the Vitim drainage. *Doklady Acad. Sci. USSR*, **35**, 2: *58—61**.

KOZHOV, M. M., 1945. On the morphology of the endemic mollusks of Lake Baikal. 1. Benedictiinae. *Zool. J.*, **24**, 5: *277—289**.

KOZHOV, M. M., 1946. Baikalian mollusks in Lake Koso Gol (Mongolia). *Doklady Acad. Sci. USSR*, **52**, 4: *369—372**.

KOZHOV, M. M., 1947. The living world of Lake Baikal. Irkutsk: *1—305**.

KOZHOV, M. M., 1949. On the history of the lake systems of the Baikal and trans-Baikal areas and their fauna. *Trudy USSR hydrobiol. Soc.*, **1**: *210—223**.

KOZHOV, M. M., 1950 a. On the morphology of the endemic mollusks of Lake Baikal. *Izvestiya Biol.-Geogr. Inst. Irk. Univ.*, **XII**, 1: *3—19**.

KOZHOV, M. M., 1950 b. The fresh waters of East Siberia. OGIZ, Irkutsk: *3—367**.

KOZHOV, M. M., 1951. On the morphology and history of the Baikalian endemic mollusks of the fam. Baicaliidae. *Trudy Baik. limnol. St. Acad. Sci. USSR*, **XIII**: *93—119**.

KOZHOV, M. M., 1954. The vertical distribution of plankton and plankton-eating fish in Lake Baikal. In *Voprosi ikhtiologii*, **2:** *7—20**.

KOZHOV, M. M., 1955 a. Seasonal and annual changes in the plankton of Lake Baikal. *Trudy USSR hydrobiol. Soc.* **VI:** *133—157**.

KOZHOV, M. M., 1955 b. New data on life in the mass of water of Lake Baikal. *Zool. J.*, **XXXIV**, 1: *17—45**.

KOZHOV, M. M., 1956. On the distribution of the modern Baikalian fauna outside of Baikal. *Trudy Karelian br. Acad. Sci. USSR*, **V**: *39—46**.

KOZHOV, M. M., 1957. The horizontal distribution of plankton and plankton-eating fish in Lake Baikal. *Trudy Baik. limnol. St. Acad. Sci. USSR*, **XV**: *337—376**.

KOZHOV, M. M., 1958 a. On the genesis of the main ecological complexes in the modern Baikalian fauna. *Izvestiya Biol.-Geogr. Inst. Irk. Univ.*, **XVII**, 1—4: *68—83**.

KOZHOV (KOSHOW), M., 1958 b. Über die Genesis des Ökologischen Hauptcomplexes in der heutigen Fauna des Baikalsees. *Verh. int. Ver. Limnol.*, **XII**: *799—804*.

KOZHOV, M. M., 1959 a. On the vertical migrations of widespread plankton species in Lake Baikal. *Trudy USSR hydrobiol. Soc.*, **IX**: *161—174**.

KOZHOV (KOSHOW), M., 1959 b. Über Richtlinien und Faktoren der Evolution der Fauna des Baikalsees. XVth int. Congr. Zool. London: *72—76*.

KOZHOV (Kožov), M. M., 1960. On species formation in the Baikal Lake. *Bull. Mos. nat. Soc. LXV* (6): *39—45**, (English summary).

KOZHOV, M. M. & TOMILOV, A. A., 1949. On the new finds of Baikalian fauna outside of Baikal. *Trudy USSR hydrobiol. Soc.* **I:** *224—227**.

KOZHOV M. M. & TJUMENZEV N. V., 1960. On the Biological sequences of the level variations in the Baikal Lake. *Bull. Mos. nat. Soc., biol.*, **LXVI**, 3: *32—39**. (English summary).

KOZHOVA, O. M., 1953. The feeding habits of the *Epischura baicalensis* Sars. (Copepoda Calanoida) of Lake Baikal. *Doklady Acad. Sci. USSR*, **90**, 2: *299—304**.

KOZHOVA, O. M., 1956 a. On the biology of the *Epischura baicalensis* Sars of Lake Baikal. *Izvestiya Biol.-Geogr. Inst. Irk. Univ.*, **XVI**, 1—4: *92—120**.

KOZHOVA, O. M., 1956 b. The phytoplankton of Lake Baikal. Synopsis of the Master's theses, Irkutsk: *3—20**.

KOZHOVA, O. M., 1957. The horizontal distribution of planktonic algae in Lake Baikal. *Izvestiya East-Sib. br., Acad. Sci. USSR*, **4—5:** *226—233**.

KOZHOVA, O. M., 1959 a. On sub-ice water bloom in Lake Baikal. *Bot. J.*, **44:** *1001—1004**.

KOZHOVA, O. M., 1959 b. The phytoplankton of the Maloye More. *Trudy Baik. limnol. St. Acad. Sci. USSR*, **XVII:** *255—274**.

KOZHOVA, O. M. & KAZANTSEVA, E. A., 1961. On the seasonal changes of bacterioplankton in the waters of Lake Baikal. In *Mikrobiologiya* 1*.

KRISHTOFOVICH, A. N., 1928. The water-chestnut *Trapa borealis* Heer. from the Tertiary deposits of the Tunka Valley in the Sayan Range. *Bull. geol. Comm.*, **III**, 9—10: *58—61**.

KRISHTOFOVICH, A. N., 1930. The main features of the development of the Tertiary flora of Asia. *Izvestiya Centr. Bot. Gard.*, **XXXIX**, 3—4: *391—401**.

KRISHTOFOVICH, A. N., 1932. A geological review of countries of the Far East, Moscow: *2—330** (English summary).

KROGIUS, F. V., 1933. Materials on the biology and systematics of the gwyniad of Lake Baikal. *Trudy Baik. limnol. St. Acad. Sci. USSR*, **V:** *5—154**.

KUZNETSOV, S. I., 1951. A comparative characteristic of the biomass of bacteria and phytoplankton in the surface layer of water of Central Baikal. *Trudy Baik. limnol. St. Acad. Sci. USSR*, **XIII:** *217—224**.

KUZNETSOV, S. I., 1955. The use of radioactive carbon dioxide C14 for establishing the comparative values of photosynthesis and chemosynthesis in a number of lakes of various types. In *Izotopy v microbiologii*. Publ. Acad. Sci. USSR: *126—135.*

KUZNETSOV, S. I., 1957. A microbiological characteristic of the waters and soils of Lake Baikal. *Trudy Baik. limnol. St. Acad. Sci. USSR*, **XV:** *388—396**.

LAMAKIN, V. V., 1950. Geological and climatic factors in the evolution of organic world in Baikal. *Bull. Comm. for the Study of the Quat. Per.*, **15,** publ. Acad. Sc. USSR*.

LAMAKIN, V. V., 1952. The Ushkany Islands and the problem of the origin of Baikal. Geografizdat, **2**, Moscow: *1—199**.

LAMAKIN, V. V., 1953. The Baikal type of Quarternary glaciation. *Izvestiya USSR geogr. Soc.*, **2:** *139—153**.

LAMAKIN, V. V., 1955. The Obruchev fault in the Baikal depression. In Voprosy geologii Azii, **2,** Moscow: *448—478**.

LAMAKIN, V. V., 1957. On the development of Baikal in the Quaternary period. *Trudy Comm. for the Study of the Quart. Per.*, **XIII:** *80—91**.

LAMAKIN, V. V., 1959. On the stratigraphical division of the Quaternary system in the coastal belt of Baikal. *Trudy Geol. Inst.*, **32:** *45—78**.

LAYIMAN, E., 1933. The parasitic worms of Baikalian fishes. *Trudy Baik. limnol. St. Acad. Sci. USSR*, **IV:** *5—93**.

LEVANIDOVA, I. M., 1948. On the causes of the immiscibility of the Baikalian and Palaearctic faunae. *Trudy Baik. limnol. St. Acad. Sci. USSR*, **XII:** *57—81**.

LINDBERG, K., 1955. Les Cyclopoids du lac Baical. *Kgl. fisiogr. Sallskkap., Lund forhangl.*, **25,** 5.

LINDHOLM, W., 1909. Die Mollusken des Baikalsees. Wiss. Erg. zool. Exp. nach d. Baikalsee, **IV**, Kiew u. Berlin: *1—104.*

LINDHOLM, W. A., 1924 a. Einige neue Gastropoda aus dem Baikalsee, *Doklady Acad. Sci.*, January-March: *22—25*.

LINDHOLM, W. A., 1924 b. Collectanea baicalica. *Arch. Molluskenk.*, **LVI**: *217—225*.

LINDHOLM, W., 1927. Kritische Studien zur Molluskenfauna des Baikalsees. *Trudy Comm. for the Study of Baikal*, **II**: *139—189*.

LINDHOLM, W., 1929. Die ersten Schnecken (Gastropoda) aus dem See Kossogol in der Nord-West Mongolei. *Doklady Acad. Sci. USSR: 315—318*.

LINEVICH, A. A., 1948. Materials to the knowledge of the larvae of the Tendipedidae of Lake Baikal. *Izvestiya Biol.-Geogr. Inst. Irk. Univ.*, **X**, 2: *100—104**.

LINEVICH, A. A., 1957. The Tendipedidae of the Angara-Baikal basin. Joint sess. Acad. Sc. USSR and Acad. Agric. Sc. USSR, Irkutsk, July 10—17. Theses of papers. Irkutsk: *1—3*.

LIVANOV (LIWANOV), N. A., 1902. Die Hirudineen Gattung *Hemiclepsis*. *Zool. Jb., Syst.*, **XVII**: *339—362*.

LIVANOV, N. A. 1962. Essays on the planarians of Baikal. *Trudy Baik. limnol. St. Acad. Sci. USSR*, **XIX***.

LOGACHOV, N. A., 1958. Cainozoic continental deposits in Baikal-type depressions. *Izvestiya Acad. Sci. USSR, geol. ser.* No. **4***.

LOPATIN, G. V., 1954. On perennial fluctuations in the level of Lake Baikal. *Doklady Acad. Sci. USSR*, **94**, 6: *1041—1043**.

LUKIN, E. I. & EPSTEIN, V. M., 1959. The 10th conference on the problems of parasitology and natural-nidus diseases, **2**: *189**.

LUKIN, E. I. & EPSTEIN, V. M., 1960 a. Endemic Baikalian leeches form the fam. Glossiphonidae. *Doklady Acad. Sci. USSR*, **131**, 2: *457—460**.

LUKIN, E. I. & EPSTEIN, V. M., 1960 b. The leeches of the subfamily Toricinae nov. subfam. and their geographical distribution. *Doklady Acad. Sci. USSR*. **134**, 2: *478—481**.

MAKUSHOK, M. E., 1925. On the problem of the origin of the sponge fauna of Lake Baikal. *Russ. zool. J.*, **5**, 4: *50—73** (German summary).

MANTEIFEL, B. P., 1959. Vertical migrations of marine organisms. 1. Vertical migrations of food zooplankton. *Trudy A. N. Severtsev Inst. Anim. Morphol.*, **13**: *62—118**.

MANTEIFEL, B. P., 1961. Vertical migrations of marine organisms. II: The adaptive significance of migrations of fishes. *Trudy A. N. Severtsev Inst. Anim. Morphol.*, **39**: *5—-46**.

MARKEVICH, A. P., 1956. Parasitic copepods of the fishes of the USSR. *Publ. Ukrainian Acad. Sci., Kiev: 151—156**.

MARTINSON, G. G., 1936. Distribution of sponge spicules in the deep borehole near the village of Posolsk on Baikal. *Doklady Acad. Sci. USSR*, **IV** (XIII), 6 (10): *261—264**.

MARTINSON, G. G., 1938. The fossil sponge fauna of Tertiary deposits in the Baikal area. *Doklady Acad. Sci. USSR*, **XXI**, 4: *212—214**.

MARTINSON, G. G., 1940. Materials for the study of the fossil fauna of the Baikal area. *Trudy Baik. limnol. St. Acad. Sci. USSR*, **X**: *425—451**.

MARTINSON, G. G., 1947. On te oscular tubes of the Baikalian sponges. *Doklady Acad. Sci. USSR*, **58**, 1: *167—168**.

MARTINSON, G. G., 1948 a. The fossil invertebrate fauna of ancient continental waters of the trans-Baikal area. *Trudy Baik. limnol. St. Acad. Sci. USSR*, **XII**: *82—106**.

MARTINSON, G. G., 1948 b. Fossil sponges from the Tunka depression in the Baikal area. *Doklady Acad. Sci. USSR*, **61**, 5: *897—900**.

MARTINSON, G. G., 1949 a. The first finds of Neogene molluska in the Irkut carboniferous basin. *Doklady Acad. Sci. USSR*, **67**, 2: *365—367**.

MARTINSON, G. G., 1949 b. New Mesozoic freshwater gastropods from the east trans-Baikal area. *Yearb. USSR paleont. Soc.*, **XIII***.

317

MARTINSON, G. G., 1951. The Tertiary mollusk fauna of the east Baikal area. *Trudy Baik. limnol. St. Acad. Sci. USSR*, **XIII**: *5—92**.

MARTINSON, G. G., 1952. Upper Mesozoic freshwater mollusks from the area of Lake Gusinoye in the west trans-Baikal area. *Doklady Acad. Sci. USSR*, **83**, 4: *131—134**.

MARTINSON, G. G., 1954 a. Some freshwater gastropod mollusks from the Neogene deposits in the Irkutsk amphitheatre. *Trudy Baik. limnol. St. Acad. Sci. USSR*, **XIV**: *108—121**.

MARTINSON, G. G., 1954 b. A zoological analysis of bottom deposits in Baikal. *Trudy Baik. limnol. St. Acad. Sci. USSR*, **XIV**: *152—168**.

MARTINSON, G. G., 1955 a. Lacustrine basins of the geological past of Asia and their fauna. In *Priroda*, **4**: *78—82**.

MARTINSON, G. G., 1955 b. Heterotypic complexes of freshwater mollusks in the Tertiary deposits of Sinkiang. *Doklady Acad. Sci. USSR*, **102**, 3: *591—593**.

MARTINSON, G. G., 1956. Determinative tables of the Mesozoic and Cainozoic mollusks of East Siberia. *Publ. Acad. Sci. USSR*: *1—91**.

MARTINSON, G. G., 1957 a. The Mesozoic freshwater mollusks of some regions of East and Central Asia. *Trudy Baik. limnol. St. Acad. Sci. USSR*, **XV**: *262—336**.

MARTINSON, G. G., 1959 a. In search of ancestors of the Baikalian fauna. Moscow: *1—112**.

MARTINSON, G. G., 1959 b. The fossil mollusks of Asia and the problem of the origin of the Baikalian fauna. *Trudy Baik. limnol. St. Acad. Sci. USSR*, **XIX**: *1—17**.

MARTINSON, G. G., 1960. The fossil mollusks of Asia and the problem of the origin of the Baikalian fauna. In Geologiya i geofizika, **2**. Novosibirsk: *47—56**.

MARTINSON, G. G. & POPOVA, S. M., 1959. Some Tertiary mollusks of the Baikalian type from the lacustrine deposits of the south of West Siberia. *Paleont. J.*, **4**: *105—109**.

MARTYNOV, A., 1909, 1910, 1914. The Trichoptera of Siberia and adjacent regions. **XIV**, 3—4: *223—225*; **XV**, 4: *351—429*; **XIX**, 2: *173—285**.

MARTYNOV, A., 1914. Die Trichopteren Sibiriens und der angrenzenden Gebiete. Th. **III**, Apataninae (Fam. Limnophilidae). *Yearb. Zool. Mus. Acad. Sci.*, **XIX**: *1—187*.

MARTYNOV, A., 1924. To the Knowledge of Baicalinini, a group of endemic Baikalian Trichoptera. *Doklady Acad. Sci. USSR*. **A**: *93—96*.

MARTYNOV, A., 1924. Caddis-flies. *Practical Ent.* **V**: *1—384**.

MARTYNOV, A., 1929. Ecological pre-requisites for the zoogeography of freshwater benthonic animals. *Russ. zool. J.*, **IX**, 3: *3—38** (German summary).

MARTYNOV, A., 1935 a. The caddis-flies of the Amur region. *Trudy Zool. Inst. Acad. Sci. USSR*, **II**, 2—3: *205—395**.

MARTYNOV, A., 1935 b. To the knowledge of the amphipods of the running waters of Turkestan. *Trudy Zool. Inst. Acad. Sci. USSR*, **II**, 2—3: *411—508**.

MAYR, E., 1944, Systematics and the Origin of Species. American Museum of Natural History. Columbia University Press.

MAYR, E., 1947. Systematics and the Origin of Species (tr. into Russian). Foreign Languages Publishing House, Moscow: *2—504*.

MAZEPOVA, G. F., 1950. To the knowledge of the Cyclopoida fauna of Lake Baikal. *Doklady Acad. Sci. USSR*, **72**, 4: *809—812**.

MAZEPOVA, G. F., 1952 a. New data on the Cyclopoida fauna of Lake Baikal. *Doklady Acad. Sci. USSR*, **82**, *805—807**.

MAZEPOVA, G. F., 1952 b. The vertical distribution of the Baikalian *Cyclops baicalensis* in Lake Baikal. *Izvestiya Biol.-Geogr. Inst. Irk. Univ.*, **XIII**, 2: *16—27**.

MAZEPOVA, G. F., 1955. A new form of *Eucyclops serrulatus* Fisch. (Copepoda Cyclopoida) from lake Baikal. *Trudy Zool. Inst. Acad. Sci. USSR*, **XVIII**: *106—111**.

318

MAZEPOVA, G. F., 1957. The cyclopes of Lake Baikal and some data on their zoogeography. Theses of papers, joint sess. Acad. Sc. USSR and Acad. Agric. Sc. USSR: *1—3**.

MAZEPOVA G. F. 1960. Morphology of the stages of metamorphosis of *Cyclops Kolensis* from Baikal Lake. *Izvestia Sibriskogo otdelenia Academii Nauk USSR*, **6**, *103—115**.

MENDELEYEV, I. D., 1935. On the anomalous density of the deep waters of Baikal. *Doklady Acad. Sci. USSR*, **III (VIII)**, 3: *105—108**.

MESYATSEV (MESSJATZEFF), I., 1926. Parasitische Copepoden aus dem Baikalsee. *Arch. Naturgesch., Berlin*, **92**, 4: *120—134*.

MEYER, K. I., 1930. An introduction to the alga flora of Lake Baikal. *Bull. Mos. nat. Soc., biol. dept.*, **39**, 3—4: *179—392**.

MICHAELSEN, W., 1901. The Oligochaeta fauna of Baikal. Fifty years of East-Sib. br. Russ. Geogr. Soc., an anniv. publ., Kiev: *67—76**.

MICHAELSEN, W., 1905. Oligochaeten d. Baikalsees. Wiss. Erg. zool. Exp. nach d. Baikalsee, **1**: Kiew u. Berlin: *1—69*.

MICHAELSEN, W., 1926 a. Zur Kenntnis d. Oligochaeten d. Baikalsees. *Russ. hydrobiol. J.*, **V**: *153—173*.

MICHAELSEN, W., 1926 b. *Agriodrilus vermivorus* aus d. Baikalsee. *Mitt. zool. Inst. zool. Mus. Hambrug*, **42**: *1—20*.

MICHAELSEN, W., 1933. Ein Panzeroligochaeta aus d. Baikalsee. *Zool. Anz.*, **102**.

MICHAELSEN, W., 1935. Eine interessante neue Tubificide aus dem Baikalsee. *Trudy Baik. limnol. St. Acad. Sci. USSR*, **VI**: *15—20*.

MICHAELSEN, W. & VERESCAGIN, G., 1930. Oligochaeten aus d. Selenga-Gebiet d. Baikalsees. *Trudy Comm. for the Study of Baikal*, **III**: *213—226*.

MIDDENDORFF, A., 1867. Sibirische Reise, **IV**, St. Petersburg: *525—783*.

MIKLASHEVSKAYA, L. G., 1935. Materials to the knowledge of the productivity of the bottom of Baikal. *Trudy Baik. limnol. St. Acad. Sci. USSR*, **VI**: *99—196** (English summary).

MISHARIN, K. I., 1947. The Baikalian gwyniads. *Izvestiya Biol. Georg. Inst. Irk. Univ.* **X**, 1: *22—65**.

MISHARIN, K. I., 1953 a. A biologo-morphological characteristic of the Posolsky race of the omul. *Trudy Irk. Univ.*, **VII**, 1—2: *39—51**.

MISHARIN, K. I., 1953 b. The natural propagation and artificial breeding of the Posolsky omul in Baikal. *Izvestiya Biol.-Geogr. Inst. Irk. Univ.*, **XIV**, 1—4: *3—133**.

MISHARIN, K. I., 1959. The Baikalian omul. In Coll. of art. on fish and fishing in the Baikal basin, Irkutsk: *130—287**.

MORDUKHAI-BOLTOVSKOI, F. D. 1960. Caspian fauna in the Azov-Black Sea basin. *Publ. Acad. Sci. USSR, Leningrad*: *5—286**.

MOZGOVOI, A. A. & RYZHIKOV, K. M., 1950. The problem of the origin of the Baikalian seal in the light of the helmintological science. *Doklady Acad. Sci. USSR*, **72**, 5: *997—999**.

MUKHOMEDIAROV, F. B., 1942. The races of the Baikalian omul. *Izvestiya Biol. Geogr. Inst. Irk. Univ.*, **IX**, 3—4: *35—96**.

NASONOV, N. V. (NASSONOW, N.) 1926. Sur quelques Turbellaria d. environs du lac Baical. *Doklady Acad. Sci. USSR*, **XI**: *203—206*.

NASONOV, N. V., 1930. About the *Acrorhynchus baicalensis* Rubtz. of Lake Baikal. *Doklady Acad. Sci. USSR*, **22**: *586—588**.

NASONOV, N. V. (NASSONOW, N.) 1930. Vertreter d. Fam. Graffilidae (Turbellaria) d. Baikalsees. *Izvestiya Acad. Sci. USSR*, **8**: *727—738*.

NASONOV, N. V. (NASSONOW, N.), 1936. Über d. Heliotropismus d. Turbellaria-Rhabdocoela d. Baikalsee. *Trudy Labor. exp. Zool. Morph. Acad. Sci. USSR*, **IV**.

NECHAYEVA, N. B. & SALIMOVSKAYA-RODINA, A.G., 1935. A microbiological analysis of the bottom deposits of Baikal. *Trudy Baik. limnol. St.*, **VI**: *5—14**.

NEIMEIER, 1886. Erdgeschichte, 1.

NIKOLAYEV, I. I., 1950. Diurnal vertical migrations of zooplankton and their protective and adaptive significance. *Zool. J.*, **XXIX, 6***.

NIKOLSKY, G. V., 1947. On the biological peculiarities of faunistic complexes and the importance of their analysis for zoogeography. *Zool. J.*, **26**, 3: *221—232***.

NIKOLSKY, G. V., 1955. On the origin of the Chinese autochthonous plainland complex in ichtyofauna. In coll. of art. in memory of L. S. Berg, Moscow: *443—448***.

NOVOZHILOV, N. I., 1954. The phylopod crustaceans of the upper Jurassic and the Cretaceous periods of Mongolia. *Trudy Paleont. Inst.*, **48**: *7—119***.

NUSBAUM, J., 1901. *Dybowcella baicalensis* n. gen. n. sp. *Biol. Zbl.*, **21**, 1: *6—18*.

OBRUCHEV, V. A., 1929. Selenga Dauria. Publ. Troitskosav. br. Geogr. Soc. *1—206***.

OBRUCHEV, V. A., 1932. A geological outline of the Baikal and Lena areas. Trudy Counc. for the Study of Product. Forces and the Geol. Inst. Essays on the Geology of Siberia. Publ. Acad. Sc. USSR: *1—128***.

OBRUCHEV, V. A., 1938. The geology of Siberia **III,** publ. Acad. Sci. USSR: *781—1357***.

OBRUCHEV, V. A., 1948. Borderland Dzhungariya. An orographical and geological essay. In Zemlevedeniye. Publ. Mos. Nat. Soc. lists. Novosibirsk, **II/XLII:** *16—37***.

OGNEV, S. I., 1935. The wild animals of the USSR and adjacent countries, **III,** Moscow: *1—752***.

PALLAS, P. S., 1776. Reise durch verschiedene Provinzen des Russischen Reiches, **III,** 1772—1773, p.p. *288—297, 707, 709, 710,* St. Petersburg.

PALLAS, P. S., 1811, Zoogeographia Rosso-asiatica, St. Petersburg, **V:** *114*.

PALIBIN, I. V., 1936. The Tertiary flora of the south-east coast of Baikal and the Tunka depression. *Trudy Oil Geol. Prosp. Inst.*, ser. A, **76**: *26—46***.

PATRIKEYEVA, G. I. 1959. The bottom deposits of the Maloye More. *Trudy Baik. limnol. St. Acad. Sci. USSR*, **XVII:** *205—254***.

PAVLOVSKY, E. V. & TSVETKOV, A. I., 1936. The north-west trans-Baikal area. *Publ. Acad. Sci. USSR:* *1—133*** (English summary).

PAVLOVSKY, E. V., 1937. The Lake Baikal depression. *Izvestiya Acad. Sci. USSR, geol. ser.*, **2:** *351—377***.

PAVLOVSKY, E. V., 1941. The problem of the origin of the Lake Baikal depression. In *Priroda*, **3:** *19—31***.

PAVLOVSKY, E. V., 1948 b. Comparative tectonics of the Meso-Cainozoic structures of East Siberia and the great rift of Africa and Arabia. *Izvestiya Acad. Sci. USSR, geol. ser.*, **5:** *25—38***.

PAVLOVSKY, E. V. & FLORENSOV, N. A., 1956. The bowels of East Siberia. In *Priroda*, **12:** *3—13***.

PELSENEER, P., 1905. L'origine des animaux d'eau douce. *Bull. Acad. R. Belgique. Cl. Sci.*, **12:** *699—740*.

PESCHEL, O., 1878. Neue Probleme der vergleichenden Erdkunde, 3. Aufl. Leipzig: *7, 117, 171*.

PIROZHNIKOV, P. L., 1937 a. Marine and Baikalian elements in the fauna of the River Yenisei. *Bull. Mosc. nat. Soc.*, **XVI**, 3: *165—172***.

PIROZHNIKOV, P. L., 1937 b. On the problem of the origin of northern elements in the Caspian fauna. *Doklady Acad. Sci. USSR*, **XV**, 8: *513—516***.

PIROZHNIKOV, P. L., 1941. The main features of the bottom population of the River Yenisei and the Yenisei Bay. *Trudy Astrakh. Inst. Fish.* **I:** *135—137***.

PIROZHNIKOV, P. L., 1955. On the problem of the enrichment of the food fauna of lakes and reservoirs. *Zool. J.*, **XXXIV**, 2: *267—277***.

PLOTNIKOV, V., 1906. Glossiphoniidae, Hirudinidae und Herpobdellidae. *Yearb. Zool. Mus. Acad. Sci. USSR*, **X:** *135—158,*

320

POMYTKIN, B. A., 1960. On the problem of wind-induced rises and falls in the level of Lake Baikal. *Trudy Baik. limnol. St. Acad. Sci. USSR*, **XVIII**: *242—263**.

POPOV, M. G., 1954. Two genera of angiosperms new to the flora of the USSR. *Botan. mater. of the herbarium of the V.L. Komarov Botan. Inst.*, **XVI**: *3—15**.

POPOV, M. G., 1957. The flora of East Siberia. *Publ. Acad. Sci. USSR: 5—553**.

POTAKUYEV, Y. G., 1954. The feeding and food relationships of the plankton-eating fishes of Lake Baikal. Synopsis of the Master's theses, Irkutsk: *3—13**.

POTAKUYEV, Y. G., 1956. On the food coefficient of the fry of the Baikalian omul. *Izvestiya Biol.-Geogr. Inst. Irk. Univ.*, **XVI**, 1—4: *139—142**.

PREOBRAZHENSKY, V. S., FADEYEVA, N. S., MUKHINA, L. I. & TOMILOV, G. M., 1959. The types of locality and natural zoning of the Buryat ASSR. *Publ. Acad. Sci. USSR: 1—215**.

PRESNYAKOV, E. A., 1940. The geological conditions of the formation of Baikal. *Trudy Baik. limnol. St. Acad. Sci. USSR*, **X**: *369—398**.

PRESTON, H. B., 1916. *Ann. Mag. nat. Hist.*, 8th ser., **XVII**: *160—161*.

PUYTORAC, P. DE, 1959. Comparaison de la fauna Infusorienne endoparasite des Oligochètes du lac d'Ochrid et de ceux du Baical. *Rec. trav. Stat. hydrobiol. Ochrid*, **VI**, 2 (18).

RAMMELMEYER, E., 1931. The fauna of the Tertiary deposits of Baikal, Part 1. *Izvestiya Acad. Sci. USSR*, **10**: *1395—1399**.

RAMMELMEYER, E. S., 1935. The fauna of mollusks from the River Vitim. *Izvestiya Acad. Sci. USSR, math. and nat. Sc. Dept.*, **3**: *450—453**.

RAMMELMEYER, E. S., 1940. Fossil mollusks from the freshwater deposits of the trans-Baikal area. *Trudy Baik. limnol. St. Acad. Sci. USSR*, **X**: *399—423**.

REIS, O., 1910. The fish slate fauna of the trans-Baikal area. In *Geol. studies and surveys along the Siberian Railway*, **XXIX**: *1—68**.

REZVOI, P., 1927. Note on sponges from lake Dzhegetai- kul in the Urjankhai region. *Doklady Acad. Sci. USSR*, A, **18**: *296—300**.

REZVOI, P. D., 1936. Freshwater sponges of the USSR. In the fauna of the USSR. New ser. Sponges, **II**, 2, *Publ. Acad. Sci. USSR: 1—104** (German summary).

RODINA, A. G., 1954. Bacteria in the productivity of the rocky littoral of Lake Baikal. *Trudy Conf. Zool. Inst. Acad. Sci. USSR*, **2**: *172—201**.

ROMANOVA, A. P., 1958. Intensity of development of bacterial flora in the littoral of Lake Baikal. In *Mikrobiologiya*, **XXVII**, 5: *634—640.**

ROSSOLIMO, L. L., 1923. On the protozoan fauna of Baikal. *Russ. hydrobiol. J.*, **II**, 3—4, Saratov: *74—81** (German summary).

ROSSOLIMO, L., 1926. Parasitische Infusorien aus dem Baikalsee. *Arch. Protistenk.* **34**, 3: *463—509*.

ROSSOLIMO, L. L., 1957. The thermal regime of Lake Baikal. *Trudy Baik. limnol. St. Acad. Sci. USSR*, **XVI**: *1—551**.

RUBTSOV, I. A., 1928. A viviparous planarian from the River Angara. *Coll. of works, Irk. Univ.*, **XIV**: *301—324**.

RUBTSOV, I. A., 1929. *Acrorhynchus baicalensis* n.g., n.sp. *Russ. hydrobiol. J.*, **VIII**, Saratov: *132—138**.

RYZHIKOV, K. M. & SUDARIKOV, V. E., 1951. The works of the 272nd joint helmintological expedition in the area of Lake Baikal in 1940. *Trudy Helmint. Exp Acad. Sci. USSR*, **V**: *270—292**.

RYZHIKOV, K. M. & SUDARIKOV, V. E., 1945. A new nematode from the cottids of Lake Baikal. *Trudy Helm. Lab. Moskow*, **195***:

SAMARINA, A. V., 1960. Some data on the production of oxygen in the photosyn-thesis processes in the mass of water of Lake Baikal. *Trudy USSR hydrobiol. Soc.*, **10**: *158—169**.

SARKISYAN, S. G., 1955. Baikal Geografizdat, Moscow: *3—79**.

SARKISYAN, S. G., 1958. The Mesozoic and Tertiary deposits of the Baikal and trans-Baikal areas and the Far East. *Publ. Acad. Sci. USSR: 3—337**.

SARS, S., 1900. On *Epischura baicalensis* a new Calanoids from Lake Baikal. *Yearb. Zool. Mus. Acad. Sci. USSR*, **V**: 226—238.

SARS, G. O., 1908. On the occurrence of a genuine Harpactid on the Lake Baikal. *Arch. Mathem. of Naturvid.*, **29**: 1—13.

SCHELLENBERG, A., 1937 a. Schlüssel und Diagnosen der Süsswasser der Gammarus nahestehenden Einheiten ausschliesslich der Arten des Baikalsees und Australiens. *Zool. Anz.*, **117**, 11/12.

SCHELLENBERG, A., 1937 b. Kritische Bemerkungen zur Systematik der Süsswassergammariden. *Zool. Jb. Syst.*, **69**: 5—6.

SCHMID, F., 1953—1954. Contribution à l'étude de la sous-famille des Apataniinae (Trichoptera, Limnophilidae). *Tijdschr. Ent.*, **96, 97**.

SEGERSTRÅLE, S. G., 1956. The Distribution of Glacial Relicts in Finland and Adjacent Russian Areas. *Soc. Sci. Fenn. Comm. biol.*, **XV**, 18: 1—36.

SEGERSTRÅLE, S. G., 1957. On the Immigration of the Glacial Relicts of Northern Europe with Remarks on their Prehistory. *Soc. Sci. Fenn. Comm. biol.*, **XVI**, 16: 1—117.

SEGERSTRÅLE, S. G., 1958. On an Isolated Finnish Population of the Relict Amphipoda *Pallasea quadrispinosa* G.O. Sars. *Soc. Sci. Fenn. Comm. biol.*, **XVII**, 5: 1—33.

SEMENKEVICH, Y. N., 1924. On the Baikalian water-bugs. *Russ. hydrobiol. J.*, Saratov, **III**, 1—2: 8—13* (German summary).

SHCHEGOLEV, G. G., 1922. A nes species of leeches from Lake Baikal. *Russ. hydrobiol. J.*, **1**, Saratov: 136—141* (German summary).

SIBIRYAKOVA, O. A., 1928. On the fauna of *Turbellaria rhabdocoela* from the River Angara. *Russ. hydrobiol. J.*, **8**, 8—9, Saratov: 237—250*.

SKABICHEVSKY, A. P., 1929. Concerning the biology of *Melosira baicalensis*. *Russ. hydrobiol. J.*, **8**, 4—5: 93—113*.

SKABICHEVSKY, A.P., 1935. Observations of the phytoplankton of the Barguzin Gulf of Lake Baikal in the summer period of 1932 and 1933. *Izvestiya Biol.-Geogr. Inst. Irk. Univ.*, **VI**, 2—4: 182—231* (German summary).

SKABICHEVSKY, A. P., 1936. New and interesting diatom algae from North Baikal. *Bot. J.*, **6**: 705—722*.

SKABICHEVSKY, A. P., 1936 a. Notes on algology of Baikal about some representatives of Cladophoraceae. *Izvestya Biol. Geogr. Inst. Irk. Univ.*, **VII**, 1—2: 32—41* (German summary).

SKABICHEVSKY, A. P., 1952. On the systematics of the Baikalian diatoms. *Bot. mat. of the cryptog. plants dept.*, **VIII**: 177—180*.

SKABICHEVSKY, A. P., 1954. The phytoplanton of the Selenga area of Baikal. *Trudy Baikol limnol. St. Acad. Sci. USSR*, **XIV**: 177—180*.

SKORIKOV, A., 1903. Three new species of Rotatoria. *Yearb. Zool. Mus. Acad. Sci.*, **VII**, 2: 19—21*.

SKVORTSOV, B. V. (SKVORTZOV, B. W.) & MEYER, K. L., 1928. A contribution to the diatoms of Baikal Lake. *Proc. Sungaree River St.*, **1**, 5: 1—55.

SKVORTSOV, B. V. (SKVORTZOV, B. W.), 1937. Bottom diatom from Olkhon Gate of Baikal Lake. Manila.

SLASTNIKOV, G. S., 1940. Concerning the discovery of the polychaete *Manayunkia* in the drainage area of the River Gyda. In *Priroda*, **7**: 75—77*.

SLASTNIKOV, G. S., 1941. New data on the distribution of the Baikalian *Manayunkia*. In *Priroda*, **7—8**: 87—88*.

SMIRNOV, S., 1929. Beiträge zur Copepoden Fauna Ostasiens. *Zool. Anz.*, **81**, 11—12.

SOKOLNIKOV, V. M., 1960. On the currents and temperature of water under the ice cover in the southern part of the Baikal near the outfall of the River Angara. *Trudy Baik. limnol. St. Acad. Sci. USSR*, **XVIII**: 265—285*.

SOKOLOV, D. S., 1936. On the Jurassic marine deposits of the east trans-Baikal area. *Bull. Mosc. nat. Soc., geol. dept.*, **XIV**, 2: 183—187*.

SOKOLOV, I., 1930. Beiträge zur Kenntnis der Hidracarinen Sibiriens. *Arch. Hydrobiol.*, **XXII**, 2: 306—305.

SOKOLOV, I. I., 1944—1945. *Baicalacarus vermiformis* n. gen. n. sp., the first repre-

322

sentative of hydrachnellas from Lake Baikal. *Trans. Leningr. St. Univ., nat. sci. ser.*, **2**: *46—53**.

Sokolov, I. I., 1948. On the systematic position of *Baicalacarus vermiformis* Sokolov and its relictary nature. *Doklady Acad. Sci. USSR*, **60**, 1: *181—183**.

Sokolov, I. I., 1952. Water mites. Part II. Halacarae In the fauna of the USSR. Arachnids, **V**: *5—201**.

Solovyov, V., 1925. The method of models and its application in the study of seiches in Lake Baikal. *Izvestiya Biol.- Geogr. Inst. Irk. Univ.*, **II**, 2: *9—25** (English summary).

Sovinsky, V. K., 1915. The Amphipoda of Lake Baikal. In *Zoological studies of Lake Baikal, Kiev*, **IX**, 1: *1—483** (German summary).

Stankovič, S., 1932. Die Fauna des Ochridsees und ihre Herkunft. *Arch. Hydrobiol.*, **XXIII**, 4: *557—617*.

Stankovič, S., 1955 a. La zone profunde du lac d'Ochrid et son Peuplement. *Mem. Ist. Idrol.*, suppl. **8**: *281—310*.

Stankovič, S., 1955 b. Sur la speciation dans le lac d'Ochrid. *Verh. int. Ver. Limnol.*, **XII**: *478—506*.

Stankovič, S., 1960. The Balkan lake Ohrid and its living world. Monographie biologicae. W. Junk, Den Haag, **IX**: *1—357*.

Starobogatov, Y., 1958. The system and philogeny of Planorbidae (Gastropoda, Pulmonata). *Bull. Mosc. nat. Soc., biol. dept.*, **XIII**: *37—53**.

Starostin, A., 1921. Zur Kenntnis d. Mollusken-Fauna des Baikalsees. *Arch. Naturgesch.*, **32**.

Stebbing, T. R., 1899. Amphipoda collected from Copenhagen Museum and other sources. *Trans. Linn. Soc. Lond.*, Zoology, **VII**, 8: *395—432*.

Sudarikov, V. E. & Ryzhikov, K. M., 1951. On the biology of *Contracoecum osculatum baicalensis*, a nematode of the Baikalian seal. *Trudy Helmint. Lab. Acad. Sci. USSR*, **V***:

Sudarikov, V. E. & Ryzhikov, K. M., 1952. Substantiation of the new family of nematodes from freshwater fishes (Spirurata, Haplonematidae). *Trudy Helmint. Lab. Acad. Sci. USSR*, **VI**: *152—157**.

Sukachov, B., 1895. Some new data on the sponges of Lake Baikal. *Trudy St. Petersb. nat. Soc.*, **25**: *1—11**.

Svarchevsky, B., 1901. Materials on the sponge fauna of Lake Baikal. *Trans. Kiev nat. Soc., Kiev*: *1—24**.

Svarchevsky (Swarczewsky), B., 1910. Beobachtungen über *Lankesteria* sp., eine in Turbellarien des Baikalsees lebende Gregarine. Festschrift zum sechzigsten Geburtstage R. Hertwig, **1**, Jena: *637—673*.

Svarchevsky, B., 1923 a. Spongiological essays. *Trudy Irk. nat. Soc.*, **I**, 6: *1—27** (German summary).

Svarchevsky, B. A., 1923 b. Essays on Hydraria. *Hydra baicalensis. Trudy Irk. Univ.* **4**: *90—102**.

Svarchevsky, B., 1925. Spongiological essays. *Izvestiya Biol.-Geogr. Inst. Irk. Univ.*, **2**: *11—28**.

Svarchevsky (Swarczewsky), B., 1928. Beobachtungen über *Spirochona elegans* n. sp. *Arch. Protistenk.* **61**: *185—222*.

Svarchevsky (Swarczewsky), B., 1928—1930. Zur Kenntniss der Baikalprotistenfauna. *Arch. Protistenk.*, **61—65, 69**.

Svatosh, Z. f., 1925. The Baikalian seal and its fishing. In Coll. of art. on nature and hunting, Kharkov: *27—49**.

Svetovidov, A. N., 1931. Materials on the systematics and biology of the graylings of Lake Baikal. *Trudy Baik. limnol. St. Acad. Sci. USSR*, **I**: *19—194** (German summary).

Taliyev, D. N., 1940. An experiment in applying the precipitation reaction in the study of the origin and history of the Baikalian fauna. *Trudy Baik. limnol. St. Acad. Sci. USSR*, **X**: *241—352** (English summary).

TALIYEV, D. N., 1941. A serologigal analysis of the races of the Baikalian omul. *Trudy Baik. limnol. St. Acad. Sci. USSR*, **VI**, 4: *68—91**.

TALIYEV, D. N., 1946. Ancestors of the Baikalian Cottoidei in the Tsipa-Tsipikan lakes. *Doklady Acad. Sci. USSR*, **52**, 8: *743—746**.

TALIYEV, D. N., 1947. Thd impact of carnivores on the divergent radiation of Cottoidei. *Doklady Acad. Sci. USSR*, **58**, 7: *1509—1512**.

TALIYEV, D. N., 1948. On the problem of the pace and causes of the divergent evolution of the Baikalian Cottoidei. *Trudy Baik. limnol. St. Acad. Sci. USSR*, **XII**: *107—158**.

TALIYEV, D. N., 1949. On the unisexual propagation of *Comephorus*. *Doklady Acad. Sci. USSR*, **69**, 1: *105—108**.

TALIYEV, D. N., 1951. On the role of fetalisation in the evolution of the endemic fauna of Baikal. *Doklady Acad. Sci. USSR*, **78**, 3: *605—608**.

TALIYEV, D. N., 1955. The miller's-thumbs (Cottoidei) of Lake Baikal. *Publ. Acad. Sci. USSR: 9—603**.

TALIYEV, D. N. & KORYAKOV, E. A., 1947. Consumption of oxygen by the Baikalian Cottoidei. *Doklady Acad. Sci. USSR*, **58**, 8: *1837—1840**.

TALIYEV, D. N. & KORYAKOV, E. A., 1948. The upper thermal limits for the Baikalian Cottoidei. *Doklady Acad. Sci. USSR*, **59**, 4: *755—758**.

TALIYEV, D. N. & KORYAKOV, E. A., 1949. The natural specific gravity of the Baikalian Cottoidei. *Doklady Acad. Sci. USSR*, **68**, 1: *169—172**.

THIELE, I., 1929—1934. Handbuch der systematischen Weichtierkunde. Jena.

TIKHOMIROV, P. V., 1927. Two new species of Rotatoria from Lake Baikal. *Russ. hydrobiol. J.*, **VI**, 6—7, Saratov: *145—147**.

TIKHOMIROV, P. V., 1929. A new species of Rotatoria from Lake Baikal. *Russ. hydrobiol. J.*, **VIII**, 6—7, Saratov: *171—173**.

TIMOFEYEV, S. I., 1928. The embriology of *Manayunkia baicalensis* Nusb. and some notes on the direct development of Polychaeta. Trudy 3rd USSR Congr. Zool., Anat. Histol., Moscow: *159—160**.

TKACHUK, V. T., YASNITSKAYA, N.V., & ANKUDINOVA, G. A., 1957. The mineral waters of the Buryat-Mongolian ASSR. *East-Sib. br. Acad. Sci. USSR, Irkutsk: 1—151**.

TOMILOV, A. A., 1954. Materials on the hydrobiology of some deep lakes of the Olekma-Vitim mountain system. *Trudy Irk. St. Univ., biol. ser.*, **XI**: *5—85**.

TSVETKOV, V. N., (ZWETKOV V. N.) 1928. Two new species of gregarines. *Doklady Acad. Sci. USSR, A*, **3***.

TUGARINA, P. Y., 1956 a. The biology of propagation of the white Baikalian grayling and ways of increasing its stock. Synopsis of the Master's theses, Irkutsk: *1—16**.

TUGARINA, P. Y., 1956 b. Some data on the propagation of the white Baikalian grayling. *Zool. J.*, **XXXV**, 6: *938—939**.

UENO, MASUZO, 1954. The Bathynellidae of Japan. *Arch. Hydrobiol.*, **49, 4**.

ULOMSKY, S. N., 1957. On the eating up of lacustrine plankton by the ripus in the lakes of the Urals. *Izvestiya USSR Res. Inst. of Lake & River Fish.*, **39**: *146—159**.

VASILYEVA, G. L., 1950. A new cyclops species from Lake Baikal. *Izvestiya Biol.-Geogr. Inst. Irk. Univ.*, **X**, 3: *3—8**.

VASILYEVA, G. L., 1956. Materials on the study of the zooplankton of the River Angara. *Izvestiya Biol.-Geogr. Inst. Irk. Univ.*, **XVI**, 1—4: *151—184**.

VERESHCHAGIN, G. Y., 1926. The systematics and biology of *Comephorus*. *Doklady Acad. Sci. USSR*, **A**: *47—50**.

VERESHCHAGIN, G. Y., 1927 a. Some data on the regime of the deep waters of Baikal in the area of Maritui. *Trudy Comm. for the Study of Baikal*, **II**: *77—138**.

324

VERESHCHAGIN, G. Y., 1927 b. An attempt to summarise the literature on Baikal and its coasts. *Trudy Comm. for the Study of Baikal.* **II**: *162—178**.

VERESHCHAGIN (WERESTSCHAGIN) G., 1928. Vorläufige Betrachtungen über den Ursprung der Fauna und Flora des Baikalsees. *Doklady Acad. Sci. USSR*, ser. **A**: *407—412.*

VERESHCHAGIN, G. Y., 1930. On the problem of the origin and history of the fauna and flora of Baikal. *Trudy Comm. for the Study of Baikal*, **III**: *77—116**.

VERESHCHAGIN, G. Y., 1933 a. The literature on Baikal in the period between 1927 and 1931. *Trudy Baik. limnol. St. Acad. Sci. USSR*, **V**: *162—178**.

VERESHCHAGIN, G. Y., 1933 b. Basic features of the vertical distribution of dynamics of the water masses in Baikal. Coll. of art. dedic. to the 50th anniv. of the scient. and pedag. activities of Acad. V.I. Vernadsky, **2**: *1207—1230** (German summary).

VERESHCHAGIN, G. Y., 1935. Two types of biological complexes of Baikal. *Trudy Baik. limnol. St. Acad. Sci. USSR*, **VI**: *199—212**.

VERESHCHAGIN, G. Y., 1937. Observations of the vertical distribution of some pelagic fishes of Baikal. *Trudy Baikal limnol. St. Acad. Sci. USSR*, **VII**: *213—218**. (German summary).

VERESHCHAGIN, G. Y., 1940 a. Theoretical questions connected with the elaboration of the problem of the origin and history of Baikal. *Trudy Baik. limnol. St. Acad. Sci. USSR*, **X**: *7—72** (French summary).

VERESHCHAGIN, G. Y., 1940 b. The origin and history of Baikal and its fauna and flora. *Trudy Baik. limnol. St. Acad. Sci. USSR*, **X**: *73—239** (French summary).

VERESHCHAGIN, G. Y., 1949. Baikal. Moscow: *7—205**.

VERESHCHAGIN, G. Y. & SIDORYCHEV, I. P., 1929. Some observations of the biology of *Comephorus. Doklady Acad. Sci. USSR*, **A**, 5: *126—130**.

VERSHININ N.V., 1960. On the problem of the origin of the relict fauna of the Norilsk group of lakes. *Trans. USSR Acad. Sci.*, **135**, 3: *753—755**.

VERTEBNAYA (WERTEBNAJA), P. I., 1929. Über eine relicte Algenflora in den Seeablagerungen Mittelrusslands. *Arch. Hydrobiol.*, **20**, 1.

VILISOVA, I. K., 1951. On the problem of the feeding of the Baikalian pelagic amphipod *Macrohectopus branizki* Dyb. *Doklady Acad. Sci. USSR.* **29**, 2: *329—331**.

VILISOVA, I. K., 1954. A comparative review of the zooplankton of the Posolsky Sor and littoral regions of open Baikal. *Trudy Baik. limnol. St. Acad. Sci. USSR*, **XIV**: *190—261**.

VILISOVA, I. K., 1959 a. The zooplankton of the Maloye More. *Trudy Baik. limnol. St. Acad. Sci. USSR*, **XVII**: *276—304**.

VILISOVA, I. K., 1959 b. To the knowledge of the microbenthos of the Maloye More. *Trudy Baik. limnol. St. Acad. Sci. USSR*, **XVII**: *305—311**.

VINBERG, G. G., 1956. Primary productivity of plankton. *J. biol. Soc.*, **XVII, 5**: *364—376**.

VISLOUKH (WISLOUCH), S. M., 1924. Beiträge zur Diatomeenflora von Asien. II. Untersuchungen über die Diatomeen des Baikalsees. *Ber. dtsch. bot. Ges.*, **42**: *1—173.*

VLASENKO, N. M., 1928. *Ankyrocotyle baicalensis* n.g., n. sp. *Russ. hydrobiol. J.*, **VII**, 10—12: *229—248**.

VORONKOV, N.V., 1925, 1927. On the geographical distribution of rotifers. Krasnoyarsk: *2—19**.

VOTINTSEV, K. K., 1948 a. On the part played by sponges in the dynamics of silicic acid in the water of Lake Baikal. *Doklady Acad. Sci. USSR*, **62**, 5: *661—663**.

VOTINTSEV, K. K., 1948 b. Observations of the regeneration of biogenous elements during the decomposition of dead *Epischura baicalensis* Sars. *Doklady Acad. Sci. USSR*, **63**, 6: *741—744**.

VOTINTSEV, K. K., 1952 a. Materials on the dynamics of biogenous elements in the waters of Lake Baikal. *Doklady Acad. Sci. USSR*, **84**, 2: *353—356**.

VOTINTSEV, K. K., 1952 b. The energy of photosynthesis and seasonal changes in the biomass of *Melosira baicalensis. Doklady Acad. Sci. USSR*, **84**, 3: *607—610**.

VOTINTSEV, K. K., 1952 c. The hydrochemistry of Lake Baikal. Synopsis of the Master's theses, Irkutsk*.

VOTINTSEV, K. K., 1953 a. The diurnal course of oxygen and primary production in the upper layer of Lake Baikal. *Doklady Acad. Sci. USSR*, **88, 7**: *149—151**.

VOTINTSEV, K. K., 1953 b. The influence of diurnal vertical migrations of zooplankton on the oxygen regime of Lake Baikal. *Doklady Acad. Sci. USSR*, **92**, 1: *157—160**.

VOTINTSEV, K. K., 1953 c. On the speed of the regeneration of biogenous elements during the decomposition of dead *Melosira baicalensis. Doklady Acad. Sci. USSR*, **92, 3**: *667—670**.

VOTINTSEV, K. K., 1955 a. The ways of the migration of silicon in Lake Baikal. *Trudy USSR hydrobiol. Soc.*, **VI**: *70—79**.

VOTINTSEV, K. K., 1955 b. The vertical distribution and seasonal dynamics of organic matter in the water of Lake Baikal. *Doklady Acad. Sci. USSR*, **101**, 2: *359—362**.

VOTINTSEV, K. K., 1956. Nitrogen and phosphorus in the waters of Lake Baikal. *Trudy USSR hydrobiol. Soc.*, **VII**: *24—35**.

VOTINTSEV, K. K., 1960. On the spread of the waters of the river Selenga in Lake Baikal in the summer-autumn period. *Doklady Acad. Sci. USSR*, **131**, 3: *620—623**.

VOTINTSEV, K. K., 1961. The hydrochemistry of Lake Baikal. *Trudy Baik. limnol. St. Acad. Sci. USSR*, **XX**: *1—312**.

VOTINTSEV, K. K., SAMARINA, A. V. 1957. The oxygen regime of Lake Baikal. *Trudy USSR hydrobiol. Soc.*, **8**: *288—304**.

WALKER, B., 1918. A Synopsis of the classification of the Freshwater Mollusca of North America, North of Mexico. University of Michigan. *Mus. Zool. Misc. Publ.*, **6**, 1: *1—213*.

WOLTERECK, R., 1931. Wie entsteht eine endemische Rasse oder Art? *Biol. Zbl.*, **51**: *231—253*.

WORTHINGTON, E. B., 1954. Speciation of fishes in African Lakes. *Nature, Lond.* **173**: *1064—1067*.

YAKHONTOV, G., 1904. A report on the excursion to Lake Baikal in the summer of 1902. *Records nat. Soc. Kaz. Univ. for 1902—1903*, suppl. to Rec. No. **212**: *1—11**.

YASHNOV, V. A., 1922. The plankton of Lake Baikal from the materials of the Baikal expedition of Moscow University's Zoological Museum in 1917. *Russ. hydrobiol. J.* **I**, 8: *225—241** (German summary).

YASNITSKY, V. N., 1923. Materials to the knowledge of the plankton of Lake Baikal. *Trudy Irk. nat. Soc.*, **1**, 1: *32—72** (English summary).

YASNITSKY, V. N., 1924. The plankton of Lake Baikal in the Kultuk and Slyudyanka area. *Izvestiya East-Sib. br. geogr. Soc.*, **47**: *147—152**.

YASNITSKY, V.N., 1926. On the problem of variability in some planktonic organisms of Lake Baikal. *Izvestiya Biol.-Geogr. Inst. Irk. Univ.*, **II**: *15—31**.

YASNITZKY, V.N., 1928. Some results of hydrobiological studies in Baikal in the summer of 1929. *Doklady Acad. Sci. USSR*, **18—19**: *353—358**.

YASNITSKY, V. N., 1930. Results of observations of the plankton of Baikal in the area of the biological station in 1926—1928. *Izvestiya Biol.-Geogr. Inst. Irk. Univ.*, **IV**, 3—4: *191—238**.

YASNITSKY, V. N., 1931. Concerning the history of the development of a new alga from Lake Baikal, *Swarzewskiella rotans* n. gen. n. sp. *Izvestiya Biol.-Geogr. Inst. Irk. Univ.*, **V**, 4: *49—58**.

326

YASNITSKY, V. N., 1934. The plankton of the northern extremity of Baikal. *Izvestiya Biol.- Geogr. Inst. Irk. Univ.*, **VI**, 1: *85—102**.

YASNITSKY, V. N., 1936. New and interesting species of diatom algae from Lake Baikal. *Bot. J.*, **6:** *689—703**.

YASNITSKY, V. N., 1952. Phenomena of gigantism in the flora of Lake Baikal. *Izvestiya Biol.-Geogr. Inst. Irk. Univ.*, **XIII**, 2: *3—11**.

YASNITSKY, V. N., 1956. The phytoplankton of the Chivyrkui Gulf. *Izvestiya Biol.-Geogr. Inst. Irk. Univ.*, **16**, 1—4: *121—138**.

YASNITSKY, V. N., & SKABICHEVSKY, V. P., 1957. The phytoplankton of Lake Baikal. *Trudy Baik. limnol. St. Acad. Sci. USSR*, **XV:** *212—261**.

YASNITSKY, V., BLANKOV, B. & GORTIKOV, V., 1927. Report on the work of the Baikal biological station in 1926—1927. *Izvestiya Biol.-Geogr. Inst. Irk. Univ.*, **III**, 3: *3—35**.

YEFREMOV, I. A., 1954. Problems of the historical development of dinosaurs. *Trudy paleont. Inst.*, **48:** *125—141**.

YEGOROV, A. G., 1947. On the fertility of the Baikalian sturgeon. *Izvestiya Biol.- Geogr. Inst. Irk. Univ.*, **X**, 1: *84—88**.

YEPOVA, N. A., 1955. Concerning the history of the habitat of *Bergenia crassifolia* L. (Fritsch.). *Izvestiya Biol.-Geogr. Inst. Irk. Univ.*, **XV**, 1—4: *5—110**.

YEPOVA, N. A., 1956. Relicts of platyphyllous forests in the fir taiga of the Khamar-Daban. *Izvestiya Biol.-Geogr. Inst. Irk. Univ.*, **XVI**, 1—4: *25—61**.

ZABUSOV, I. P., 1901 a. Notes on the morphology and systematics of Triclada. On the anatomy of *Rimacephalus pulvinar* Grube. from Lake Baikal. *Trudy Kaz. Univ. nat. Soc.*, **XXXVI**, 1: *1—76** (German summary).

ZABUSOV, I. P., 1901 b. To the data on the planarians of Lake Baikal. Coll. of art. dedic. to the 50th anniv. of the East-Sib. br. of the Geogr. Soc. The fauna of Baikal, 1, Kiev: *43—49**.

ZABUSOV, I. P., 1903. Notes on the morphology of Triclada. The first preliminary report on the Baikalian planarians collected by V. P. Garyayev. *Trudy Kaz. Univ. nat. Soc.*, **XXXVI**, 6: *1—58**.

ZABUSOV, I. P., 1906. The second preliminary report on the Baikalian planarians collected by V. P. Garyayev. *Kaz. Univ. nat. Soc.*, **XXXVII**, 6: *1—28**.

ZABUSOV, I. P., 1911. Studies of the morphology and systematics of the Baikalian planarians. 1, The genus *Sorocelis* Grube. *Trudy Kaz. Univ. nat. Soc.*, **XLIII**, IV: *1—422**.

ZAITSEV, N. S., 1947. On Pliocene deposits and young movements in the Tannu-Ola Range. *Doklady Acad. Sci. USSR*, **57**, 9: *931—938**.

ZAKHVATKIN, A. A., 1932. To the knowledge of the diurnal migrations of zooplankton. *Trudy Baik. limnol. St. Acad. Sci. USSR*, **II**, 55—106**.

ZENKEVICH, L. A., 1922 a. The new parasitic rotifer *Albertia voronkovi* from Lake Baikal. *Russ. hydrobiol. J.*, **I**, 4: *134—136** (German summary).

ZENKEVICH, L. A., 1922 b. New data on the zoogeography of Lake Baikal. *Russ. hydrobiol. J.*, **I**, 5—6, Saratov: *159—163**.

ZENKEVICH (ZENKEWITSCH), L., 1925. Biologie, Anatomie u. Systematik d. Süsswasser-polychaeten d. Baikalsees. *Zool. Jb.*, **50:** 1—60.

ZENKEVICH (ZENKEWITSCH), L., 1935. Über d. *Manayunkia (M. polaris)* an den Murman-Küsten. *Zool. Anz.*, **199**, 7/8.

ZENKEVICH, L. A., 1947, 1951. The fauna and biological productivity of the sea. **1, 2**, Moscow.

ZHADIN, V. I., 1937. The mollusks of the mountain bodies of water of the Baikal Range. *Trudy Baik. limnol. St. Acad. Sci. USSR*, **VII:** *97—101*.

ORGANISM INDEX

The page, on which an organism is described at some length, is indicated by bold type.
The page, on which an organism is pictured, is indicated by italics.

338

SUBJECT INDEX

340

342

Light period and intensity 256, 260, 263, 265, 266
Limnocottus, parasites 74
Limnophiles 285
Limnophilidae, origin 122
Limnorheophiles 285
Littoral group community 302
— sor complex 302
— — zone, 158, 159, 170
— — —, fauna and flora 170
— zone 181
— —, benthos 181
— —, fauna 183
— —, vegetation 182
Lubomirskiidae, occurrence 67
—, skeletal spicules 65
—, skeleton 66

Macrohectopus branicki, parasites 77
Malacostraca 106
Maloye More, composition of plankton 242
— —, fauna 212
— —, morphology 210
— —, vegetation 210
Mammalia 145
Manayunkia, development 81
—, distribution 83, 222, 225, 227
—, origin 83
—, relative species 82
— *baicalensis* 81
— —, development 83
— —, distribution 83
— —, origin and migration 81
Manganese, annual changes in water 48
Marine origin of fauna 282
Melania, parasites 90
Melosira, seasonal and annual fluctuations 232
— *baicalensis*, vertical distribution 259
Metalimnion 43
Migrations of fish, influence of chemical and thermal factors 166
— — pelagic fish 228
—, vertical, of mass pelagic dwellers 251
Miller's thumbs, parasites 56, 59
Minnow, parasites 99
Mollusca 126
—, depths 129
—, fossil 134, 273
—, genitalia 131, 132
—, horizontal division in open Baikal 176
—, influence of temperature on genital products 166
—, nervous system 131
—, on sands 186

—, parasites 61
—, relationships with other faunae 132
—, systematics 127
—, tertiary Ponto-Caspian (Balkan) faunae 135
—, thinness of shell 130
Monogena 72
Morphology of Baikal basin 19

Nanism 298
Nematodes 76
—, parasitic 76, 77, 78
Neotenic species 298
Nitrates, annual changes in water 48
—, changes with depth 49
Norilsk lake, fauna 221
Nyasa lake, 292, 293
— —, speciation 293

Ohrid lake 62, 68, 70, 71, 84, 103, 104, 181, 193, 243, 281
Oligochaeta 84
—, parasites 61, 62, 79
—, systematics 85, 86
Omul s.a. *Coregonus autumnalis migratorius*
—, annual migrations 142
—, horizontal feeding migrations 264
—, influence of temperature on feeding 266
—, life cycle 142
—, migrations 243
—, parasites 75, 76, 77, 98, 99
—, shoals 142
—, speed of movement 263
Ostracoda 102
Oxidizability of water 49, 50
Oxygen, diurnal changes in water 52
—, seasonal changes in water 51
—, vertical distribution in water 49, 50

P/B factor (Production/Biomass) 217
pH 54
Palaeontological research 273
Paracottus 139
—, parasites 75, 76, 219
Parasites 56, 57, 59, 60, 61, 62, 72, 74, 75, 76, 77, 78, 79, 89, 90, 98, 99, 100, 126, 147, 219, 226, 289
Pelagic Cottoidei, horizontal feeding migrations 264
— fish, seasonal migrations 254
— —, vertical distribution 254
— group community 302
Perca fluviatilis, parasites 75, 77, 78, 98, 100
Phoca 145

344

ERRATA

Page	1,	line 35:	for	specialisation	read	speciation
,,	3,	,, 34:	,,	specialisation	,,	speciation
,,	68,	,, 9:	,,	*Pachydictym*	,,	*Pachydictyum*
,,	79,	,, 12:	,,	*labsis*	,,	*labis*
,,	84,	,, 27:	,,	*buthius*	,,	*bythius*
,,	86,	,, 2:	,,	*pygmaeus*	,,	*pygmeus*
,,	116,	,, 24:	,,	*Grangonyx*	,,	*Crangonyx*
,,	157,	,, 7:	,,	*Geminiphora*	,,	*Gemmiphora*
,,	213,	,, 38:	,,	*tenucosta*	,,	*tenuicosta*